D0908976

Ecological Studies

Analysis and Synthesis

edited by

J. Jacobs
München

O. L. Lange
Würzburg

J. S. Olson
Oak Ridge

W. Wieser
Innsbruck

Volume 14

with contributions by

Elgene Box
Institut für Physikalische Chemie, KFA, Jülich, F.R.G., and
Curriculum in Ecology, University of North Carolina
Chapel Hill, North Carolina   27514

John S. Bunt
Australian Institute of Marine Science
Townsville, Queensland, 4810 Australia

Charles A. S. Hall
Marine Biological Laboratory, Woods Hole, Massachusetts, and
Section of Ecology and Systematics
Cornell University, Ithaca, New York   14850

Helmut Lieth
Department of Botany
University of North Carolina
Chapel Hill, North Carolina   27514

Gene E. Likens
Section of Ecology and Systematics
Langmuir Laboratory, Cornell University
Ithaca, New York   14850

Peter L. Marks
Section of Ecology and Systematics
Langmuir Laboratory, Cornell University
Ithaca, New York   14850

Russell Moll
Great Lakes Research Division
University of Michigan
Ann Arbor, Michigan   48104

Peter G. Murphy
Department of Botany and Plant Pathology
Michigan State University
East Lansing, Michigan   48824

Douglas D. Sharp
Department of Botany
University of North Carolina
Chapel Hill, North Carolina   27514

David Sharpe
Department of Geography
Southern Illinois University at Carbondale
Carbondale, Illinois   62901

Dennis Whigham
Department of Biology, Rider College
Trenton, New Jersey   08602

Robert H. Whittaker
Section of Ecology and Systematics
Langmuir Laboratory, Cornell University
Ithaca, New York   14850

Primary Productivity of the Biosphere

edited by Helmut Lieth
and Robert H. Whittaker

with 67 figures

Springer-Verlag

New York Heidelberg Berlin

1975

Library of Congress Cataloging in Publication Data

Lieth, Helmut.
    Primary productivity of the biosphere.

    (Ecological studies; v. 14)
    Includes index.
    1. Primary productivity (Biology) I. Whittaker,
Robert Harding, 1920–   joint author. II. Title.
III. Series.
QH541.3.L5   574   74-26627

ISBN 0-387-07083-4   Springer-Verlag   New York Heidelberg Berlin
ISBN 3-540-07083-4   Springer-Verlag   Berlin Heidelberg New York

Preface

The period since World War II, and especially the last decade influenced by the International Biological Program, has seen enormous growth in research on the function of ecosystems. The same period has seen an exponential rise in environmental problems including the capacity of the Earth to support man's population. The concern extends to man's effects on the "biosphere"—the film of living organisms on the Earth's surface that supports man. The common theme of ecologic research and environmental concerns is primary production—the binding of sunlight energy into organic matter by plants that supports all life. Many results from the IBP remain to be synthesized, but enough data are available from that program and other research to develop a convincing summary of the primary production of the biosphere—the purpose of this book.

The book had its origin in the parallel interests of the two editors and Gene E. Likens, which led them to prepare a symposium on the topic at the Second Biological Congress of the American Institute of Biological Sciences in Miami, Florida, October 24, 1971. Revisions of the papers presented at that symposium appear as Chapters 2, 8, 9, 10, and 15 in this book. We have added other chapters that complement this core; these include discussion and evaluation of methods for measuring productivity and regional production, current findings on tropical productivity, and models of primary productivity. The book is directed toward the interests of a range of readers, from those seeking summaries of research techniques to those concerned with our synthesis of global production.

Several institutions and people have helped to complete this work in its present form. The chapters contributed or coauthored by Lieth and Sharpe were supported in part by the Eastern Deciduous Forest Biome US-IBP. The chapters contributed by Whittaker and Hall were supported in part by Brookhaven National Laboratory; the contributions by Likens and Whittaker were supported in part by the National Science Foundation. During the final stage of editing this volume, one of the editors (HL) worked as guest researcher at the Nuclear Research Center (KFA) in Jülich, West Germany. We gratefully acknowledge the financial and logistic help received at the KFA through Prof. Dr. K. Wagener and his staff at the Institut für Physikalische Chemie. The index was compiled by Margot Lieth and Cyndi Grossman. We thank them both for their assistance. We gladly give credit to the staff of Springer-Verlag New York for excellent assistance in improving the book.

We hope this book will be of value for its characterization of the biosphere as a productive system. We are not confident of man's ability to control the future of the world or even his own existence. Nevertheless, we should be gratified if a focal point of the book—the net primary production of the biosphere—is one day seen as a figure of real significance to man. If in the future man's population and industry are stabilized, then to biosphere production as a steady-state flow of biological energy in the world will be related two other steady-state flows—of food energy from the biosphere to man and of industrial energy—that will support a human world society living in a durable balance with its environment.

Helmut Lieth
Robert H. Whittaker

Contents

Part 1   Introduction

Preamble   2

1 Scope and Purpose of This Volume   3
    *Robert H. Whittaker, Gene E. Likens, and
    Helmut Lieth*

2 Historical Survey of Primary Productivity
    Research   7
    Helmut Lieth

Part 2   Methods of Productivity Measurements

3 Methods of Assessing Aquatic Primary
    Productivity   19
    Charles A. S. Hall and Russell Moll

4 Methods of Assessing Terrestrial
    Productivity   55
    Robert H. Whittaker and Peter L. Marks

5 The Measurement of Caloric Values   119
    Helmut Lieth

6 Assessment of Regional Productivity in
    North Carolina   131
    *Douglas D. Sharp, Helmut Lieth, and
    Dennis Whigham*

7 Methods of Assessing the
Primary Production of Regions   147
David M. Sharpe

Part 3   Global Productivity Patterns

8 Primary Productivity of Marine
Ecosystems   169
John S. Bunt

9 Primary Productivity of Inland Aquatic
Ecosystems   185
Gene E. Likens

10 Primary Productivity of the Major
Vegetation Units of the World   203
Helmut Lieth

11 Net Primary Productivity in Tropical
Terrestrial Ecosystems   217
Peter G. Murphy

Part 4   Utilizing the Knowledge of Primary Productivity

12 Modeling the Primary Productivity of
the World   237
Helmut Lieth

13 Quantitative Evaluation of Global
Primary Productivity Models
Generated by Computers   265
Elgene Box

14 Some Prospects beyond Productivity
Measurement   285
Helmut Lieth

15 The Biosphere and Man   305
Robert H. Whittaker and Gene E. Likens

Index   329

Part 1

Introduction

PREAMBLE

The last decades of biologic, and especially ecologic, research have made it clear that

1. The notion that man's population and wealth can increase without limit is self-deception and an invitation to self-destruction

2. The unregulated increase of the human population beyond the world's sustainable carrying capacity must be considered a moral crime

3. The relentless increase in the gross national products of the industrial nations, at the expense of the world population, must be considered a social crime

4. The reckless exploitation of our fossil fuel sources for short-term profit and growth, rather than careful planning for a reasonable use for a long-term future, is a crime against our own children

Helmut Lieth

1

Scope and Purpose of This Volume

Robert H. Whittaker, Gene E. Likens, and Helmut Lieth

Some commonplace ideas of our time are that the surface of the earth is occupied by a film of living organisms, the "biosphere"; that the life of man and all other heterotrophic organisms is dependent on the primary production of the biosphere; and that the growth of man's population and industry affects the biosphere with increasing pressures, particularly those of harvest and chemical influence. These ideas are familiar, but some of the quantitative characteristics of the biosphere and man's relationship to it are not. Only in the last decade have sufficient data become available so that productive dimensions of the biosphere can be characterized by something better than educated guesses. Only in the last two or three decades has the unstable character of man's relationship to the biosphere become apparent to more than a small circle of scholars.

The word biosphere is used to mean either the global film of organisms or the surface environments of the world in which these organisms live and with which they interact (Hutchinson, 1970). This volume refers to "biosphere" in the first sense and expresses the second meaning as the "ecosphere" (Cole, 1958). The basis of all biosphere function is *primary productivity*, the creation by photosynthetic plants of organic matter incorporating sunlight energy. (This volume does not deal with the much smaller contribution of chemosynthetic autotrophic organisms.) The purpose of this volume is to synthesize current knowledge of world primary productivity in terms of methods of measurement, environmental determinants, the quantities for different communities and for the biosphere as a whole, the relationship to other biosphere characteristics, and the implications for man.

KEYWORDS: Primary productivity; ecology; phytogeography; biosphere.
Primary Productivity of the Biosphere, edited by Helmut Lieth and Robert H. Whittaker.
© 1975 by Springer-Verlag New York Inc.

Our concern centers on *net* primary productivity, which is that part of the total or gross primary productivity of photosynthetic plants that remains after some of this material is used in the respiration of those plants. The remaining portion, net productivity, is available for harvest by animals and for reduction by saprobes. Net primary productivity provides the energetic and material basis for the life of all organisms besides the plants themselves. Net primary *productivity* is most commonly measured as dry organic matter synthesized per unit area of the Earth's surface per unit time, and is expressed as grams per square meter per year (g/m²/year $\times$ 8.92 = lb/acre/year).[1] Net *production* of ecosystem types in the world is expressed as metric tons (t = 10^6 g) of dry matter per year (metric tons $\times$ 1.1023 = English short tons). *Biomass* is the dry matter of living organisms present at a given time per unit of the Earth's surface, and may be expressed as kilograms per square meter (kg/m² $\times$ 10 = t/ha, $\times$ 8922 = lb/acre). Productivity may also be expressed as grams of carbon or calories of energy in the dry matter formed per unit area and time. The relationship of carbon to dry matter is variable, but 2.2 is a reasonable average by which carbon production may be multiplied to obtain dry matter. The energy content of plant biomass (in kilocalories per dry gram of tissue) is also variable, with a world average of about 4.25 for land plants, but with values around 4.9 for plankton and coniferous forest (see Table 7–2).

One of the purposes of this book is to summarize available data into an estimate of the world's total net primary production, for which we obtain 172 $\times$ 10^9 t/year. The pattern of production relationships in different kinds of communities that underlies this value has some complexity. In the three realms, the land, oceans, and freshwaters, net primary productivities range downward from 2000 to 3000 g/m²/year or more to near zero in desert conditions. Great contrasts in productivity are determined by water availability on land and nutrient availability in fresh and salt water, whereas temperature affects productivity everywhere. Over all, land communities are much more productive than are those of the oceans because land makes possible extensive community structure that retains nutrients and supports leaf surfaces. Marine plankton communities are far smaller in biomass, chlorophyll, and content of critical nutrients, as well as in the productivity that depends on these. Efficiency in use of light energy for productivity is generally correlated with primary productivity itself, but efficiency in productivity per unit chlorophyll is higher in marine plankton than it is in much more productive forests. Fractions of gross primary productivity spent in plant respiration vary with temperature and community biomass from 75% in tropical rain forest to probably 20–30% in some plankton communities. The energy content of plant biomass from different land communities varies in a

---

[1] As a way of expressing productivity we prefer g/m²/year for its direct translation into English as grams per meter square per year, and in particular prefer it to the cumbersome g·m⁻²·year⁻¹. The g/m²/year form is potentially ambiguous, since it is possible to interpret it so that the year would go into the numerator. We have never encountered anyone who has thus misinterpreted it and doubt that the potential ambiguity is a real problem, but g/m²/year should of course be interpreted as g/(m²·year) or (g/m²)/year.

definite pattern, from low values in tropical rain forest to high values in boreal forest.

At the moment it seems that man will not be able to restrain the growth of his population and industry before serious damage is done to the biosphere. If he is to do so, he must set limits on himself and plan for wise long-term use and conservation of the biosphere, based on knowledge of its characteristics. This book contributes to the understanding of the biosphere on which man's life and the healthfulness and attractiveness of his environment depends.

References

Cole, L. C. 1958. The ecosphere. *Sci. Amer.* 198(4):83–92.

Hutchinson, G. E. 1970. The biosphere. *Sci. Amer.* 223(3):45–53.

2

Historical Survey of Primary Productivity Research

Helmut Lieth

From a recent paper on the history of the discovery of photosynthesis (Rabinovitch, 1971), it appears that many biologists equate photosynthesis with productivity and identify the raw materials of photosynthesis (water, carbon dioxide, and sunlight energy) as the direct controls of productivity. Photosynthesis and primary productivity are not so simply identical. Indeed, primary productivity—the actual energy bound into organic matter—is the product of photosynthesis. Yet primary productivity requires more than photosynthesis alone. The uptake and incorporation of inorganic nutrients into the diverse organic compounds of protoplasm are essential to the photosynthesizing organism. Temperatures govern annual productivity in various ways that do not result from temperature dependence of the photosynthetic process. On land, productivity is strongly affected by the availability of water, not primarily for use in the photosynthetic process itself, but to replace the water lost through the stomata that are open to allow carbon dioxide uptake.

This chapter compiles the key sources in the historical understanding of plant productivity as distinguished from photosynthesis. These include the gradual assessment of the global amounts and, to a limited degree, the understanding of the importance of primary productivity for man and environment.

In this history there are at least three major periods: (1) before Liebig, (2) from Liebig to the International Biological Program (IBP), and (3) the IBP and its consequences. Let us follow this sequence to see how the modern viewpoints and methods have developed.

KEYWORDS: Primary productivity; history; ecology.

Primary Productivity of the Biosphere, edited by Helmut Lieth and Robert H. Whittaker.

From Aristotle to Liebig

384–322 B.C.   Aristotle taught that soil, in a manner comparable to that of the intestinal tract of animals, provides predigested food for the plants to take up through their roots. Thus he rightly emphasized the relationship between plant and soil while wrongly interpreting plant nutrition with an idea that was held generally for 1800 years.

1450 A.D.   Nicolai de Cusa expressed the almost revolutionary idea that "the water thickens within the soil, sucks off soil substances and becomes then condensed to herb by the action of the sun."

A reading of the entire paper "Ydiote de staticis experimentis" (the Ydiote here meant is layman, most likely a practitioner with high technical skill) in Nicolaus de Cusa (Cusanus) *Werke* (1967) gives the impression that the "agricultural engineers" of his time held this plant–water relationship as a general consensus. Nicolai's view emphasized this relationship between plant and water. This paper appears to be the design for van Helmont's experiment about 150 years later.

ca. 1600   van Helmont, besides performing odd experiments to find meth-
(1577–1644)   ods of obtaining mice from junk and sawdust, did one rather intelligent experiment. He grew a willow twig weighing 5 lb in a large clay pot containing 300 lb of soil, and irrigated it with rainwater. After 5 years, he harvested a willow tree of 164 lb of wood with a loss of only 2 oz of soil. van Helmont concluded from this that water was condensed to form plants.

1772–1777   Priestley, Scheele, and Ingenhousz were the first to discuss the
or 1779   interaction between plants and air. They spoke about "melioration" and the "spoiling" of the air by plants in light or darkness.

1804   de Saussure studied the gas exchange of plants and gave the correct equation for photosynthesis:

$$\text{Carbon dioxide} + \text{water} = \text{plant matter} + \text{oxygen}$$

Following Rabinovitch's (1971) manner of indicating persons whose work led up to the primary production equation (not the photosynthetic equation), we have added the names of those who were instrumental in first evaluating the importance or necessity or both of each of the elements. Entries from Rabinovitch are in parentheses; our entries are in brackets [ ].

$$CO_2 \quad + \quad H_2O \quad + \quad \text{light} \quad + \quad \begin{array}{c}\text{inorganic}\\\text{nutrients}\end{array}$$

<table>
<tr><td>(Senebier)</td><td>(de Saussure)</td><td>(Ingenhousz)</td><td></td></tr>
<tr><td>[Priestley–Scheele]</td><td>[van Helmont]</td><td></td><td>[Liebig]</td></tr>
</table>

YIELD

$$= \quad \text{Oxygen} \quad + \quad \text{organic matter} \quad + \quad \text{chemical energy}$$

<table>
<tr><td>(Priestley)</td><td>(Ingenhousz)</td><td>(Mayer)</td></tr>
<tr><td></td><td>[van Helmont]</td><td>[Boltzmann]</td></tr>
</table>

Following the development of this equation, plant production was subjected to widespread, serious investigation, although not on the scale of present-day studies. The newly founded Colleges of Agriculture and Forestry dealt with various aspects of such questions.

From Liebig to the IBP

1840   The development of analytical chemistry enabled Liebig to show the importance of minerals for plant nutrition. He fought intensely against the generally accepted humus theory, which was based on the assumption that plants lived from organic matter only. While studying the relationship between dry-matter production and nutrient supply, Liebig formulated the well-known Law of the Minimum.

1850–1900   Plant chemistry uncovered the major relationships among plants, mineral nutrients, soil, water, and air. The importance of humus was investigated for all physical and chemical parameters significant in agriculture and forestry. The principles of matter cycles were widely discussed all over Europe; today it is difficult to determine who had the original ideas or evidence for primary productivity. These results were summarized in a few books that were cited frequently up to the early twentieth century (Boussingault, 1851; Liebig, 1862; and Ebermayer, 1876, 1882).

1862   Liebig was the first to think quantitatively about the impact of vegetation on the atmosphere. In 1862 he said, "If we think of the surface of the earth as being entirely covered with a green meadow yielding annually 5000 kg/ha, the total CO_2 content of the atmosphere would be used up within 21–22 years if the CO_2 were not replaced," (230–240×10^9 metric tons CO_2 consumption per year, according to Liebig). This sentence marked the beginning of the geochemical treatment of productivity.

1882   Yield studies were easy to do with agricultural plants in laboratories and in the field, but forests presented special difficulties. The first dry-matter productivity figure for forests was not presented until 1882 when Ebermayer compared matter productivity of forests in Bavaria (from his own measurements) and field crops in France (data of Boussingault). Of course, the forests were more productive. His figures in kilograms per hectare of dry matter ($= 10$ times grams per square meter) are as follows:

Beech	Wood	3163 kg/ha	Potatoes	4080–4340 kg/ha
	Litter	3334	Clover	4200
	Total	6497	Wheat	4500
			Oats	4250
Spruce	Wood	3435 kg/ha		
	Litter	3007		
	Total	6442		
Pine	Wood	3233		
	Litter	3186		
	Total	6419		

These remained the key figures for about 50 years and were used again and again by geochemists in calculations of chemical elements in the biosphere. Forty years later, similar measurements were made by Boysen Jensen, Burger, Harper (see Lieth, 1962). Ebermayer presented the first estimation of world carbon binding of vegetation based on field measurements restricted to land areas. From his calculations for Bavaria he extrapolated that the annual consumption of CO_2 for the entire world was 90×10^9 t.

1900–1930   More than 60 years after Liebig's Law of Minimum, E. A. Mitscherlich developed this into the Law of Yield. This delay is rather surprising because the measurement of yield and dry-matter production had become very popular during Liebig's time. Mitscherlich's yield law is the first attempt to model productivity (Mitscherlich, 1954).

1908–1913   Figures similar to Ebermayer's (100×10^9 t) for CO_2 consumption were given by Arrhenius in 1908, and Cimacian in 1913, but neither gave additional biologic information (see Noddack and Komor, 1937).

1919   Schroeder (1919) provided the next major contribution to the knowledge of dry-matter production from the land. He based his calculations primarily upon Ebermayer's studies, but utilized more reliable information regarding the surface areas of forests, steppes, and cultivated land. Schroeder's calculations gave the following figures for the total land area of the earth:

Carbon	Carbon dioxide	Dry matter
13×10^9 t	48×10^9 t	28×10^9 t

He had based his figures on crude, superficial, geographic classification. The next refinement of the production figures could be expected when the plant geographers developed their first vegetation maps.

1930   This probably began with Drude at the end of the nineteenth century, and later led to the widely used physiognomic map of Brockmann-Jerosch (1930). Production calculations could be made from these vegetation maps as soon as information from the different vegetation units became available. Most later calculations for the production of the world, including recent ones, were based upon areas of physiognomically established vegetation units [see Lieth, 1964; Whittaker and Likens (in Whittaker, 1970); Golley, 1972].

1937   Schroeder's (1919) land-production figures were refined by the geochemist Noddack (1937) and then were used in reviews and textbooks until 1965. Schroeder was apparently the first to offer some information about the benthic algae, but he did not venture to say anything about the plankton. The first estimate of total aquatic carbon binding, 28.6×10^9 t/year, was made by Noddack and Komor (1937). This was more an opinion than a solidly based figure.

1944–1959   Only during the last 30 years has aquatic production received much interest. Within a short period of time, figures were presented by Riley (1944), Steemann Nielsen (1954), Steemann Nielsen and Aabye Jensen (1957), Fleming (1957), Fogg (1958), and Ryther (1959), and summarized in Gessner (1959) (see also Chapter 8, this volume).

1960   Müller (1960) summarized the estimates of world production up to that time in an extensive review and gave his own estimates of 10.3×10^9 t of carbon net production on land, and 25×10^9 t for land and sea. From 1960 on, there have been a large number of estimates that need not be reviewed in detail. Whereas Table 2–1A summarizes contributions to knowledge of the requirements for productivity, Table 2–1B summarizes some estimates of most significance.

1964   Figure 2–1 represents the first attempt to combine all productivity pattern information for the earth in one map (Lieth, 1964).

Table 2–1   Historical development of knowledge of primary productivity

A. Matter and Energy Metabolism of the Plants

Years	Sources[a]	Metabolic categories			
		Soil	Water	Air	Energy
384–322 B.C.	Aristotle	×	—	—	—
1450 A.D.	Nicolai de Cusa (Cusanus)	×	×	—	—
1577–1644	van Helmont	—	×	—	—
1772–1779	Priestley, Scheele, and Ingenhousz	—	—	×	×
1804	de Saussure	×	×	×	×
1840–1890	Liebig, Boussingault, and Ebermayer	Mineral cycling	×	×	
1886	Boltzmann	—	—	—	Radiant energy conversion
Today's problems in the various categories:		Mineral cycling	Water balance	Gaseous metabolism	Energy function

B. Estimates of the Annual World Primary Production

Years	Sources	Number of vegetation units distinguished	Value for annual production (10^9 t)	Category assessed
1862	Liebig	1	230–240	CO_2
1882	Ebermayer	3, on land only	90	CO_2
1919	Schroeder	4, on land only	13	C
			48	CO_2
			28	Dry matter
1937	Noddack	5	64	C
1960	Müller	5	25	C
1969	Whittaker and Likens	15	164	Dry matter
1964, 1968	Lieth; Junge, and Czeplak	none [b], land only	32–44	C
1972	and this volume	20	155–180	Dry matter
			687	Calories
			70	C
			257	CO_2
1975	Rodin, Bazilevich, and Rozov	106, land only	172	Dry matter

C. Mapping of World Productivity

Years	Sources	Area mapped	Category mapped
1955	Sverdrup	Ocean	C
1957	Fleming	Ocean	C
1957	FAO team (in Gessner, 1959)	Ocean	C
1956	Patterson (in Lieth, 1962)	Forested areas	C (dry matter, calories, CO_2)
1964	Lieth	1st world map	Volume of wood
1965 (1966)	Rodin and Bazilevich	Land	Mineral cycling, dry matter
1968, 1970	Bazilevich et al.	Land	Mineral cycling, dry matter
1971	Lieth, Zaehringer, and Berryhill (in Lieth, 1972)	Land, 1st computer map	C (dry matter, calories)

[a] Literature before 1900 usually is quoted from secondary literature, and is not included in the reference list.
[b] Pattern of Lieth's 1964 map was used.

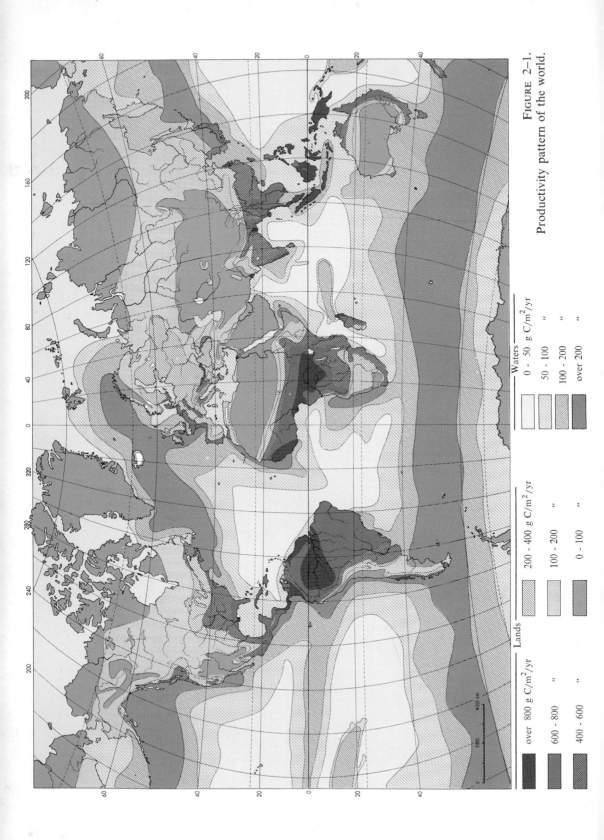

Lands

▨ over 800 g C/m²/yr
▨ 600 - 800   "
▨ 400 - 600   "

▨ 200 - 400 g C/m²/yr
▨ 100 - 200   "
▨ 0 - 100   "

Waters

▨ 0 - 50 g C/m²/yr
▨ 50 - 100   "
▨ 100 - 200   "
▨ over 200   "

FIGURE 2–1.
Productivity pattern of the world.

The IBP and Its Consequences

This review of the history of productivity studies has emphasized three trends: First, the development of the production equation—its relation to photosynthesis and the controls on primary productivity—has been outlined. The production equation was completed in the last century, but the factors affecting productivity are still under quantitative investigation today. Second, the increasing refinement and convergence of world production estimates have been observed. Liebig (1882) extrapolated from a single community, which he assumed to be a representative unit of world production. Schroeder (1919) distinguished four land community types, but current estimates are based on 20 or more vegetation types. There is encouraging agreement among these results. I hope I am not unjustified in suggesting that the major estimates offered here—of 100 to 125 $\times$ 10^9 t of dry matter and 425 to 530 $\times$ 10^{18} cal of organic energy as the net primary production for the land vegetation of the earth—will be subject to refinement in detail but not to major revision. The assessment of these figures is described in Chapter 8 for the ocean, Chapter 9 for inland waterbodies, and in Chapter 10 for the world's major vegetation formation classes. These chapters also describe the history of world productivity assessment after the establishment of the IBP.

References

Bazilevich, N.I., A.V. Drozdov, and L. E. Rodin. 1968. Productivity of the plant cover of the earth, general regularities of its distribution and relation to climatic factors. [In Russian with Engl. summ.] *Zh. Obshch. Biol.* 29:267–271.

———, L. E. Rodin, and N. N. Rozov. 1970. Untersuchungen der biologischen Produktivität in geographischer Sicht. (In Russian.) *5th Tagung Geogr. Ges. USSR Leningrad.*

Boussingault, J. B. 1851. *Die Landwirtschaft in ihren Beziehungen zur Chemie, Physik und Meteorologie.* (German transl. by N. Graeger), Vol. 1, 399 pp. Halle: Graeger Verlag.

Cusanus, N. (Nicolai de Cusa). 1450. De statics experimentatis.

——— (Nicolaus von Cues). 1967. *Schriften des-Heft 5. Den Laie über Versuche mit der Waage* (Transl. by Hildegund Menzel-Rogner), Leipzig: Meiner Verlag, 85 pp.

Ebermayer, E. 1876. *Die gesamte Lehre der Waldstreu mit Rücksicht auf die chemische Statik des Waldbaues,* 300+116 pp. Berlin-Heidelberg: Springer-Verlag.

———. 1882. Naturgesetzliche Grundlagen des Wald- und Ackerbaues, Vol. 1: *Die Bestandteile der Pflanzen,* Part I *Physiologische Chemie der Pflanzen.* 861 pp. Berlin-Heidelberg: Springer-Verlag.

Fleming, R. H. 1957. General features of the oceans. In *Treatise on Marine Ecology and Paleoecology,* J. W. Hedgepeth, ed. Vol. 2: *Ecology,* pp. 87–108. Geological Society of American Mem. 67(1), New York.

Fogg, G. E. 1958. Actual and potential yields in photosynthesis. *Advan. Sci.* 14:359–400.

16

Part 1: Introduction

Gessner, F. 1959. *Hydrobotanik* Vol. 2, 701 pp. Berlin: Deutscher Verlag der Wissenschaften.

Golley, F. B. 1972. Energy flux in ecosystems, In *Ecosystem Structure and Function*, J. A. Wiens, ed. Corvallis, Oregon: *Oregon State Univ. Ann. Biol. Colloq.* 31: 69–90.

Junge, C. E., and Czeplak, G. 1968. Some aspects of the seasonal variation of carbon dioxide and ozone. *Tellus* 20: 422–434.

Liebig, J. von. 1840. *Organic Chemistry and Its Applications to Agriculture and Physiology* (Engl. ed.: L. Playfair and W. Gregory.) 387 pp. London: Taylor and Walton.

———. 1862. *Die Naturgesetze des Feldbaues*, 467 pp. Braunschweig: Vieweg.

Lieth, H. 1962. *Die Stoffproduktion der Pflanzendecke*, 156 pp. Stuttgart: Fischer.

———. 1964. Versuch einer kartographischen Darstellung der Produktivität der Pflanzendecke auf der Erde. *Geographisches Taschenbuch*, 1964/65, pp. 72–80. Wiesbaden: Max Steiner Verlag.

———. 1972. Über die Primärproduktion der Pflanzendecke der Erde. *Z. Angew. Bot.* 46:1–37.

Mitscherlich, E. A. 1954. *Bodenkunde für Landwirte, Forstwirte und Gärtner*, 7th ed., 168 pp. Berlin and Hamburg: Parey.

Müller, D. 1960. Kreislauf des Kohlenstoffs. *Handbuch Pflanzenphysiol.* 12:934–948.

Noddack, W. 1937. Der Kohlenstoff im Haushalt der Natur. *Angew. Chem.* 50:505–510.

———, and J. Komor. 1937. Über die Ausnutzung des Sonnenlichtes beim Wachstum der grünen Pflanzen unter natürlichen Bedingungen. *Angew. Chem.* 50:271–277.

Rabinowitch, E. 1971. An unfolding discovery. *Proc. Natl. Acad. Sci. US* 68:2875–2876.

Riley, G. A. 1944. The carbon metabolism and photosynthetic efficiency of the earth as a whole. *Amer. Sci.* 32:129–134.

Rodin, L. E., and N. I. Bazilevich. 1966. *Production and Mineral Cycling in Terrestrial Vegetation.* 288 pp. Edinburgh: Oliver and Boyd.

———, ———, and Rozov, N. N. 1975. Productivity of the world's main ecosystems. In *Proceedings of the Seattle Symposium.* Washington, D. C.: National Academy of Science. (*In press.*)

Ryther, J. H. 1959. Potential productivity of the sea. *Science* 130:602–608.

Schroeder, H. 1919. Die jährliche Gesamtproduktion der grünen Pflanzendecke der Erde. *Naturwissenschaften* 7:8–12.

Steemann Nielsen, E. 1954. On organic production in the oceans. *J. Cons. Perm. Int. Explor. Mer.* 19:309–328.

———, and E. Aabye Jensen. 1957. Primary oceanic production. The autotrophic production of organic matter in the oceans. *Galathea Rep.* 1:49–136.

Sverdrup, H. U. 1955. The place of physical oceanography in oceanographic research. *J. Marine Res.* 14:287–294.

Whittaker, R. H. 1970. *Communities and Ecosystems*, 162 pp. New York: Macmillan.

Part 2

Methods of

Productivity Measurement

During the last decade the methods for assessing primary productivity have developed gradually, starting with direct harvest of the material produced and advancing to elaborate procedures for measuring gas exchange and even correlation models linked to parameters of the production process. The measurement of primary productivity has never assumed accuracy better than $\pm 10\%$, and comparisons and summaries are sometimes affected by different results from different techniques. The major methods used in compiling the productivity data reported in Chapters 3 and 4 are as follows:

To assess aquatic productivity
>   Gas exchange (light and dark bottle)
> ^{14}C uptake
>   Diurnal curves of O_2 or CO_2
>   Chlorophyll and light relations

To assess terrestrial productivity
>   Harvest technique
>   Structural and compositional analysis (dimension analysis)
>   Gas exchange
>   Vegetation-type correlation

Used for both
>   Yield statistics
>   Environmental parameter correlation
>   Determination of combustion value

The productivity tables in Chapter 3 refer to the techniques most frequently used to assess the productivity of a given unit.

The techniques are described so that the reader will be able to judge the limitations of different approaches. Readers interested only in the results of production research are invited to scan or skip the Methods section. We have included extensive details in this section both to indicate the basis of our results and to render it useful as a summary and reference work.

3

Methods of Assessing
Aquatic Primary Productivity

Charles A. S. Hall and Russell Moll

This chapter is a concise description and comparison of the most commonly used methods in measuring "primary productivity" in water. The reader may also wish to consult reviews by Doty (1961); Goldman (1969); Vollenweider (1969b); Wrobel (1972); and Wetzel (1973).

Often the various methods used in the determination of primary productivity measure different processes, the most obvious case of which is the difference between gross and net productivity. However, there are also more subtle problems; for example, the "net productivity" measured by ^{14}C assimilation and the "net productivity" measured by free-water oxygen techniques are very different, although the same descriptive name is used for each. Figure 3–1 attempts to clarify this situation with a model of the photosynthetic process; the labeled pathways are used throughout this chapter. The symbols used are those developed by H. T. Odum (1967).

We might follow the pathway of the energy contained in an incident photon as follows: first the incident photon (A in Fig. 3–1) strikes the surface of the leaf or wall of the plant cell. Some of the energy contained in the photon is reflected from the surface of the leaf (B), and much of the remainder is trapped by chlorophyll in the chloroplasts. The incident energy is captured in the interior of the chlorophyll molecule as an orbiting electron is shifted to an orbit with a higher energy. This energy is then used to split a water molecule. Free oxygen is liberated (C) and ionic hydrogen is made available for the eventual production of sugar. The original energy captured from the photon is used to reduce (hence energize) a cycling series of organic compounds, and the donor chemical

KEYWORDS: Primary productivity; methods; aquatic
ecosystems; ecology.

Primary Productivity of the Biosphere, edited by
Helmut Lieth and Robert H. Whittaker.
© 1975 by Springer-Verlag New York Inc.

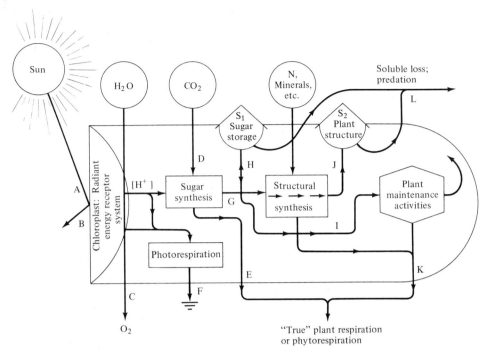

FIGURE 3–1. Diagrammatic representation of metabolism and productivity in aquatic ecosystems. A, incident sunlight; B, reflected sunlight and sunlight converted directly to heat at leaf surface; C, gross photosynthesis as measured by oxygen evolution; O, total carbon fixed (often considered gross photosynthesis); E, energy lost in transfer of energy in ATP to energy in sugar; F, photorespiration (unavoidable oxidation in C_3 plants); G, gross energy fixed as sugar; H, energy stored as sugar (S_1); I, energy previously fixed by photosynthesis and used for plant maintenance activities (left-pointing arrow indicates that some of this occurs "upstream" in diagram); J, energy stored as structural materials (S_2); K, energy changed to heat during processes of structural synthesis and plant maintenance activities; L, energy lost from plant as soluble organic "exudates" or by predation by herbivores; M, change in biomass $= \Delta S_1 + \Delta S_2 = $ Net production of standing crop; and N, net production $= L + M$.

is, in turn, oxidized. Eventually the energy is stored as ATP and it is in this form that the energy is used to produce high-energy organic compounds.

The energy trapped as ATP then is used for synthesizing sugars and other organics; the cost of this process in calories is represented as pathway E. Photorespiration, in which some of the Calvin cycle intermediates (glycolic acid) are oxidized in the peroxisomes, is indicated by pathway F. This oxidation is reduced or prevented in C_4 plants, which have separate chloroplasts in the mesophyll (where 4-carbon acids are produced and oxygen released) and in the bundle sheaths (where the Calvin cycle occurs with little or no photorespiration). The occurrence of photorespiration in ordinary, C_3 plants complicates the definition of gross primary productivity for these. However, since no energy useful to the plant is released by photorespiration, which appears to be inadvertant oxidation, we have excluded it from our definition of gross primary productivity. Production measurements in general fail to measure photorespiration because this involves rapid turnover of both CO_2 and O_2 without accumulation of fixed carbon.

Gross primary productivity normally is measured as the observed release of O_2, or uptake of CO_2, resulting from photosynthesis. Because the first step in the process of photosynthesis that is measurable readily in the field is the production of O_2, this, with appropriate corrections for simultaneous respiration, is commonly considered the best available measurement of gross productivity. Uptake of CO_2 is also readily measured in the field, and, with respiration corrections, it is justifiably considered gross productivity. The ratio of oxygen produced to carbon dioxide used is the photosynthetic quotient and is a reflection of the relative quantities of sugars, fats, and proteins produced by the plant.

Once the energy and carbon are fixed as sugar, the sugar may be stored (pathway H), used for constructing plant structural tissue (J), or used for various plant maintenance functions (K), including the energy expended for making sugar and plant structure (E and I). We may consider the summation of all these oxygen-using activities, excluding photorespiration, as "true" plant respiration (which we like to call *phytorespiration*). Net production is then defined as gross plant production minus true plant respiration. Net production includes changes in plant biomass, but it also includes particulate or soluble organics lost to the surrounding environment as well as organic material eaten by herbivores. The last two categories, which are missed by some production measurements, are represented by pathway L. Table 3–1 gives some useful conversions for productivity studies.

Methods of Measurement

Free water versus in situ *incubation versus shipboard incubation*

Where metabolism is sufficiently concentrated, normally in waters less than 5 or 10 m deep or in deeper waters rich in mineral nutrients, O_2 and CO_2 changes in the aquatic environment itself will be sufficient to be measured readily by Winkler oxygen determinations, by gas chromatograph measurements of CO_2, or by a regular-scale pH meter. In this method "free water" samples are taken directly in the aquatic medium and the O_2 or CO_2 concentrations are compared over time. Normally, O_2 concentrations will increase during the daylight hours owing to the photosynthetic activities of autotrophs, and will drop during the night owing to the respiration of all aerobic organisms. Changes in CO_2 concentration will be the opposite of those for O_2, falling during the day and rising at night. As pH is inversely related to CO_2 (CO_2 in an aqueous environment will combine with water to form carbonic acid, which will slightly dissociate and lower the pH) the pH, like the O_2, will rise during the day and fall at night. Thus the aquatic medium acts as a giant and completely natural bell jar trapping

Table 3–1   Some useful approximations for studies of primary production[a,b]

To convert from (g)	To	Multiply by	Reference
O_2 metabolized	kcal metabolized	3.5	Brody (1945)
Dry weight	kcal	3.3–4.9	Cummins and Wuycheck (1971)
			E. P. Odum (1971)
Dry weight	g carbon	~0.5	E. P. Odum (1971)
		~0.45	Chapter 15, this volume
O_2 metabolized	g organics metabolized	~1.0 (depends on PQ)	H. T. Odum and Hoskin (1958)
O_2 metabolized	g CO_2 metabolized	1.38 (if PQ = 1.0)	By molecular weights

[a] For more specific treatment, consult the referenced primary literature or, for accurate work, run the appropriate field and laboratory studies. See also E. P. Odum (1971), and Table 5–3, this volume.

[b] Basic equation of photosynthesis and respiration:

$$6\ CO_2 + 6\ H_2O \rightleftharpoons C_6H_{12}O_6 + 6\ O_2$$

or as elaborated in E. P. Odum (1971):

1.3 · 10^6 kcal radiant energy + 106 CO_2 + 90 H_2O + 16 NO_3 + 1 PO_4 + mineral elements =

13,000 kcal potential energy in 3258 g protoplasm (106 C, 180 H, 46 O, 16 N, 1 P, 815 g mineral ash) + 154 O_2 + 1,287,000 kcal heat energy dispersed (99%)

This formulation is an average based on element ratios of plant protoplasm, and accounts for the fact that the eventual product of photosynthesis is not simply sugar but a variety of organics.

the metabolic gases released, and providing a convenient measurement of the quantities absorbed.

Free-water methods are free of errors that result from enclosing the organisms; however, free-water methods are subject to errors arising from diffusion (gas interchange between air and water). Reasonable corrections may be made for this, but the techniques for determining diffusion tend to be substantially more difficult than those for gas concentration determination. Free-water methods may measure the metabolism of the entire aquatic community including planktonic, benthic, and pelagic components, whereas bottle methods measure only the planktonic or benthic communities. Free-water methods have been developed for O_2, CO_2, chlorophyll, and enumeration techniques but not for ^{14}C. In principle, any of these methods can be used for bottle measurements, but in practice, bottle methods generally have been restricted to ^{14}C and O_2.

In situ methods, based in large measure on the work of Gaarder and Gran (1927) provide greater control over the experiment, allow more precise measurements, and are useful for ^{14}C methods as well as metabolic gases. Several errors may arise when organisms are isolated in bottles. The turbulence in nature has a stimulatory effect on metabolism (Westlake, 1967; Olinger, 1968; Mann *et al.* 1972), either by supplying raw materials that may be depleted locally, by carrying off wastes, or by an unknown process. Mann found that by rotating his bottles in the water with a device similar to a paddle wheel, the productivity of his samples was increased by 30%. This may or may not be the case in general. Schindler *et al.* (1973b) found that CO_2 diffusion limitations gave erroneous estimates with bottle methods.

Another error introduced by isolating planktonic communities in bottles is caused by glass surface effects (see Jannasch and Pritchard, 1972). A variety of largely undetermined effects may arise because the glass is a good substrate for bacteria or for some species of algae. This difficulty could be investigated easily by running a series of productivity estimates in bottles with different numbers of glass slides suspended inside. Then, if differences in rates were detected, the original estimates for bottles without glass slides could be corrected for the area of glass on the inside of the bottle alone. This, we suspect, would turn out to be trivial, but, to our knowledge, it has not been measured. Acid-sterilizing or cleaning bottles before each day's sampling should eliminate most problems with bacteria.

If bottle methods are used, the samples must be incubated for some period of time under conditions that approximate the conditions of the natural communities. This is normally done by the *in situ* method, that is, by resuspending the samples in the water from a float at the depth from which they were taken. This has the advantage of, in theory, maintaining the conditions that the natural communities are experiencing; however, it eliminates the vertical mixing into different light regimes. Another disadvantage, according to Fee (1973a, b), is that *in situ* methods have virtually no predictive value because there are many unmeasured, changing parameters, such as light.

Ship time is very expensive, and because much more information may be gained by taking many samples on the same day under controlled conditions of light, temperature, or nutrients, an alternative method to the *in situ* procedures is to construct a shipboard or laboratory incubator. These bottles are maintained for the incubation period in a water bath at the same temperature and light intensity as was measured *in situ*, or at some other standard conditions (see Goldman, 1967; Fee, 1971, 1973a, b). If the latter is maintained, samples from different regions or under different intensities of insolation can be compared for their potential productivity. Saunders *et al.* (1962) and Fee (1973a) found *in situ* and shipboard productivity estimates to be closely related. Plastic bags (McAllister *et al.*, 1961) and flexible plastic columns are an interesting variation on this theme and allow some of the environmental turbulence to be transferred into the incubation chamber (Bender & Jordan, 1970). It would be interesting to float plastic bags down a river, taking oxygen samples at hourly intervals.

For both bags and bottles, some error results from the absorption of light by the enclosing material. Perhaps this could be corrected for by incubating a representative sample in two layers of glass or plastic, followed by subtracting the difference between incubations with one and two layers from the estimates made with only one layer. Alternatively, and more simply, a measurement can be made of the fraction of photosynthetically important light attenuated by a layer of glass and by correcting the original measurement upward by this amount. In any case, whether using bottles or free-water methods, application of statistical methods is essential (Cassie, 1961).

Gas-exchange methods

LIGHT-AND-DARK BOTTLES. Light-and-dark-bottle oxygen methods are simple and versatile, and give a measure of both photosynthesis and respiration at the same time. They may be used in conjunction with free-water O_2 estimates to partition the relative importance of planktonic communities relative to the metabolism of the whole system (Day *et al.*, 1973).

To run a light-and-dark-bottle O_2 series, prepare three glass biochemical oxygen demand (BOD) bottles, normally of 300 ml each. Two are left clear and one is made into a "dark" bottle by painting it black, wrapping it with black electrician's tape, and then wrapping the black bottle in aluminum foil. The three bottles are filled with the water (containing plankton communities) of interest by overflowing each three times to eliminate atmospheric contamination. Two of the bottles, one clear and one black, are resuspended in the water from a nonshading float at the depth from which the samples were taken (or alternately in a shipboard incubator) for a period of 1 hr to half a day. In general, shorter time periods give more accurate estimates but increase the importance of analytic error. The third bottle (initial bottle, IB) is immediately fixed with Winkler reagents. After the incubation period the other bottles are fixed with Winkler reagents, and the three bottles are titrated for O_2. Community metabolism can be calculated as follows (sample data from Czaplewski and Parker, 1973, slightly modified for clarity; incubation: 1 hr):

Net photosynthesis $= LB - IB = 8.40 - 8.05 = 0.35$ ppm O_2 per hour
Respiration $= IB - DB = 8.05 - 8.00 = -0.05$ ppm O_2 per hour
Gross photosynthesis $= LB - DB = 8.40 - 8.00 = 0.40$ ppm O_2 per hour

Of course, if a true picture of photosynthesis over an entire day is desired, it is necessary to repeat this process several times and at several depths. Czaplewski and Parker suggest that by using an oxygen probe instead of Winkler titrations many more samples can be processed each time with no loss in accuracy.

FREE-WATER OXYGEN OR DIURNAL CURVE. Free-water oxygen methods are useful in waters of moderate to high productivity, or in some cases in waters of low productivity where the water depth is not great. The existence of diel (often called diurnal) oxygen changes in water has been known since the early 1900s, but the methodology was not developed or used extensively until the studies by H. T. Odum and co-workers in the mid-1950s. The method is based on the daily rise and fall of oxygen content in natural waters, a rise and fall that is proportional, with appropriate corrections, to the photosynthetic and respiratory activities of the organisms residing within the water mass. In effect, the water mass itself acts as a giant incubation bottle, and the many problems of "bottle" estimates discussed previously are eliminated. The principal disadvantages here are (1) lack of sensitivity in oligotrophic waters, (2) the uncertain diffusion corrections, which are particularly important in waters that are shallow or turbulent or both, and (3) the fact that the method measures the gross metabolism of the entire community (a drawback if only the planktonic component is of interest). Another disadvantage is that you have to stay up all night if you do not have automatic recording equipment.

MEASURING OXYGEN CONCENTRATION IN WATER. The concentration of oxygen in water is measured normally by the Winkler method (APHA, 1971), or with a galvanic probe, supplied by most scientific supply companies. Winkler bottle methods are inexpensive and probably less susceptible to error, particularly by inexperienced investigators. They should be used only with an "antidiffusion cork" or similar device made from a two-holed stopper and two pieces of glass tubing, one that will reach the bottom of the BOD bottle and one that just penetrates the top. This will eliminate oxygen exchange with the atmosphere as the bottle is filled. If diffusion errors during filling are still suspected, a series of these bottles with a suction pump will eliminate the problem. Alternately, a limnologic sampling bottle, such as the Van Dorn, may be used. When filling a Winkler bottle from a Van Dorn, the water sampler should overflow the Winkler bottle three times. For other considerations of oxygen sampling, see Carpenter (1965a, b), Efford (1968), and Strickland and Parsons (1972).

Galvanic probes tend to be expensive ($500–$1000 or more), may be subject to drift and gremlins, and because all galvanic probes are also temperature probes, the results must be algebraically temperature-compensated unless a probe with an electronic compensation feature is used. Compensation can be determined (and the automatic temperature compensation checked) by immersing the probes (temperature, oxygen, and, if present, temperature compensation) in a sealed bottle with water of known oxygen concentration at room temperature. The bottle is then cooled in an ice bath and the temperature and oxygen recorded at intervals. If the oxygen probe is correctly temperature-compensated, there should be no change in the oxygen reading as the temperature changes. If there is change, the probe is either incorrectly compensated or not compensated at all. Corrections may be made with the following formula:

$$O_c = O_u + \{t \cdot r\,[-c \cdot (100 - S)]\}$$

where O_c is the corrected oxygen value; O_u is the uncorrected oxygen value; t is the centigrade difference between the temperature at which O_u was determined and the temperature of calibration; r is the rate-difference change of the oxygen probe per degree at 100% oxygen saturation; c is a constant that expresses the rate at which the true temperature compensation diminishes as the reading departs from saturation; and S is percent oxygen saturation of the water. In other words, the slope of the correction line is greater near oxygen saturation than at 2 ppm O_2. So approximate calculations of dissolved oxygen must be made in order to correct fully for temperature compensation. These corrections may or may not be important but must at least be considered for accurate work. A correction curve derived for a Gulton non-temperature-compensated meter was

$$O_c = O_u + \{t \cdot 1.14 - [0.0086\,(100 - S)]\}$$

This probably would serve as an approximation for many noncompensated instruments, but several datum points should be checked. In addition, the correction curves should be checked for nonlinearity. In practice, these corrections should be done by a computer or by drawing a family of curves for oxygen reading versus temperature at different saturation values, and interpolating on the graph. In contrast, it becomes apparent that good electronic temperature compensation, such as is available with the Martek Model DOA and other modern equipment, is well worth the money [see also corrections for temperature and salinity given in APHA (1971)]. If all samples are taken within a degree or two of the temperature at which the calibration was made, no corrections are needed except for extremely accurate work.

TWO-STATION ANALYSIS. The most accurate estimates of photosynthesis and community respiration can be obtained for flowing water by the two-station method of oxygen analysis (H. T. Odum, 1956; Owens, 1969). The two-station analysis is based on the actual change in dissolved oxygen as a parcel of water flows from one region of the stream to another one that is ideally about a 1- or 2-hr flow time through a homogeneous environment. We have found considerable oxygen-rate differences over distances of only a few hundred meters in a small, partially forested stream in upstate New York owing to the shading effects of forests. Thus changes over a clearly defined area can be measured and rates of change determined from differences in oxygen concentration between a water mass and that same mass some 1 or 2 hr later. According to H. T. Odum (1956), the area-based change in oxygen concentration is given as follows:

$$Q \quad = \quad P \quad - \quad R \quad + \quad D \quad + \quad A$$

Q	P	R	D	A
(Rate of change of oxygen per area)	(Rate of gross primary productivity per area)	(Rate of respiration per area)	(Rate of oxygen diffusion per area)	(Rate of drainage accrual)

Ideally a stretch of river is picked where A is trivial. By dividing the above units by depth z, the relationships are expressed as volume or concentration units, and are normally given as lower-case letters. Hence

$$q = p - r + d = \frac{Q}{z}$$

Diffusion constants, K, are often given per square meter, but the corrections applied to the data are normally expressed per cubic meter, making conversion of diffusion constants necessary: $k = K/z$. The use of diffusion constants will be explained on pages 28 and 33. It is best to pick a section of stream without waterfalls or rapids, as they make diffusion corrections more difficult.

The analysis of metabolism by the two-curve method is identical to the single-station method presented in the next section, except the rate of change for graph *d* is obtained by subtracting oxygen values at the downstream station from oxygen values at the upstream station but at an earlier time, the time interval being equal to the time it takes for the water to flow from one station to the next. This may be measured with dye, a handful of salt and a conductivity meter, or a dozen oranges. In each case the mean (integrated) time is what is needed.

SINGLE-CURVE METHOD. The two-station method is often impractical in streams owing to logistics or time considerations, and it is not necessary in standing waters except in studying spatial variations, which may be large in very productive and/or poorly mixed ponds. In flowing waters, where upstream and downstream diurnal curves are similar, one may use a single-station curve as an approximation to the two-station analysis (H. T. Odum, 1956). This assumes that the oxygen concentration measured at the one station would be the same as a sample taken upstream a distance equal to the distance used for calculations (about 1–2 hr), at the same time. Hornberger (*personal communication*, 1974) has found that single- and double-curve analyses generally give similar results for a Virginia stream. The basic procedure in estimating stream metabolism by this method is to measure oxygen and temperature in the field every 2 or 3 hr beginning at midnight, using either Winkler oxygen methods or a galvanic probe, either with or without a recorder. The data then are plotted (Fig. 3–2a, b).

If there were no biotic or chemical activity in the water under study, there would be only the change in the oxygen concentrations over the day due to temperature changes affecting saturation values. The oxygen concentration would be at virtual saturation for the entire day. However, biotic respiration tends to lower the oxygen in the water throughout the day and night, and the photosynthesis of green plants raises the oxygen during the day. Thus, a characteristic oxygen curve is produced, rising during daylight and falling at night, and often dropping less rapidly later at night as the amount of oxygen that diffuses into the stream equals the amount of oxygen being used by respiring organisms (Fig. 3–2).

The following is a sample calculation for one point on the corrected rate-of-change curve (Fig. 3–2d):

For the 9:00 A.M. rate-of-change point in Figure 3–2:

1	Oxygen concentration at 10:00 A.M.	= 12.25	From Figure 3–2a
2	Oxygen concentration at 8:00 A.M.	= 11.70	From Figure 3–2a
3 = 1–2	Difference	= 0.55	For 2 hr
4	Rate of change per hour =	0.27	Plotted in Figure 3–2d
5 = 4 + 6	Rate of change corrected for diffusion	= 0.18	Plotted in Figure 3–2d (with triangles)

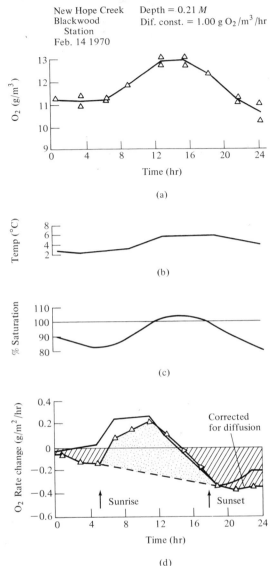

FIGURE 3–2. Representative analysis of community metabolism for single upstream station in New Hope Creek, North Carolina. (*a*) Mean oxygen concentration based on Winkler determinations. Triangles represent single samples. (*b*) Temperature at 3-hr intervals. (*c*) Percentage saturation of average of two Winklers at temperatures of sample. (*d*) Unmarked line is rate of change of oxygen concentration (g/m³/hr). Curve with triangles depicts rates of change after correction for diffusion of oxygen across air–water interface. Stippled area represents gross photosynthesis of water mass represented by these water samples. Cross-hatched area is estimated gross community respiration. Planimetry is used to measure these areas, or a computer program may be used. Depth is average depth in meters for 1-hr flow distance above sampling station. (1 mg/l=g/m³=1 ppm.)

Diffusion correction for the above point:

$$d = k \times SD$$

Total oxygen diffused equals diffusion constant times saturation deficit.

Sample: 0.09 $= 1.00 \times (100 - 91) \times 0.01$
Units: g O₂/m³/hr $= (g\ O_2/m^3 \cdot hr \cdot atm) \times (atm)$

In other words, at 9:00 A.M. the amount of oxygen that diffused into the water is the quantity that would have moved in if there were no oxygen in the water times the saturation deficit (100% saturation minus the observed saturation). This must be multiplied by 0.01 to change from percent to a proportion. Percent saturation is determined as observed oxygen/oxygen saturation concentration for that temperature, as determined from standard tables (i.e., Carpenter, 1966; Churchill *et al.*, 1962; APHA, 1971). We prefer the numbers of Churchill *et al.* (1962) as these values, if nothing else, are intermediate with respect to other published values.

When the above procedure is repeated and plotted for each 2-hr interval over 24 hr, starting at 1:00 A.M. (the 12:00 midnight to 2:00 A.M. interval), the metabolism for the entire day then may be calculated by integrating the slanted and stippled regions given in Figure 3–2d. See Manny and Hall (1969) for applications in large lakes. This is obviously an extremely laborious task, and if many of these curves are to be run, an appropriate computer program is invaluable.

Alternately, an abbreviated method has been developed by Welch (1968), which uses only two Winkler determinations per day instead of nine or more. It should be noted that the times for the two Winklers suggested by Welch may not be optimal for all situations. It is recommended that the daily maximum and minimum oxygen concentrations are first determined and then these times used for the calculation. Diffusion may be estimated crudely using the mean oxygen saturation of the two points. We prefer the method using nine points per day.

DAYTIME RESPIRATION CORRECTION. Actual measurement of respiration during the daytime is impossible by this method, or for that matter, any other field technique. In order to overcome this difficulty, it is assumed that daytime respiration equals nighttime respiration. A line drawn on the rate-of-change curve at the average nighttime respiration rate as in Figure 3–3a was suggested initially as an approximation of daytime respiration. Further refinement of this method (H. T. Odum and Wilson, 1962) takes into account the varying nature of daytime respiration, which is greater toward the end of the day when temperatures and oxygen levels are higher. Therefore a sloping line drawn from the predawn low point on the rate-of-change curve to the postsunset minimum (Fig. 3–3b) is probably a more accurate representation of what occurs in nature. Most 24-hr oxygen curves, based on many environments examined by the authors, have the postsunset rate-of-change point lower than the predawn point, indicating greater respiration during the latter part of the day.

Further studies (Sollins, 1969; H. T. Odum *et al.*, 1969) have indicated that daytime respiration may be considerably higher due to higher oxygen levels and photorespiration. Therefore the actual daytime respiration curve may dip considerably as suggested in Figure 3–3c and d. This oxygen consumption obviously is compensated for by a greater amount of oxygen being concurrently produced by photosynthesis, as the oxygen level in the water rises during the day. Thus community metabolism during the day may be considerably greater than during the night. However, until some adequate means for measuring photorespiration in the field becomes available, the method of connecting the predawn

FIGURE 3–3. Curves representing daytime respiration based on data from New Hope Creek, North Carolina. Curve *a* shows constant daytime respiration at level of average nighttime rates (H. T. Odum and Hoskins, 1958); curve *b* represents varying nighttime rates (H. T. Odum and Wilson, 1962); curve *c* shows hypothetical assumption of respiration proportional to oxygen concentration (Sollins, 1969); and curve *d* is hypothetical plot correcting for photorespiration (H. T. Odum *et al.*, 1969). Corrected rate-of-change curve from Wood Bridge Station, New Hope Creek, Oct. 4, 1968. Daytime respiration as represented by curve *b* is suggested as a reasonable approximation to use.

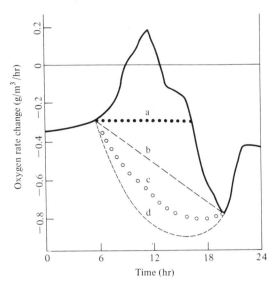

point by a straight line to the postsunset point is at least objective and is probably a minimal, but reasonable, estimate of all community respiration.

A recent refinement of Odum's basic free-water method has been suggested by Kelly *et al.* (1974). This method uses a continuous function derived from differences in oxygen between two stations fitted to a Fourier series and daytime respiration estimates derived from the zero intercept of a light versus net productivity plot. Although it is early for a complete analysis of this method, the preliminary results appear to be very promising. The same paper gives a rather strong argument indicating that daytime respiration does not dip substantially as in Figure 3–3c or d, and consequently that lines such as Figure 3–2a or b give a good approximation of daytime respiration. Free water oxygen methods give gross photosynthesis (pathway C of Figure 3–1).

ESTIMATE OF METABOLISM FROM pH CHANGES. Measurements of community metabolism may be made easily and elegantly using the diurnal pH method (Beyers *et al.*, 1963) or by applying the basic principles to light-and-dark bottles. The development of expanded-scale field pH meters makes this easily applicable to oligotrophic waters as, in general, unproductive waters are also poorly buffered. The production of carbon dioxide by the respiration of living organisms produces carbonic acid by the following formula:

$$\text{Sugar, etc.} \underset{\text{photosynthesis}}{\overset{\text{respiration}}{\rightleftharpoons}} CO_2 + H_2O \rightleftharpoons H_2CO_3 \rightleftharpoons \text{(Other carbon compounds, bicarbonate, carbonate, etc.)}$$

Thus, respiration lowers the pH of the water, and photosynthesis raises the pH. Over 24 hr, O_2 and pH curves have a similar shape. As the interaction of

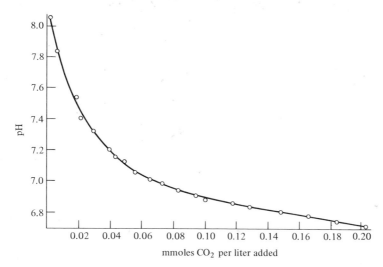

FIGURE 3–4. Carbon dioxide titration of New Hope Creek water for metabolic studies. Abscissa represents carbon dioxide in water sample added to that present at start of titration.

various carbon compounds in natural waters is extremely complicated and subject to unknown buffering, *a priori* coordination of pH and amounts of CO_2 produced or utilized is extremely difficult. However, this relation can be determined empirically by titrating the water of interest with distilled water of known CO_2 concentration (Beyers *et al.*, 1963). A sample titration of New Hope Creek water with CO_2-saturated distilled water is supplied (Fig. 3–4). As long as the buffering capacity of the water does not change, one pH–CO_2 curve can be used for many days of metabolic measurements.

The change in relative amounts of CO_2 in the water can be determined by reading the pH–CO_2 graph or by using the computer program supplied by Beyers *et al.* (1963). To determine absolute values of CO_2 in the water requires separate determinations of total CO_2 at the start of titration. However, this is not necessary because the metabolic determinations are based on changes in CO_2, not on absolute values.

Total inorganic carbon, and hence CO_2 changes, also may be determined by the syringe–gas chromatograph technique given by Stainton (1973). In this method, a 50-cm³ syringe is filled with 20 cm³ of sample water, 1 cm³ of dilute sulfuric acid to change all inorganic carbonates to CO_2, and 29 cm³ of helium. Once equilibrium is reached, the helium is analyzed for CO_2. This method is preferable in humic waters (Schindler, *personal communication*, 1974) and should be checked against the pH method.

Once CO_2 changes are known, estimates of total production and respiration are made by a procedure similar to that used for oxygen. Plots are made of relative amounts of CO_2 in the water over 24 hr (Fig. 3–5). The first derivative of CO_2 concentration is plotted as a negative function to make the results com-

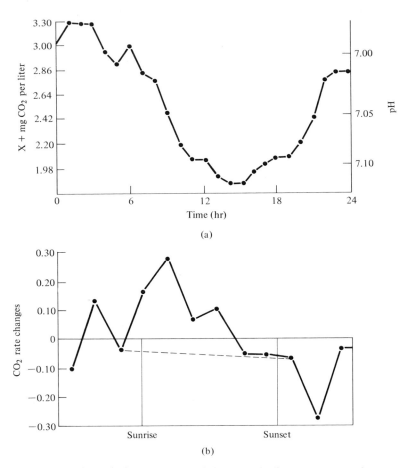

FIGURE 3–5. Technique of determining metabolism of an aquatic ecosystem (New Hope Creek, North Carolina) from changes in CO_2 showing pH and CO_2 plotted over 24 hr (*a*) and showing rate of change (*b*).

patible with oxygen data, which, of course, behave in reverse fashion. Daytime respiration is estimated according to the method discussed in the previous section, and total photosynthesis and respiration are determined by integrating the same areas discussed for oxygen. A sample determination is included (Fig. 3–5). No corrections for diffusion were made in this example, but the results (gross production = 1.33 g/m^3/day; respiration = 2.0 g/m^3/day) agree fairly well with oxygen estimates for the same day (1.5 and 2.2 g/m^3/day). Diffusion corrections for this method can be made in the same way as for oxygen and diffusion constants can be estimated from those determined for oxygen by the application of Graham's law. The determination of partial pressure of CO_2 in water is considered under the ^{14}C method. This method gives gross photosynthesis (pathway D of Figure 3–1).

DIFFUSION CORRECTIONS. In order to use the free-water oxygen, CO_2, or pH methods accurately, it is absolutely necessary to make accurate diffusion corrections. The following methods are recommended: (1) for lakes and ponds, the radioactive radon method (Emerson *et al.*, 1973); direct ΣCO_2 method (Schindler and Fee, 1973); the vertical column method (Juliano, 1969) or the dome method (Copeland and Duffer, 1964; Hall *et al., unpublished observations*); (2) for streams and rivers, the predictive equation method (see Thomann, 1971); or the displaced equilibrium method (Owens, 1969—incidently, nitrogen gas appears to be preferable to sodium sulfite for deoxygenation). The diurnal curve method for diffusion corrections (H. T. Odum, 1956; H. T. Odum and Hoskin, 1958) appears to give erroneously high values because nighttime respiration is not constant, but is higher in the evening (see H. T. Odum and Wilson, 1962, for a critique). However, this method may be useful for establishing upper limits. It should be noted that diffusion per volume is not linearly related to depth as has been assumed in a number of earlier studies. Where possible, both diffusion determinations and corrections should be done on a per-volume basis. At this time, a comprehensive review of all diffusion methods and values is needed.

^{14}C method

DESCRIPTION OF TECHNIQUE. The ^{14}C technique is very useful for measuring the productivity of plankton communities in waters of low productivity such as those of the open ocean (Steemann Nielsen, 1952, 1963; Steemann Nielsen and Aabye Jensen, 1957). The essential concept of this technique is that a small amount of radioactive carbon (^{14}C) relative to the entire sample is added to a sample in the form of bicarbonate. The total amount of CO_2 in the sample is measured before incubation and thus the ratio of normal $^{12}CO_2$ to $^{14}CO_2$ can be determined. After incubation for a suitable time period, the amount of ^{14}C incorporated into the plankton by photosynthesis is measured. Knowing the $^{12}C/^{14}C$ ratio then allows the determination of the actual amount of carbon taken up by the photosynthetic organisms during the incubation period. Developments by Morris *et al.* (1971) and Schindler *et al.* (1972) have enabled measurement of the production of both particulate and dissolved carbon by modifications of the basic ^{14}C method.

One important advantage of the ^{14}C method is that there is no reliance on the evolution of oxygen for its results, a process that can be confounded notoriously by production–respiration interactions in a closed incubation chamber (Steemann Nielsen, 1963).

A major drawback in determining productivity with ^{14}C is that a long list of expensive equipment is needed for even the crudest estimates of photosynthesis. This list, which includes the use of radioactive materials, precludes most investigators from using the technique on a one-time or instructional basis. However, once the proper equipment is amassed, many experiments can be easily set up and incubated with reasonable reproducibility (see Table 3–2).

A prerequisite for ^{14}C incubations is proper-strength solutions of radioactive carbonate in airtight containers. The original charge of radioactive carbon is

Table 3–2   Equipment normally used in investigations of phytoplankton productivity using the ^{14}C technique

---

Incubation bottles 125–300 ml: Two light to one dark, dark bottle blackened with paint and vinyl electrical tape

Spreaders: Metal rod or flat stock to hang bottles apart (at least 15 cm) during incubation

Chains, cable, or ropes: To hang bottles on spreaders on floats

    (*All of the above can be replaced with incubation racks, neutral density filters, and cooling apparatus for lab or shipboard incubations or both*)

Carbon-14: Prediluted into correct-strength working solutions

Dispensing syringe: Preferably Cornwall automatic type

Field thermometer

Plastic jars: 250-ml darkened, for alkalinity, pH, and salinity

pH meter: Field or laboratory if samples can be measured quickly

100 ml and 25 ml volumetric pipets in conjunction with sufficient 0.01 N HCl for alkalinity determinations. See Appendix, page 45.

Glass vials: 4–8-g size, for holding DOC samples

Compressed N_2 or air: To bubble through sample (DOC)

25-mm-diameter Millipore filters (0.45) or similar membrane filters

Filtering apparatus for the above filters (vacuum pump, side-arm flask, tubing from pump to flask, fritted glass filter holder and cork, spring clamp and filter reservoir: This is essentially one filter set up as obtained from Millipore, it is preferable to do several filtrations at once in which case much of the apparatus may be custom made)

Scintillation vials: Glass, type to fit into automatic counters

Scintillation cocktails: Water-soluble or toluene based or both

Scintillation counter

---

obtained commercially (generally in 1–10 mCi lots) and diluted into a standard (usually 50–100 $\mu Ci/ml$), which is rediluted into proper strength working solutions as needed, usually 1–10 $\mu Ci/ml$. An alternative is to buy premade glass ampules containing the proper strength of ^{14}C desired, although this method is more costly and less flexible. The strength of the working solution depends on several variables including the expected photosynthetic activity of the plankton crop. If a study of long duration is planned, a fairly active standard should be maintained and from this dilutions can be made in accordance with the appropriate needs of the season. The strength of the radioactive carbon added to the incubation chamber can be ascertained readily by the simple formula (Strickland and Parsons, 1972):

$$\text{Strength of } ^{14}C \text{ added in microcuries} = R_s/(E \times U \times N)$$

where R_s is the counts per minute (cpm) of the sample and should be greater than 1000, E is the efficiency of the counting machine (this varies, but machine specifications should give a reasonable estimate), U is the anticipated uptake of C in milligrams of carbon per cubic meter per hour (mg $C/m^3/hr$), and N is hours incubated. Therefore it is possible to increase the accuracy of the productivity determination merely by adding a higher initial charge of radioactive carbon. However, two considerations must be kept in mind. First, keep radiation levels low enough for safe usage in the environment (use no more than 1 mCi in one spot during the same day). Second, the volume of the bicarbonate solution added to the incubation sample should not exceed 2–3% of the total sample volume (Strickland and Parsons, 1972). The solution employed in making up both the standard and the working solutions is a simple, basic salt fluid (pH $\sim$ 11.3) of 5% w/v NaCl solution made from analytic quality salt and distilled water. To this solution add 0.3 g sodium carbonate (Na_2CO_3 anhydrous) and one pellet ($\sim$ 0.2 g) of NaOH to each liter of solution (Strickland and Parsons, 1972). Considerably less carbonate and hydroxide should be used in making solutions for use in fresh waters (Vollenweider, 1969b). This solution is then used to dilute the radioactive standard into the appropriate strength working solution based on results calculated from the above formula. The working solution is then dispensed accurately (with a syringe or automatic pipet filler) into glass ampules of the appropriate size (usually 2 ml Neutraglas® with prescored breakable necks). The ampules are then quickly sealed with an oxyacetylene or oxyhydrogen torch and allowed to cool. The ampules are then placed in distilled water containing phenolphthalein and placed into an autoclave for ½ hr. Reject any ampules showing pink discoloration caused by the seepage of phenolphthalein through the seal.

After proper preparation of the glass ampules and the working solution, the ^{14}C is injected into the incubation chamber in a precise and known amount.[1] The essence of the procedure dictates that the same amount of radioactive carbon reach the plankton every time a new experiment is started (Steemann Nielsen, 1952). The introduction of a precise amount of ^{14}C into the incubation chamber is normally achieved in one of two ways: (1) an accurate volume syringe (preferably automatic Cornwall type), or (2) the entire contents of a small prefilled ampule emptied into the sample. The drawback with the latter technique is that the ampule must be filled accurately and sealed without any loss of fluid. However, in either case the advent of water-soluble scintillation cocktails permits a check (Pugh, 1973) on the amount of radioactivity reaching the sample.[2] Drawing off a small amount of incubation fluid before incubation begins and counting the activity of this fluid makes it feasible to determine how much ^{14}C reaches the sample. In any method for measuring productivity with bottle incubations, care must be taken not to "light shock" the plankton crop by exposing it to direct solar radiation (Mahler and Cordes, 1966; Wallen and Green, 1971; Morris *et al.*, 1971). This "light shock" will cause a disruption in the normal photosynthetic process and result in a greatly reduced estimate of

[1] An alternative method to making heat-sealed glass ampules is that of using small screwtop vials to hold the working solution. After filling the screwtop vials are treated identically to glass ampules, that is, autoclave for leaks.

[2] A liquid scintillation cocktail is a mixture of aqueous chemical reagents, that, when mixed with small amounts of radioactive material create a fluorescent reaction. The scintillation cocktail and one sample are mixed together in a small glass vial and the radioactivity is measured in a scintillation counter. A water-miscible cocktail is a scintillation cocktail in which water is soluble allowing the determination of radioactivity in solutions.

production (Wallen and Green, 1971). For this and other reasons, it is normally suggested that incubations be carried out in replicate if at all possible.

While a water sample is taken for incubation, 200 ml of additional sample must be taken for the determination of the CO_2 present in the sample. The total weight of CO_2 in the sample is determined in some cases by an alkalinity titration (see Table 3–2). However, a better determination of total CO_2 would be by means of a direct method, such as gas chromatography (Stainton, 1973). Additionally, a determination of the pH of the water must be made *in situ*, as well as a temperature reading, if possible, and an approximate determination of the salinity, if applicable.

The samples then are incubated for a measured amount of time and the photosynthetic process is arrested by an injection of 1 ml of neutral Formalin, or by placing the samples in the dark. Holding the samples in the dark has the drawback that some photosynthesis could occur later, while the samples are being filtered in the light. However, problems associated with the injection of Formalin include (1) absorption of Formalin into the glass walls of the incubation jar, which may interfere with subsequent incubations, and (2) killing the phytoplankton can break cells apart and liberate large amounts of fixed dissolved organics that should normally be counted as particulate production. In all cases of bottle incubations (both light–dark oxygen and ^{14}C), Pyrex-type glass jars must be used, as normal glass will slough off silicon that may stimulate high levels of production.

Length of incubation period is apparently critical to the final results in any study employing chambers (Steemann Nielsen, 1963; Hobbie *et al.*, 1972). Previous studies often have used classic incubation periods of 6, 12, and 24 hr. However, given a sufficient substrate or a reduced oxygen level, bacteria in the sample will grow until they represent a larger biomass in the incubation chamber than in the natural environment. This enlarged bacterial crop will have a greater respiratory demand inside the chamber than in the same volume of free water. In most cases a short incubation period (2–4 hr) will not allow the bacteria to achieve a significant gain in biomass (Hobbie *et al.*, 1972). This is an advantage of the ^{14}C technique because the light–dark oxygen method may require too lengthy an incubation period (Strickland and Parsons, 1972). Particular care must be given to this bacteria–phytoplankton interaction in areas of normally high standing crops of bacteria in marshes, salt marshes, small ponds, and stagnant waters.

Additional care must be taken in waters low in dissolved CO_2 in which the plankton may become CO_2-limited (Schindler and Fee, 1973). Again this problem can be minimized by a short incubation period so that the phytoplankton do not use up the CO_2 available in the incubation chamber. In all cases of ^{14}C productivity measurements, an initial measure of both the pH and alkalinity must be made to determine the total available CO_2 before incubation (Strickland and Parsons, 1972). In areas of suspected CO_2 limitation, a second determination of CO_2 remaining in the incubation chamber will indicate whether potential CO_2 limitation occurred. In these cases of low levels of initial CO_2, special techniques must be employed such that the bicarbonate carrier with the radioactive carbon will not stimulate extra photosynthesis by the introduction of additional carbon dioxide (Nygaard, 1968; Schindler and Fee, 1973).

After arresting photosynthesis the sample then can be treated in one of two ways. In most uses of the technique, the sample is split into two fractions by filtering out the particulate production on membrane filters (usually 0.45-μm HA Millipore type) and counting ^{14}C disintegrations from the filter. In some cases the filters are known to retain a sizable percentage of the dissolved organics, thus the filter gives an estimate of production higher than particulate production alone (Nalewajko and Lean, 1972). It is also the case, however, that this technique gives an underestimation of productivity because some of this has been lost from cells into the water as dissolved organics (Schindler *et al.*, 1972; Saunders, 1972; Stephens and North, 1971; Wetzel and Otsuki, 1973). In a current modification of the technique the unused carbon dioxide in the sample is removed by acidifying the sample ($\sim$ pH 3–4) with a small drop of dilute phosphoric or hydrochloric acid and by bubbling N_2 or air through the sample for about 5 min (Schindler *et al.*, 1972). The sample is then filtered; ^{14}C is counted for both the filtered (and primarily particulate) material and the filtrate water (containing dissolved organics tagged with ^{14}C). The sum of these expresses total productivity, and the filtrate count indicates rate of loss of organics from cells into the water (see Stull *et al.*, 1972 for applications to individual species). For practical reasons, the older method of filtering the crop and counting the filter only may have to be used in areas of low production, as there may be insufficient counts in the dissolved fraction.

In determining the radioactivity of the particulate fraction some effort should be made to remove the small amount of unused ^{14}C (as carbonate) dried on the filter pad. This is done readily by "fuming" the filters over HCl, that is, placing the filters in a desiccator for 2 hr with an open beaker of HCl. The HCl fumes in the desiccator convert all carbonates to $^{14}CO_2$, which diffuses off the filter pads. The unused ^{14}C in the liquid samples is removed by the method of "stripping" (acidifying the sample and bubbling N_2 gas through it) mentioned before. This process of stripping off the unused CO_2 must be applied to any liquid ^{14}C sample.

The samples are now ready for determination of their specific radioactivity (see Jitts, 1961; Ward and Nakanishi, 1971; Schindler and Holmgren, 1971). The particulate samples on the filter pads are placed in any suitable scintillation cocktail. These cocktails are either made from reagents or purchased in concentrated form and diluted with toluene. Currently, toluene-based scintillation cocktails give the best results for dry material, dissolving the filter completely and suspending the plankton in the cocktail (Pugh, 1973). Liquid samples require a different type of scintillation medium which allows the radioactive water to become completely mixed in the scintillation cocktail. These water-miscible cocktails, such as Aquasol®, also are purchased either premade or in concentrated form. The purpose of the scintillation cocktail is to suspend the radioactive material in a fluid so that when the scintillation counter passes a special beam of light through the sample, the sample fluoresces. The amount of fluorescence is, in turn, dependent on the amount of radioactivity in the sample. The sample, to which 10.0 ml of cocktail is added, is placed in a small scintillation vial designed for use in scintillation counters. In all cases, glass scintillation vials should be used instead of the plastic counterparts, as it has been found that after 48 hr the plastic vials will readily absorb toluene from the scintillation cocktail including the ^{14}C in it.

Scintillation counters, like any other spectrophotometric device, have a small inherent variation with each use. For this reason, a blank (container of scintillation cocktail without radioactivity) and a standard of predetermined radioactivity should be counted along with the samples each day a new batch of samples is counted. The standard of known activity usually is obtained commercially along with purchase of the scintillation counter.

Two other aspects must be considered before the final results can be calculated. Because every batch of ^{14}C tracer varies, and dilutions are not exactly the same, each time the working solutions are made the exact activity of ^{14}C reaching the sample must be determined. This is accomplished by adding the ^{14}C to incubation chambers in the normal manner immediately followed by taking a small sample from the incubation chamber and counting the activity in the incubation fluid without stripping off the carbon dioxide. This substandard is counted along with the known standard and the true activity reaching the incubation chamber is determined. This procedure should be carried out on four to five randomly chosen ampules each time a new working solution is made up. The total amount of activity reaching a sample is determined and the counted value of the known standard is measured on the day the activity is determined. Thus each day a new batch of samples is counted, the known standard is counted and any change in the value of the known standard is used to adjust the value of the activity that reaches the sample:

$$\text{Activity reaching the sample} = R_s/R_c \times S$$

where S is the value of activity reaching the sample (determined once earlier), R_s is the value of known standard on day that activity of sample was determined, R_c is the value of known standard during present counting of samples. In essence, this is merely an adjustment in the value of the activity reaching the sample owing to daily variation in machine-counting efficiency.

The second correction, given to the counts from ^{14}C, is due to the ^{14}C being a "soft" emitter in that the radiation given off is in the form of low-energy β-particles (Patterson and Greene, 1965). These β-particles, because they are low in energy, are absorbed, in part, by the scintillation cocktail in which they are suspended. This effect, called "quenching," results in the underestimation of the amount of radioactivity in any scintillation vial. However, this problem can be corrected by counting one or more standards of known radioactive content and then adjusting the observed value from the scintillation counter to match the true known value of the standard. This correction will vary slightly depending on the exact mixture of the scintillation cocktail, and in practice a complete quench series must be established for every scintillation counter (Patterson and Greene, 1965). The amount of quenching in any given sample may be determined by a channels ratio, the ratio of high-energy emitters to low-energy emitters, and the output from the scintillation corrected accordingly. The entire problem of quenching is of considerable complexity involving the physics of radioisotopes. Other references (e.g., Patterson and Greene, 1965), may be consulted for details of quench correction.

Given the corrected activity of a sample, the calculation of productivity is achieved by the following formula (Strickland and Parsons, 1972; see Table 3–3 for a complete set of calculations):

$$\text{mg C/m}^3/\text{hr} = [(R_a - R_b) \times W \times 1.05]/(R_s \times H)$$

where R_a is the corrected activity of sample (corrected for quench), R_b is the blank value obtained from counting a vial with no radioactivity, W is the original total weight of carbon dioxide in the sample calculated from pH readings, 1.05 is a correction factor for the difference in the use of ^{14}C to that of normal ^{12}C, R_s is the amount of known activity added to the sample, and H is the incubation time of the sample in hours. Current considerations of ^{14}C-measured productivity do not subtract dark-bottle activity from light-bottle activity, but rather consider the two processes of light-and-dark uptake as somewhat interdependent (Morris *et al.*, 1971). The values obtained from light-bottle fixation are usually considered as net production (pathway N, Fig. 3–1). The reasoning behind this latter consideration is best explained in the words of Ruttner (1960):

> The ^{14}C method does not measure the oxygen given off in assimilation or used up in respiration, but rather the *carbon* incorporated in the photosynthate. Hence, it is important to know whether the respiration of the plant cell is based on the carbon presently being assimilated, part of which is labeled with ^{14}C, or whether older carbohydrate deposits are being drawn upon for this purpose. In the former instance the quantities of assimilated labeled C will be reduced, whereas in the latter instance they will remain unchanged. In other words, ^{14}C determination would give net production in the first instance (related to plant cells) and gross production in the second.

Although the ^{14}C method has a number of inherent disadvantages such as cost and possible bottle effects, it is very sensitive and flexible; and it measures, in theory, the net productivity of phytoplankton (for a critque see Sheldon *et al.*, 1973) available to higher trophic levels. Recently there have been many minor corrections in the method, and anyone interested in using the method would be well advised to check the most recent literature. A summary of our presentation is given in Figure 3–6, and some results in Figure 3–7; Figure 3–7 also emphasizes the importance of giving results as both volume and area values.

Chlorophyll and biomass methods

The determination of chlorophyll is a simple analytic procedure explained in most standard texts (Strickland and Parsons, 1972). In practice, a sample of water ranging from 0.05 to 5.0 liters is filtered through a membrane (Millipore-type AA) or glass fiber filter (Whatman GF/C) as soon as possible after collection. Filters are then frozen for no longer than 2 weeks, at which time the chlorophyll is extracted. The extraction of chlorophyll from membrane filters

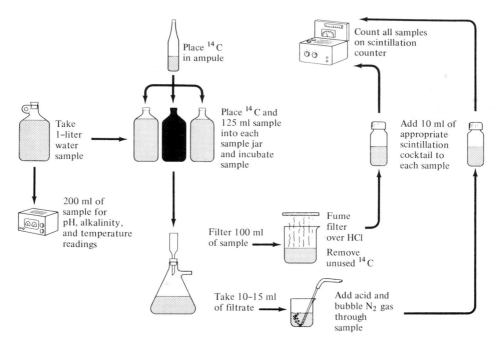

FIGURE 3–6. Flow chart for measurement of production with ^{14}C technique.

FIGURE 3–7. Radiation curve and primary productivity according to ^{14}C method in Lake Erken based on short-term exposures (I–V) and long-term exposure (VI); Σ is sum of I–V (from Vollenweider, 1969a).

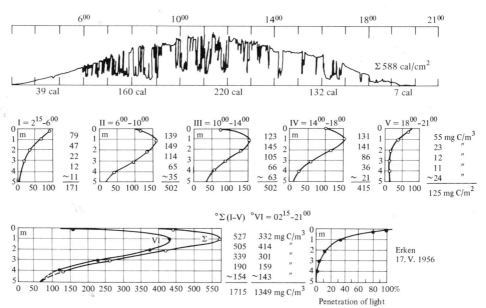

is simple. The filter is placed in 10.0 ml of 90% acetone where both the filter and the chlorophyll dissolve.

Membrane filters are relatively easy to use but they have two drawbacks: (1) they are very expensive, and (2) they impart a small amount of turbidity to the sample, which interferes with spectrophotometric measurements (Strickland and Parsons, 1972). Glass fiber filters are cheaper and produce no significant turbidity when placed in 90% acetone. However, fiber filters, as well as membrane filters when the algae crop is composed mostly of green algae, must be ground up in a tissue grinder with about 5 ml of 90% acetone to extract the chlorophyll. Once the fiber filters are ground up, they are placed into a glass centrifuge tube along with the acetone used in grinding, and the acetone level in the tube is made up to 12.0 ml. At this point, samples collected on either membrane or glass fiber filters are treated identically. Samples are centrifuged from 15 to 30 min and then placed in a dark refrigerator for 20 hr. The chlorophyll is allowed to rewarm to room temperature, decanted into a 6-ml, 10.0-cm spectrophotometer cell, and the extinctions are read at 4800, 6300, 6450, and 7500 Å on a spectrophotometer. The 7500 reading serves as a "blank" or a wavelength for which chlorophyll should not cause any extinction, but only turbidity, which is subtracted from all other readings. In all cases of cholorophyll extraction, work must be carried out in either subdued or green light as normal white light will rapidly change chlorophyll and will give erroneous results. Chlorophyll concentrations are determined from extinction readings by formula (consult Strickland and Parsons, 1972 for complete details; see also Lorenzen, 1966). It should be mentioned that a very popular alternative to the spectrophotometer for measuring chlorophyll is the fluorometer. Fluorometers are simple to use, are reasonably accurate, and require a much smaller sample than do spectrophotometers. However, a fluorometer requires constant standardization against a spectrophotometer to correct for drift (Strickland and Parsons, 1972). An additional correction for phaeophytin (dead chlorophyll) can be made to either spectrophotomically or fluorometrically measured chlorophyll values by acidifying and rereading the sample (Yentsch, 1967).

The advent of simple and accurate measurements of standing crop of chlorophyll led researchers to speculate that photosynthesis may be predicted from chlorophyll data. This idea was developed initially by Ryther and Yentsch (1957) who predicted photosynthesis from chlorophyll and light data. The success of the prediction of net photosynthesis is based on the assumption that chlorophyll is a uniformly photosynthetic pigment. An initial calibration of chlorophyll readings against empirically derived photsynthetic measurements must be made for each set of conditions. However, the inherent advantage of the technique is obvious. It is considerably simpler to evaluate chlorophyll and light data to predict photosynthesis than to measure photosynthesis directly. Efforts have been made to improve the accuracy and usefulness of the initial predictions made by Ryther and Yentsch (H. T. Odum *et al.*, 1958; Wright, 1959; Aruga and Monsi, 1963; Small, 1963; Aruga, 1966; Williams and Murdoch, 1966; Dally *et al.*, 1973). Other developments have included refinements in estimates of standing-crop production by measuring changes in numbers and size of particles (McAlice, 1971; Strickland and Parsons, 1972; Sheldon *et al.*, 1973), or grazing rates (Haney, 1971).

However, a comparison of photosynthetic estimates from chlorophyll and associated data with empirically derived estimates of photosynthesis (^{14}C, O_2, free water, etc.) indicates that the chlorophyll technique leaves much to be desired. The results obtained with the chlorophyll technique can be no better than the equations used to predict photosynthesis, and to date most researchers do not agree on these equations. Furthermore, the chlorophyll must accomplish a uniform amount of photosynthesis per unit weight and light, or the estimation of production is impossible. There is good reason to believe that chlorophyll can become more or less photosynthetically active depending on the age of the plankton crop, water temperature, and season of the year (Thomas, 1961; Talling and Driver, 1961; Margalef, 1968). In general, the use of chlorophyll data to estimate aquatic primary productivity is not recommended.

Periphyton methods

Periphyton is a term that describes the small plant communities that coat the surfaces of mud, rocks, logs, and macrophytes of aquatic environments. An indication of net production of standing crop (pathway M, Fig. 3–1) can be obtained from glass slides or other artificial (or precleaned natural) substrate inserted in the water and periodically measured for the accumulation of periphyton biomass (Margalef, 1949; Blum, 1957). Wetzel (1964a) and Allen (1971) describe a method for periphyton production estimates using a cylinder placed over the substrate into which a sample of ^{14}C is introduced. After a suitable incubation period, the substrate was removed and stored, to be counted later. Another chamber technique that uses benthic trays colonized over an extended period, metabolic gases, and an enclosed flowing system has been described by Hansmann *et al.* (1971). These latter methods are preferable to the artificial substrate methods because they use the natural system intact. Benthic biomass may be estimated by scraping the bottom within a special cylinder (Ertl, 1971) or by radioactive apportionment (Pomeroy, 1961; Nelson *et al.*, 1969). For a review of periphyton productivity methods, see Wetzel (1964a). Recent developments of chamber techniques using changes in CO_2 or ^{14}C in the water of the chamber, as well as critiques of earlier methods, are given in Schindler *et al.* (1973a).

Macrophyte methods

Where there are large growths of macrophytes that die back each winter, sequential harvesting will give a minimum estimate of net primary production (see, e.g., Kaul and Vass, 1972). Corrections must be made, as with all aquatic plants, for the amount of productivity that is eaten by herbivores or lost by dissolution (exudation) of organics. This is difficult to do but may be estimated by enclosing the macrophytes in some sort of chamber and determining the rate of organic buildup in the water surrounding the plants. Belowground production also must be measured. This may be done by planting the macrophytes in some sort of box with no roots in it to start with, then harvesting the box at the end of the growing season or by isolating root sections with aluminum tubes to measure decomposition rates, hence production from biomass changes, or by other means (see Gorham and Pearsall, 1956). None is considered to be particularly satisfactory.

In addition, there may be loss of leaves from the plant during the growing season. A very simple method for determining this for *Spartina*, for example, and probably for many other macrophytes, is to drop a small plastic ring over each new leaf as it appears. As the lower leaves drop off, the plastic rings will fall to the bottom of the stem. The sum of these plastic rings at the end of the year is the number of leaves that have fallen off. Mann (1972) has determined productivity of large seaweeds by punching holes along the length of the blades. The holes serve as markers that move away from the base as the blade elongates; the holes thus permit measurement of rate of elongation (and, hence of dry matter growth) of the blade.

Given growth rates for individual macrophytes, one can extrapolate to a square-meter basis for net primary productivity. This was done also for peat communities by Clymo (1970). Wetzel (1964b) has studied the rate of ^{14}C productivity of macrophytes by enclosing individual plants in Plexiglas containers, injecting the ^{14}C, and then proceeding as with ^{14}C estimates for plankton. In another study, Mathews and Westlake (1969) made an interesting application of the "Allen curve" method of fish production to aquatic macrophytes, in which the number of plants in a population is plotted against the mean weight. Production corrected for grazing or other sources of mortality then can be determined by integrating the proper sections of this curve.

Estimation of Primary Production from Models

It has been found that the aquatic primary production tends to be a function of such variables as depth, sunlight intensity, temperature, species present, chlorophyll concentration, and the degree of opaqueness of the water. With this information quantified from a series of field correlations or laboratory determinations, primary production then can be estimated for a given body of water from some or many of these parameters, which often are easier to obtain than measurements of primary production itself.

Consequently there has been a development of a number of models (i.e., simplifications that aid in understanding and prediction) for primary production in water. In general, models may be characterized by different degrees of precision, generality, and realism (Levins, 1966; see application for phytoplankton modeling in O'Connor and Patten, 1968). For example, models that are rather general tend to sacrifice precision, and models that are very precise for one lake may not be generally applicable to other lakes.

It is not the purpose of this chapter to detail the procedures by which it might be possible to model primary production in waters. Interested readers are referred to the classic papers of Steemann Nielsen (1952), Ryther (1956), Ryther and Yentsch (1957), Talling (1957), Rodhe *et al.* (1958), and Vollenweider (1969), Patten (1968), as well as more recent papers by DiToro *et al.*

(1971), Fee (1973a, b), Bannister (1974), Kelly and Spofford (*in press*), and Nixon and Kremer (*in press*). Fee's papers are recommended for a fresh approach to the problem. Some models of marine production are discussed in Chapter 8.

The data for models are generated usually in one of two ways: by isolation, that is, by doing carefully controlled laboratory experiments in which only one variable at a time is changed (e.g., Fee, 1973a); or by correlation, that is, by comparing a dependent variable such as photosynthesis, with a number of independent variables under a variety of natural conditions, and then attempting to determine the effects of each of the independent variables by statistical regression or some other means (e.g., Brylinsky and Mann, 1973). The former method often has problems associated with the potentially unnatural behavior of any natural system once brought into the laboratory. For example, Fee's (1973a) studies, although elegant, are potentially subject to the same errors of CO_2 limitation elucidated in the same journal issue by Schindler *et al.* (1973b). Statistical regressions can deal with natural systems, but they do not demonstrate cause and effect, only co-occurrence. Ideally, an effective model would be based on coefficients determined from a combination of both methods.

Many of these models are based on the concept of assessing what the rate of photosynthesis would be under the most favorable conditions, and then determining what percentage of the most favorable conditions exists, in turn, for each potentially limiting parameter (i.e., for light, temperature, and limiting nutrients). The integration of these factors would predict photosynthesis. This approach has been criticized by Fee (1973a) because changing physiologic characteristics of the algae over time alter the optimum conditions, and significant variations in the conditions arise during the incubation times that form the basis for the model coefficients. Fee (1973a) offers an alternate scheme that eliminates some, but not all problems. In a second paper, Fee (1973b) extends the concept to three dimensions. These approaches appear to be very promising, but also indicate the many problems yet to be understood.

A remarkable new approach is that of Lehman *et al.* (*in press*). They have constructed a computer library of the response of individual algae species to sunlight intensity, nutrients, competition parameters, and so forth. When all the relevant environmental parameters are fed into the model it predicts species composition and productivity. As the characteristics of more and more species are added to the data bank, this approach could lead to a very robust model, capable of predicting photosynthesis under a wide range of conditions.

Concluding Remarks

We have reviewed the various methods that would be helpful in measuring primary production in water. The techniques we have discussed all have limitations and sources of uncertainty, and although we would like to present a clear choice of a single, simple, and reliable method for use in general, we are unable to offer one. It seems to us that for some ecosystems in which diffusion is not a problem or can be measured adequately, the free-water methods have advantages that have not been given sufficient recognition. Diffusion problems and research practicality may incline choice toward other methods, notably (1) the ^{14}C technique for low-productivity plankton, (2) the oxygen light-and-dark-bottle technique for work with high-productivity plankton at less expense and effort than with the ^{14}C technique, and (3) growth measurements or other techniques, as appropriate to circumstance, for algae and attached aquatic plants. Beyond the observation that each of these methods has wide utility and problems, it is difficult to generalize about choice of techniques. Our suggestions are summarized in Figure 3–8.

Other problems go beyond the techniques of measuring primary productivity, however, and we would like to touch upon them in conclusion. The mere measurement of primary productivty does not, in itself, constitute worthwhile science, unless it is done in the context of a class exercise. We should ask such questions as: Why are we interested in the measurement of primary productivity? What relationship does it have with other parameters of the ecosystem in question? How is it related to environmental variables? What are the patterns among different ecosystems? What are desirable levels of primary productivity in waters that we might be able to manage? What is the relationship between global primary production and man's welfare? Other chapters in this book touch upon some of these problems, and the readers are encouraged to view the measurement of primary productivity not as an end in itself, but as a means to the multiple ends of better understanding of ecosystems and more intelligent management of the biosphere.

Appendix (to Table 3–2): Alkalinity Calculations

1. Calculate total alkalinity:

$$\text{Total alkalinity} = 2.500 - (1250\text{-}a_H/f)$$

   where $a_H = 10 - \text{pH}$ (correct pH for *in situ* to measurement temperature differences; pH refers to the pH of the water sample) and f is the factor ranging from 0.890 to 0.753 depending on the salinity and acid–sample mixture pH.

2. Calculate carbonate alkalinity:

$$\text{Carbonate alkalinity} = \text{total alkalinity} - A$$

   where A is a conversion factor 0.0 to 0.29 depending on the salinity, temperature, and pH of the initial water sample.

3. Calculate the total weight of carbon dioxide in the sample:

$$\text{Total CO}_2 = \text{carbonate alkalinity} \times F_t$$

   where F_t is a conversion factor of 1.07 to 0.77 interpolated from tables depending on the salinity, temperature, and pH of the sample.

For the appropriate tables needed in all three steps to find the total weight of carbon dioxide in a sample, see any standard text on productivity analysis (e.g., Vollenweider, 1969a; Strickland and Parsons, 1972). Salinity determinations need only be approximate and generally can be assumed in a certain range if yearly variations are small. The standard techniques used in determination

Figure 3–8   Best methods for measuring aquatic primary production. Special conditions may require special techniques.

Community type	Flowing water	Shallow productive still water	Shallow unproductive still water	Deeper waters
Whole autotrophic communities	Free-water O_2, CO_2, pH (2-station or 1-station)	1 station Free-water O_2, CO_2, pH	Free water pH or CO_2	Free water pH or CO_2
Plankton communities	Drifting plastic bags or bottles w/O_2, CO_2, pH, or ^{14}C	Light–dark bottle O_2, CO_2, or pH	Bottles w/pH, CO_2, or ^{14}C	Bottles w/^{14}C
Benthic communities	Special chambers, slides	Chambers w/O_2, pH, CO_2, or ^{14}C	Chambers w/pH, O_2 or ^{14}C	Not significant

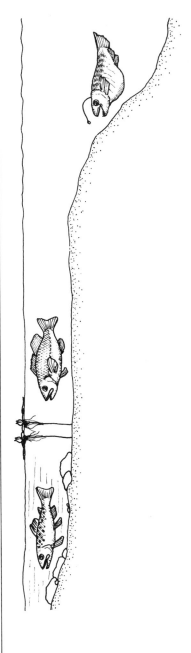

of total alkalinity vary with the type of water in which the sampling is carried out (i.e., fresh, brackish, open-sea water, etc.). It should be noted that measurement of total carbon dioxide in soft, humic, acidic, or polluted water is very difficult. It is recommended that a direct technique such as gas chromatography or infrared analysis be used when sampling in such water (Stainton, 1973).

References

Allen, H. L. 1971. Primary productivity, chemo-organotrophy, and nutritional interactions of epiphytic algae and bacteria on macrophytes in the littoral of a lake. *Ecol. Monogr.* 41:97–127.

American Public Health Association (APHA). 1971. Standard methods for the examination of water and waste water, 874 pp. Washington, D.C.

Aruga, Y. 1966. Ecological studies of photosynthesis and matter production of phytoplankton. III. Relationship between chlorophyll amount in water and primary productivity. *Bot. Mag. Tokyo* 79:20–27.

————, and M. Monsi. 1963. Chlorophyll amount as an indicator of matter productivity in bio-communities. *Plant Cell Physiol.* 4:29–39.

Bannister, T. T. 1974. Production equations in terms of chlorophyll concentration, quantum yield and upper limit to production. *Limnol. Oceanogr.* 19:1–12.

Bender, M. E., and R. A. Jordan. 1970. Plastic enclosure versus open lake productivity measurements. *Trans. Amer. Fish. Soc.* 99:607–610.

Beyers, R. J., J. Larimer, H. T. Odum, R. B. Parker, and N. E. Armstrong. 1963. Instructions for the determination of changes in carbon dioxide concentrations from changes in pH. *Publ. Inst. Marine Sci. Univ. Tex.* 9:454–489.

Blum, J. L. 1957. An ecological study of the algae of the Saline River, Michigan. *Hydrobiologia* 9:361–408.

Brody, S. 1945. *Bioenergetics and Growth.* 1023 pp. New York: Reinhold.

Brylinsky, M., and K. H. Mann. 1973. An analysis of factors governing productivity in lakes and reservoirs. *Limnol. Oceanogr.* 18:1–14.

Carpenter, J. H. 1965a. The accuracy of the Winkler method for dissolved oxygen analysis. *Limnol. Oceanogr.* 10:135–140.

————. 1965b. The Chesapeake Bay Institute technique for the Winkler dissolved oxygen method. *Limnol. Oceanogr.* 10:141–143.

————. 1966. New measurements of oxygen solubility in pure and natural water. *Limnol. Oceanogr.* 11:264–277.

Cassie, R. M. 1961. Statistical and sampling problems in primary production. In *Proc. Conf. Primary Productivity Measurement, Marine and Freshwater, Hawaii, 1961*, M. S. Doty, ed., pp. 163–171. Washington, D.C.: U.S. Atomic Energy Commission, Div. Technical Information.

Churchill, M. A., R. A. Buckingham, and H. L. Elmore. 1962. The prediction of stream reaeration rates, 98 pp. Chattanooga, Tennessee: Tennessee Valley Authority, Div. of Health and Safety, Environmental Hygiene Branch.

Clymo, R. S. 1970. The growth of *Sphagnum:* Methods of measurement. *J. Ecol.* 58:13–49.

Copeland, B. J., and W. R. Duffer. 1964. The use of a clear plastic dome to measure diffusion of natural waters. *Limnol. Oceanogr.* 9:494–495.

Cummins, K. W., and J. C. Wuycheck. 1971. Caloric equivalents for investigations in ecological energetics. *Mitt. Int. Ver. Limnol.* 18:1–158.

48

Part 2: Methods of Productivity Measurement

Czaplewski, R. L., and M. Parker. 1973. Use of a BOD oxygen probe for estimating primary productivity. *Limnol. Oceanogr.* 18:152–154.

Dally, R. J., C. B. J. Gray, and S. R. Brown. 1973. A quantitative, semiroutine method for determining algal and sedimentary chlorophyll derivatives. *J. Fish. Res. Bd. Can.* 30:345–356.

Day, J. W., W. G. Smith, P. R. Wagner, and W. C. Stowe. 1973. Community structure and carbon budget of a salt marsh and shallow bay estuarine system in Louisiana, 79 pp. Baton Rouge, Louisiana: Louisiana State Univ.

Di Toro, D. M., D. J. O'Connor, and R. V. Thomann. 1971. A dynamic model of phytoplankton populations in the Sacramento–San Joaquin delta. *Advan. Chem. Ser.* 106:131–180.

Doty, M. S. (ed.) *Proc. Conf. Primary Productivity Measurement, Marine and Freshwater, Hawaii, 1961.* Washington, D.C.: U.S. Atomic Energy Commission, Div. Technical Information.

Efford, I. E. 1968. Winkler titration for oxygen. (mimeogr.) Vancouver, British Columbia: Institute of Animal Ecology, Univ. of British Columbia.

Emerson, S., W. S. Broecker, and D. W. Schindler. 1973. Gas-exchange rates in a small lake as determined by the radon method. *J. Fish. Res. Bd. Can* 30:1475–1484.

Ertl, M. 1971. A quantitative method of sampling periphyton from rough substrates. *Limnol. Oceanogr.* 16:576–577.

Fee, E. J. 1971. Digital computer programs for estimating primary production, integrated over depth and time, in water bodies. Special Rep. 14, Center for Great Lakes Studies, 43 pp. Milwaukee, Wisconsin: Univ. of Wisconsin.

———. 1973a. A numerical model for determining integral primary production and its application to Lake Michigan. *J. Fish. Res. Bd. Can.* 30:1447–1468.

———. 1973b. Modelling primary production in water bodies: A numerical approach that allows vertical inhomogeneities. *J. Fish. Res. Bd. Can.* 30:1469–1473.

Gaarder, T., and H. Gran. 1927. Investigations of the production of plankton in the Oslo Fjord. *Rapp. Cons. Explor. Mer* 42:1–48.

Goldman, C. R. 1967. Integration of field and laboratory experiments in productivity studies. *Estuaries*, G. H. Lauff, ed., Special AAAS Publ., 83:346–352.

———. 1968. Aquatic primary production. *Am. Zool.* 8:31–42.

———. (ed.) 1969. *Primary Productivity in Aquatic Environments*, 464 pp. Berkeley, California: Univ. of California Press.

Gorham, E., and W. H. Pearsall. 1956. Production ecology. III. Shoot production in *Phragmites* in relation to habitat. *Oikos* 7:206–214.

Hall, C. A. S. 1972. Migration and metabolism in a temperate stream ecosystem. *Ecology* 53:585–604.

Haney, J. F. 1971. An *in situ* method for the measurement of zooplankton grazing rates. *Limnol. Oceanogr.* 16:970–977.

Hansmann, E. W., C. B. Lane and J. D. Hall. 1971. A direct method of measuring benthic primary production in streams. *Limnol. Oceanogr.* 16:822–826.

Hobbie, J. E., O. Holm-Hansen, T. T. Packard, L. R. Pomeroy, R. W. Sheldon, J. P. Thomas, and W. J. Wiebe. 1972. A study of the distribution and activity of microorganisms in ocean water. *Limnol. Oceanogr.* 17:544–555.

Jannasch, H. W., and P. H. Pritchard. 1972. The role of inert particulate matter in the activity of aquatic micro-organisms. *Mem. Inst. Ital. Idrobiol.* 29 (Suppl.): 289–308.

Jitts, H. R. 1961. The standardization and comparison of measurements of primary production by the carbon-14 technique. In *Proc. Conf. Primary Productivity Measurement, Marine and Freshwater, Hawaii, 1961*, M. S. Doty, ed. pp. 114–120, Washington, D.C.: U.S. Atomic Energy Commission, Div. of Technical Information.

Juliano, D. W. 1969. Reaeration measurements in an estuary. *J. Sanit. Eng. Div., ASCE* 95 (SA6; Proc. Paper 6987): 1165–1178.

Kaul, V., and K. K. Vass. 1972. Production studies of some macrophytes of Srinagar lakes. In *Productivity Problems of Freshwaters: Proc. IBP-UNESCO Symp.*, Z. Kajak and A. Hillbricht-Ilkowska, eds. pp. 725–731. Warsaw and Krakow: Polish Scientific Publ.

Kelly, M. G., G. M. Hornberger, and B. J. Cosby. 1974. Continuous automated measurement of rates of photosynthesis and respiration in an undisturbed river community. *Limnol Oceanogr.* 19:305–312.

Kelly, R. S., and W. Spofford. 1975. Application of an ecosystem model to water quality management: The Delaware estuary. In *Models as Ecological Tools: Theory and Case History*, C. Hall and J. Day, eds. New York: Wiley (Interscience). *(In press.)*

Lehman, J. T., D. B. Botkin, and G. E. Likens. 1975. The assumptions and rationales of a computer model of phytoplankton population dynamics. *Limnol. Oceanogr.* *(In press.)*

Levins, R. 1966. Strategy of model building in population biology. *Am. Sci.* 54:420–431.

Lorenzen, C. S. 1966. A method for the continuous measurement of *in vivo* chlorophyll concentrations. *Deep Sea Res.* 13:223–227.

Mahler, H. R., and E. H. Cordes. 1966. *Biological Chemistry*. New York: Harper and Row.

Mann, K. H. 1972. Ecological energetics of the seaweed zone in a marine bay on the Atlantic Coast of Canada. II. Productivity of the seaweeds. *Marine Biol.* 14:199–209.

———, R. H. Britton, A. Kowalczewski, T. J. Lack, C. P. Mathews, and I. McDonald. 1972. Productivity and energy flow at all trophic levels in the River Thames, England. In *Productivity Problems of Freshwaters: Proc. IBP-UNESCO Symp.*, Z. Kajak and A. Hillbricht-Ilkowska, eds., pp. 579–596. Warsaw–Krakow: Polish Scientific Publ.

Manny, B. A., and C. A. S. Hall. 1969. Diurnal changes in stratification and dissolved oxygen in the surface waters of Lake Michigan. *Conf. Great Lakes Res. Proc. Int. Ass. Great Lakes Res.* 12:622–634.

Margalef, D. R. 1949. A new limnological method for the investigation of thin-layered epilithic communities. *Hydrobiologia* 1:215–216.

———. 1968. *Perspectives in Ecological Theory*, 111 pp. Chicago, Illinois: Univ. of Chicago Press.

Mathews, C. P., and D. F. Westlake. 1969. Estimation of production by populations of higher plants subject to high mortality. *Oikos* 20:156–160.

McAlice, B. J. 1971. Phytoplankton sampling with the Sedgwick-Rafter cell. *Limnol. Oceanogr.* 16:19–28.

McAllister, C. D., T. R. Parsons, K. Stephens, and J. D. H. Strickland. 1961. Measurements of primary production in coastal sea water using a large-volume plastic sphere. *Limnol. Oceanogr.* 6:237–258.

Morris, I., C. M. Yentsch, and C. S. Yentsch. 1971. Relationship between light carbon dioxide fixation and dark carbon dioxide fixation by marine algae. *Limnol. Oceanogr.* 16:854–858.

Nalewajko, C., and D. R. S. Lean. 1972. Retention of dissolved compounds by membrane filters as an error in the ^{14}C method of primary production measurement. *J. Phycol.* 8:37–43.

Nelson, D. J., N. R. Kevern, J. L. Wilhm, and N. A. Griffith. 1969. Estimates of periphyton mass and stream bottom area using phosphorus-32. *Water Res.* 3:367–373.

Nixon, S., and J. Kremer. 1975. Narragansett Bay—The development of a composite simulation model for a New England estuary. In *Models as Ecological Tools: Theory and Case Histories*, C. Hall and J. Day, eds. New York: Wiley (Interscience). (*In press.*)

Nygaard, G. 1968. On the significance of the carrier carbon dioxide in determinations of the primary production in soft-water lakes by the radiocarbon technique. *Mitt. Int. Ver. Limnol.* 14:111–121.

O'Connor, J. S., and B. C. Patten. 1968. Mathematical models of plankton productivity. *Proc. Reservoir Fishery Resources Symp., April 5–7, 1967*, pp. 207–228. Athens, Georgia: Univ. Georgia.

Odum, E. P. 1971. *Principles of Ecology*, 574 pp. Philadelphia, Pennsylvania: Saunders.

Odum, H. T. 1956. Primary production of flowing waters. *Limnol. Oceanogr.* 2:85–97.

———. 1967. The energetics of world food production. In *The World Food Problem*. Vol. 3:55–94. Report of the president's science advisory committee panel on world food supply. White House, Washington, D.C.

———, and C. M. Hoskin. 1958. Comparative studies on the metabolism of marine waters. *Publ. Inst. Marine Sci. Univ. Tex.* 5:159–170.

———, and F. R. Wilson. 1962. Further studies on reaeration and metabolism of Texas bays, 1958–1960. *Publ. Inst. Marine Sci. Univ. Tex.* 8:159–170.

———, W. McConnell, and W. Abbott. 1958. The chlorophyll "A" of communities. *Publ. Inst. Marine Sci. Univ. Tex.* 5:65–96.

———, S. Nixon, and L. Di Salvo. 1969. Adaptations for photoregenerative cycling. In *The Structure and Function of Fresh Water Microbial Systems*, J. Cairnes, ed., pp. 1–29. Blacksburg, Virginia: Virginia Polytechnic Institute.

Olinger, L. W. 1968. *The Effect of Induced Turbulence on the Growth of Algae*, 58 pp. Atlanta, Georgia: Georgia Institute of Technology.

Owens, M. 1969. Some factors involved in the use of dissolved-oxygen distributions in streams to determine productivity. In *Primary Productivity in Aquatic Environments*, C. R. Goldman, ed., pp. 209–224. Berkeley, California: Univ. of California Press.

Patten, B. C. 1968. Mathematical models of plankton production. *Int. Revue ges. Hydrobiol.* 53:357–408.

Patterson, M. S., and R. C. Greene. 1965. Measurement of low energy beta-emitters in aqueous solution by liquid scintillation counting of emulsions. *Anal. Chem.* 37:854–857.

Pomeroy, L. R. 1961. Isotopic and other techniques for measuring benthic primary production. In *Proc. Conf. Primary Productivity Measurement, Marine and Freshwater, Hawaii, 1961*, M. S. Doty, ed., pp. 97–102. Washington, D.C.: U.S. Atomic Energy Commission, Div. of Technical Information.

Pugh, P. R. 1973. An evaluation of liquid scintillation counting techniques for use in aquatic primary production studies. *Limnol. Oceanogr.* 18:310–318.

Rodhe, W., R. Vollenweider, and A. Nauwerk. 1958. The primary production and standing crop of phytoplankton. In *Perspectives in Marine Biology*, A. A. Buzzati-Traverso, ed., pp. 299–322. Berkeley, California: Univ. of California Press.

Ruttner, F. 1960. *Fundamentals of Limnology*, 295 pp. Toronto: Univ. of Toronto Press.

Ryther, J. H. 1956. Photosynthesis in the ocean as a function of light intensity. *Limnol. Oceanogr.* 1:61–70.

———, and C. S. Yentsch. 1957. The estimation of phytoplankton production in the ocean from chlorophyll and light data. *Limnol. Oceanogr.* 2:281–286.

Saunders, G. W., Jr. 1972. The kinetics of extracellular release of soluble organic matter by plankton. *Verhandl. Int. Ver. Limnol.* 18:140–146.

———, F. B. Trama, and R. W. Bachmann. 1962. Evaluation of a modified C-14 technique for shipboard estimates of photosynthesis in large lakes. *Univ. Michigan, Great Lakes Res. Div. Publ.* 8:1–61.

Schindler, D. W., and E. J. Fee. 1973. Diurnal variation of dissolved inorganic carbon and its use in estimating primary production and CO_2 invasion in lake 227. *J. Fish. Res. Bd. Can.* 30:1501–1510.

———, and S. K. Holmgren. 1971. Primary production and phytoplankton in the Fisheries Research Board Experimental Lakes Area, northwestern Ontario, and and other low-carbonate waters, and a liquid scintillation method for determining ^{14}C activity in photosynthesis. *J Fish. Res. Bd. Can.* 28: 189–202.

———, R. V. Schmidt, and R. A. Reid. 1972. Acidification and bubbling as an alternative to filtration in determining phytoplankton production by the ^{14}C method. *J. Fish. Res. Bd. Can.* 29:1627–1631.

———, V. E. Frost, and R. V. Schmidt. 1973a. Production of epilithiphyton in two lakes of the Experimental Lakes Area, northwestern Ontario. *J. Fish. Res. Bd. Can.* 30:1511–1524.

———, H. Kling, R. V. Schmidt, J. Prokopowich, V. E. Frost, R. A. Reid, and M. Capel. 1973b. Eutrophication of lake 227 by addition of phosphate and nitrate: The second, third and fourth years of enrichment, 1970, 1971, and 1972. *J. Fish. Res. Bd. Can.* 30:1415–1440.

Sheldon, R. W., W. H. Sutcliffe, and A. Prakish. 1973. The production of particles in the surface waters of the ocean with particular reference to the Sargasso Sea. *Limnol. Oceanogr.* 18:719–733.

Small, L. F. 1963. Effect of wind on the distribution of chlorophyll *a* in Clear Lake, Iowa. *Limnol. Oceanogr.* 8:426–432.

Sollins, P. 1969. Measurements and simulation of oxygen flows and storage in a laboratory blue-green algal mat ecosystem. Masters thesis, Chapel Hill, North Carolina: Univ. of North Carolina.

Stainton, M. P. 1973. A syringe gas-stripping procedure for gas-chromatographic determination of dissolved inorganic and organic carbon in fresh water and carbonates in sediments. *J. Fish Res. Bd. Can.* 30:1441–1445.

Steemann Nielsen, E. 1952. The use of radioactive carbon (^{14}C) for measuring organic production in the sea. *J. Cons. Perm. Int. Explor. Mer.* 18:117–140.

———. 1963. Fertility of the oceans: Productivity, definition and measurement. In *The Sea*, Vol. 2, M. N. Hill, ed., pp. 129–164. New York: Wiley.

————, and E. Aabye Jensen. 1957. Primary oceanic production. The autotrophic production of organic matter in the oceans. *Galathea Rep.* 1:49–136.

Stephens, G. C., and B. B. North. 1971. Extrusion of carbon accompanying uptake of amino acids by marine phytoplankters. *Limnol. Oceanogr.* 16:752–757.

Strickland, J. D. H., and T. R. Parsons. 1972. A practical handbook of seawater analysis. *J. Fish. Res. Bd. Can.* 167: 311 pp.

Stull, E. A., E. deAmezaga, and C. R. Goldman. 1972. The contribution of individual species of algae to primary productivity of Castle Lake, California. *Verhandl. Int. Ver. Limnol.* 18:1776–1783.

Talling, J. F. 1957. Photosynthetic characteristics of some freshwater plankton diatoms in relation to underwater radiation. *New Phytol.* 56:1–132.

————, and D. Driver. 1961. Some problems in the estimation of chlorophyll-*A* in phytoplankton. In *Proc. Conf. Primary Productivity Measurement, Marine and Freshwater, Hawaii, 1961*, M. S. Doty, ed., pp. 142–146. Washington, D.C.: U.S. Atomic Energy Commission, Div. Technical Information.

Thomann, R. V. 1971. *Systems Analysis and Water Quality Management*, 286 pp. New York: Environmental Research and Applications.

Thomas, W. H. 1961. Physiological factors affecting the interpretation of phytoplankton production measurements. In *Proc. Conf. Primary Productivity Measurement, Marine and Freshwater, Hawaii, 1961*, M. S. Doty, ed., pp. 147–162. Washington, D.C.: U.S. Atomic Energy Commission, Div. of Technical Information.

Vollenweider, R. A. 1969a. Calculation models of photosynthesis-depth curves and some implications regarding day rate estimates in primary production measurements. In *Primary Production in Aquatic Environments*, C. Goldman, ed., pp. 428–457. Berkeley, California: Univ. of California Press.

————. 1969b. Methods for measuring production rates. In *A Manual on Methods for Measuring Primary Production in Aquatic Environments*, R. A. Vollenweider, ed. *International Biological Programme Handbook* No. 12, 41–127. Oxford and Edinburgh: Blackwell Scientific Publ.

Wallen, D. G., and G. H. Green, 1971. The nature of the photosynthate in natural phytoplankton populations in relation to light quality. *J. Marine Biol.* 10:157–168.

Ward, F. J., and M. Nakanishi. 1971. A comparison of Geiger-Mueller and liquid scintillation counting methods in estimating primary productivity. *Limnol. Oceanogr.* 16:560–563.

Welch, H. C. 1968. Use of modified diurnal curves for the measurement of metabolism in standing water. *Limnol. Oceanogr.* 13:679–687.

Westlake, D. F. 1967. Some effects of low-velocity currents on the metabolism of aquatic macrophytes. *J. Exp. Bot.* 18:187–205.

Wetzel, R. G. 1964a. A comparative study of the primary productivity of higher aquatic plants, periphyton, and phytoplankton in a large, shallow lake. *Int. Rev. Ges. Hydrobiol.* 49:1–61.

————. 1964b. Primary productivity of aquatic macrophytes. *Verhandl. Int. Ver. Limnol.* 15:426–436.

————. 1973. Primary production. In *River Ecology*, M. Owens, and B. Whitten, eds. Oxford: Blackwell Scientific Publ.

————, and A. Otsuki. 1975. Allochthonous organic carbon of a Marl Lake. *Arch. Hydrobiol.* (*In press.*)

Williams, R. B., and M. B. Murdock. 1966. Phytoplankton production and chlorophyll concentration in the Beaufort Channel, North Carolina. *Limnol. Oceanogr.* 11: 73–82.

Wright, J. C. 1959. Limonology of Canyon Ferry Reservoir. II. Phytoplankton standing crop and primary production. *Limnol. Oceanogr.* 4:235–245.

Wrobel, S. 1972. Comparison of some methods of determining the primary production of phytoplankton in ponds. In *Productivity Problems of Freshwaters: Proc. IBP–UNESCO Symp.*, Z. Kajak, and A. Hillbricht-Ilkowska, eds., pp. 733–737. Warsaw and Krakow: Polish Scientific Publ.

Yentsch, C. 1967. The relationship between chlorophyll and photosynthetic carbon production with reference to the measurement of decomposition products of chloroplastic pigments. In *Primary Production in Aquatic Environments*, C. R. Goldman, ed., pp. 323–346. Berkeley, California: Univ. of California Press.

4

Methods of Assessing
Terrestrial Productivty

Robert H. Whittaker and Peter L. Marks

Major contrasts between aquatic and terrestrial communities result from the short life spans and small accumulation of biomass in aquatic plants, and the longer life spans and substantial accumulation of biomass in land plants. The relationship can be expressed as the biomass accumulation ratio: the ratio of the standing crop or biomass present, to the annual net primary productivity. Such ratios are fractions of one in most aquatic communities, but range from one up to 50 or more in terrestrial communities. Significant differences in structure, function, and diversity of communities are related to this contrast between rapid turnover of the community's organic matter in short-lived organisms, and accumulation of the productivity from several years in the complex structure of woody organisms (Whittaker and Woodwell, 1971b). The contrasts extend to the prevalent means of measuring productivity.

As Chapter 3 has described, measurement of plankton productivity is based primarily on measuring oxygen and carbon dioxide exchange by small samples of these communities enclosed in bottles. Measurement of productivity through accumulation of biomass is not generally feasible in aquatic communities, although it has been applied to attached algae and submerged vascular plants. Because of the large size of dominant plants and complexity of structure in land communities, gas-exchange measurements for communities are difficult and demand extensive and usually expensive effort. Most production measurements

---

KEYWORDS: Allometry, biomass, gas exchange, primary productivity; productivity methods; terrestrial ecosystems; ecology.

Primary Productivity of the Biosphere, edited by Helmut Lieth and Robert H. Whittaker.
© 1975 by Springer-Verlag New York Inc.

on land are based instead on measurement of growth, and the corresponding accumulation of biomass, by individual plants of the community. Gas-exchange techniques are less widely applied, although they are an essential complement to the study of net primary productivity if gross primary productivity is to be known.

Different aspects of growth and biomass accumulation should be distinguished. The net production by an individual plant is the amount of organic matter it synthesizes and accumulates in tissues per unit time; it is the profit remaining from the photosynthesis of the plant, or its gross production minus its respiration. Some part of the net production of the plant may be lost with the death and loss of tissues, and in production measurement this loss must be taken into account. Hence the growth by a tree through a year may appear as a 10-kg increase in its weight, but the net production by the tree includes this growth plus net production expended in leaves, fruits, flowers, bud scales, branches, and roots that were lost during the course of the year, and loss by leaching and exudation of organic substances.

The sum of the net productions by all individual plants in a unit area of the Earth's surface is net primary productivity. Measurement of net primary productivity may be affected by the death and loss of whole individual plants, as well as by loss of tissues from living plants. Net primary productivity is thus more than the increase in mass of the plants in a study plot, from one year to the next. It is this increase, plus losses of net production in the death and loss of plant tissues, plus the losses of net production in the deaths (if any) of individual plants. The increase of plant mass in a study plot is net community growth, or net ecosystem production (Woodwell and Whittaker, 1968; Whittaker and Woodwell, 1969; Duvigneaud, 1971). Net ecosystem production and net primary productivity bear no necessary relationship to one another. In a young, fast-growing forest a sizable fraction (30–60%) of net primary productivity may accumulate from one year to the next as net ecosystem production. In a mature climax community, net primary productivity may be equaled by the death and loss of tissues and individuals; it is then possible for net primary productivity to be high, whereas net ecosystem production is zero.

We shall consider several approaches to measurement of primary productivity on land: (1) Harvest techniques are based on harvesting the plants from sample plots and determining their growth (with correction for loss). The approach is appropriate to simple communities of shorter-lived plants. (2) Forest and shrubland productivity techniques are based on more complex measurements of the growth of different tissues in trees and other plants. Such techniques are, in some respects, an elaboration of the harvest approach to deal with the complex structure of forests; but they usually rely on mathematical treatment of plant growth in relation to plant size. (3) Gas-exchange techniques for land communities are discussed along with some results from these. (4) Relationships of net primary productivity to light, leaf-surface area, chlorophyll, and other community dimensions or indices are considered as of interest, even though they are not used for productivity measurement. (5) Problems of measuring root productivity are discussed separately from these four approaches.

Harvest Techniques

The simplest approach to net shoot production for communities of modest stature (annual plants, crop plants, old fields, grasslands, tundra, bogs, marshes, and some shrub-dominated communities) is through sequential harvest of aboveground plant parts during the course of the growing season (E. P. Odum, 1960; Ovington *et al.*, 1963; Wiegert and Evans, 1964; Golley, 1965; Milner and Hughes, 1968; Bliss, 1966, 1969; Boyd, 1970; Forrest, 1971; and Singh and Yadava, 1974). For some annuals, including many crop plants, biomass and net production are nearly equivalent, making it necessary to harvest only once for each species, provided the time of harvest coincides with peak biomass accumulation for each species. Even in communities of annuals, loss of early leaves or other plant parts before the time of harvest often must be corrected. In more diverse perennial communities several harvests may be needed to determine the time of peak biomass for each species and to correct for losses.

Wiegert and Evans (1964) have proposed a somewhat more complex scheme for estimating aboveground net production. Their technique uses the rate of decomposition of dead plant material in conjunction with estimates of standing crop of live and dead plant material, based on a paired-plot sampling scheme (one harvested at the beginning, the other at the end of a time interval of a month or so). A modification of this (Lomnicki *et al.*, 1968) estimates net production from the mass of dead material at the outset and the mass of live and dead material about a month later (the time interval must be kept short to minimize decomposition of dead material). Clymo (1970) and Reader and Stewart (1972) discuss modifications of harvest technique for *Sphagnum* bogs.

Production relationships in most woody successional communities and tall shrub communities probably are best approached through a detailed dimension analysis (Whittaker, 1962; Kestemont, 1971; Marks, 1971, 1972, Zavitkovski and Stevens, 1972; see below). In very young (1–3 years) dense stands, however, sequential harvests may be preferred to the more time-consuming dimension analysis. For determination of biomass, complete harvests (above- and belowground) of randomly selected plots are sufficient (Zavitkovski, 1971; Young, 1971). Kimura (1969) and Nemeth (1973) have estimated production relations through sequential harvests of young conifers; Ford and Newbould (1970) and Kestemont (1971) have developed techniques for deciduous broadleaf coppices. Once stands contain appreciable biomass, as would be true of most woody, successional stands after the first 2–5 years, we believe it makes better sense to examine intensively the growth relationships of a relatively small number of sample trees as is done in dimension analysis, than to process more superficially a large amount of biomass as must be done when entire quadrats are harvested (see Lieth *et al.*, 1965).

If total community productivity in successional stands is desired, then dimension analysis can be applied to the trees, whereas herb and shrub productions can be determined from sequential harvests. For many woody communities, both successional and climax, the contribution of the undergrowth to community productivity is small (from several percent to less than 1%). Measurement of undergrowth production may then be justified more by interest than by increase in accuracy of the community productivity value. Chew and Chew (1965) used a modified harvest method relating cumulative dry weight to shoot age in a creosote bush (*Larrea divaricata*) desert. A considerable fraction of the net productivity of this desert (130 $g/m^2/year$ aboveground) was in subordinate plants, about 15% in shrubs other than creosote bush and 14% in herbs. Results from harvest approaches to a few representative communities are given in Table 4–1.

Forests and Dimension Analysis

Those who study forests have no choice about one characteristic of their concerns—the complexity of the living systems they deal with. In forests, this complexity, which is also present in ecosystems dominated by smaller organisms, is expanded and made conspicuous in community structure. One can see in the forest the layering of species and the staging in depth of foliage, the intricate, branching pattern of bark surfaces, and the subtle mosaic of undergrowth. For our work in analysis of forest net production we offer this thesis: The manifest complexity in which forests exceed other communities is not a handicap but an opportunity. It makes possible measurements within the forest community by which we can learn much about its functional design, while we also obtain more satisfactory estimates of net productivity than might otherwise be possible.

Forest productivity, both gross and net, can be approached in some circumstances through gas-exchange measurements on the forest as a whole as discussed in the following section. Most studies of net primary productivity of forests, however, are based on direct measurements of sizes and weights of plants and plant parts (Newbould, 1967; Whittaker and Woodwell, 1971a). There are at least three ways of synthesizing such measurements into production estimates: mean-tree, production ratio, and regression analysis approaches. The three approaches tend to correspond to three subjects of production measurement: single-age plantation stands, forest undergrowths and shrub communities, and the trees of mixed-age forests, although it is possible to use other approaches to each of these subjects.

Plantation and mean-tree approaches

Plantations of known ages were used by Ovington and his colleagues (1956–1957) in early studies of forest production and nutrient cycling (Ovington and Pearsall, 1956; see also Boysen-Jensen, 1932; Burger, 1940; Möller, 1945, 1947; Möller *et al.*, 1954b). These workers took advantage of the simplification that occurs if several plantations with stands of different ages but each comprised of single-age trees are available on similar sites (Ovington, 1962). Dividing the woody mass of a 50-year-old plantation by 50 gives an indication of the relative production rate, but this does not give an effective estimate of current productivity. However, when plantations of different ages, say 40 and 50 years, in closely similar environments can be compared, current net production can be estimated with reasonable precision.

The difference in stem-wood weight in the two stands represents 10 years'

Table 4–1 Net primary productivity of selected communities by harvest methods

Communities and species	Total (g/m²/year)	Stem and branch wood (%)	Leaves and twigs (%)	Fruit and flower (%)	Root system Rhizomes (%)	Roots (%)	Reference
Wheat	294	—	53.0	29.4		17.6	Filzer (1951)
Barley	242	—	46.6	35.5		17.9	Filzer (1951)
Zea mays (maize, high-yield)	1935	16.8	17.1	61.4		4.6	Lieth (1968)
Helianthus annuus (sunflower)	3213	37.5	17.6	36.0		8.9	Lieth (1968)
Arctic tundra, all species							Rodin and Bazilevich (1967)
Biomass (g/m²)	500	10	20	—		70.	
Production	100	2	28	—		70.	
Populus (7-year-old poplars)							Lieth, Osswald, and Martens (1965)
Biomass (g/m²)	467	61.2	18.8	—		20.0	
Production	226	48.6	36.7	—		14.7	
Blanket bog							Forrest (1971)
Biomass (g/m²)							
Calluna vulgaris	1547	28.4	19.4	—	—	52.2	
Eriophorum vaginatum	482	—	72.2	0.2	11.2	16.4	
Empetrum and others	93	28	18	—	—	(54)	
Sphagnum, other bryophytes	103	—	100	—	—	—	
Lichens	43	—	100	—	—	—	
Total:	2268	20.5	35.8	0.1		43.6	
Net production							
Calluna vulgaris	351	10.8	37.0	—	—	52.2	
Eriophorum vaginatum	221	—	78.2	1.4	8.6	11.8	
Empetrum and others	26	12	38	—	—	(50)	
Sphagnum, other bryophytes	47	—	100	—	—	—	
Lichens	3	—	100	—	—	—	
Total:	648	6.3	56.0	0.5		37.2	

accumulation; from it (assuming no death of trees) average annual stem wood and bark production for the period can be calculated. Similarly the difference in branch weights and root-system weights in the two stands may give first estimates of wood and bark production by branches and roots. Both values must be corrected for loss of dead branches and roots. Collection of fallen dead branches in litter trays may supply the basis for correcting the branch estimate (Möller *et al.*, 1954a); no secure basis for correcting the root-production estimate is available. Leaf production can be approached either through litter-trap collections, or by measurements of the mass of current leaves on living trees. As measurements of leaf production, leaf collections in litter traps are incomplete (Bray and Gorham, 1964). Separate measurements may be needed to correct for leaching loss, translocation from leaves before they fall, and insect consumption (Rothacher *et al.*, 1954; Bray, 1961, 1964; Bray and Dudkiewicz, 1963; Whittaker and Woodwell, 1968; Reichle *et al.*, 1972, 1973a). Flower, fruit, and bud-scale productions also may be measured either on sample trees or in litter collections; these are smaller, but not insignificant fractions (Ovington, 1963; Bray and Gorham, 1964; Whittaker and Woodwell, 1968; Gosz *et al.*, 1972; Whittaker *et al.*, 1972). Much of our knowledge of forest production comes from summing these or related measurements and estimates for plantations (Ovington, 1962, 1965; Art and Marks, 1971).

A crucial problem is the conversion from measurements on individual trees to biomass and production of stands. In a single-age plantation, the trees may be consistent in size; their dimensions may form a bell-shaped frequency distribution of small or moderate dispersion. It then seems reasonable to multiply the mass of an average tree (or one of its tissues) by the number of trees per unit area to obtain a biomass value for the stand (Ovington and Pearsall, 1956; Ovington, 1957; Ovington and Madgwick, 1959a; Peterken and Newbould, 1966). However, the complex geometry of trees implies complex relationships among the frequency distributions and means of different measurements, particularly those with different numbers of dimensional components: (1) *DBH* and height; (2) basal area, foliage area; and (3) stem volume, wood mass, etc. Use of different dimensions leads to choice of different trees as "average," and consequently to different estimates of community biomass (Ovington and Madgwick, 1959b; Baskerville, 1965b; Attiwill, 1966; Attiwill and Ovington, 1968; Ovington *et al.*, 1968). Baskerville (1965b) found biomass estimates for a stand of *Abies balsamea* to be in error by 25–45% when based on trees of average *DBH*. Errors may be smaller in plantations of evenly spaced trees of more consistent size; and errors are smaller and of tolerable magnitudes (within 5% or 10%) when based on trees of mean stem volume or basal area (Baskerville, 1965a; Crow, 1971). Even for single-age plantations, however, there may be advantage in the approach through regressions to be discussed in the section on dimension analysis, which follows. Results of mean-tree and regression estimates have been compared by Satoo and Senda (1966), Satoo (1968b, 1970), Kira and Shidei (1967), Ovington *et al.* (1968), Crow (1971), and Madgwick (1971).

Undergrowth and production ratios

Limitations of the mean-tree approach to production also can be escaped in part by choosing to measure those characteristics of the forest stand that are most expressive of production rate—as distinguished from biomass—and by using ratios to estimate net primary production from these measurements.

When dealing with forest undergrowth, clipping dry weights (of current twigs with leaves of shrubs and tree seedlings, aboveground current growth of herbs) best combine relative ease of measurement with effective expression of undergrowth production. In work in the Great Smoky Mountains (Whittaker 1961, 1963, 1966) clipping weights by species were obtained for 20, 0.5×2.0 m subquadrats, randomly located within a 0.1-ha sample quadrat. Dispersions of the clipping weights for subquadrats are high, but relative errors (coefficients of variation) tend to be lower the higher the undergrowth production being measured (Whittaker, 1966). Dimension analyses were carried out on major shrub populations to determine mean ratios relating net production by different fractions of the plants (as well as biomass of these fractions and leaf surface and chlorophyll) to clipping dry weights (Whittaker 1962, Table 4–2). Dimension-analysis procedures were as described below, except that the analyses were used to obtain mean ratios for sets of 10 mature or shrub-canopy plants (and for 10 subordinate shrubs) of a species rather than to obtain regressions. From clipping measurements and these ratios, shrub stratum productions for a series of forests, forest heaths, and heath balds were estimated (Table 4–3, Whittaker, 1962, 1963, 1966).

Production of larger trees, which cannot be approached through clippings, may be measured through estimated volume increments (*EVI* is one-half annual wood area increment at breast height times plant height). Because it includes a measurement of wood growth rate, *EVI* expresses production in a way that diameter, basal area, and stem volume do not. *EVI* was computed by species for 0.1-ha forest samples as described below. Sets of 10 canopy trees for each of three species from an oak–pine forest were subjected to dimension analysis (Whittaker *et al.*, 1963). Ratios of aboveground production for different parts of the trees to *EVI* were applied to the *EVI* values for a pine–oak forest and an oak heath in the Great Smoky Mountains to give preliminary estimates of aboveground tree production (Whittaker *et al.*, 1963; Whittaker, 1966). Biomass values for aboveground fractions of the trees were similarly related to parabolic volume (*VP* is one-half basal area at breast height $\times$ tree height) to obtain ratios that could be applied to parabolic volume measurements in forest samples (Whittaker, 1966). Table 4–3 illustrates the approach for three communities in the Great Smoky Mountains: a chestnut oak (*Quercus prinus*) heath with a dense shrub canopy at 2–3 m and scattered taller trees on an open west-facing slope at 970 m, a mixed heath bald on a northeast slope at 1500 m, and a subalpine heath with a single-species shrub stratum on a northeast slope at 2010 m. The production values for trees and arborescent shrubs in the heath are based on ratios to estimated volume increment, the

Table 4-2 Mean dimensions of shrub shoots, Great Smoky Mountains, Tennessee, based on sets of 10[a]

1	2	3	4	5	6	7	8
	Rhododendron maximum			*Rhododendron catawbiense*	*Clethra acuminata*	*Viburnum alnifolium*	*Gaylussacia baccata*
	Cove forest:	Oak heath:		Subalpine heath bald:	Oak health:	Spruce–fir forest:	Pine heath:
Mean dimension	Large	Dominant	Subordinate	Dominant	Dominant	Large	Large
Basal diameter (cm)	6.2	4.2	1.35	1.83	1.95	1.97	0.59
Height (m)	4.22	3.05	1.08	1.98	2.69	2.34	0.83
Age, years	40	37	35	23	24	21	9
Radial increment (mm/year)	0.52	0.25	0.12	0.20	0.18	0.25	0.24
Above ground biomass (dry g)	6529	1966	65	258	298	418	12.2
stem wood (%)	44.1	49.4	51.7	53.9	61.4	40.8	43.0
stem bark	3.9	6.3	7.8	8.1	8.6	9.8	12.0
branches	31.1	28.4	14.9	22.2	22.2	37.2	18.8
older leaves	16.7	11.2	21.0	5.1	—	—	—
curr. twigs and leaves	4.0	4.5	4.7	10.5	7.4	11.8	23.3
fruits	0.2	0.2	—	0.2	0.5	0.4	2.9
Aboveground production (dry g/year)	607	214	7.5	50.5	46.4	84.	5.1
stem wood (%)	24.2	20.8	19.9	18.2	23.7	16.3	19.6
stem bark	2.7	2.9	3.5	3.4	2.9	3.8	5.2
branches	22.1	20.4	9.4	18.8	23.4	20.6	11.8
older leaves	15.7	12.7	26.5	5.9	—	—	—
curr. twigs and leaves	33.7	41.4	40.8	53.8	47.1	57.5	56.4
fruits	1.6	1.7	—	0.1	2.9	1.8	7.0
Biomass accumulation ratio	8.9	9.3	8.6	4.3	6.4	7.1	2.4
Aboveground production ratio to:							
Curr. twigs and leaves (g/g)	2.99	2.57	2.45	1.87	2.31	1.79	1.67
Standard error	± 0.11	± 0.19	—	± 0.08	± 0.28	± 0.05	—
Estim. vol. incr. (g/cm³)	3.23	3.87	—	5.16	4.21	5.68	—
Standard error	± 0.42	± 0.45	—	± 0.44	± 0.59	± 0.53	—
Leaf-blade area (g/m²)	65.4	96.	63.5	179.	63.3	67.1	115
Leaf-blade chlorophyll (g/g)	133	217	136	577	287	161	414

[a] Data from Whittaker (1962).

production values for the other shrub strata on ratios to clipping dry weight (Whittaker, 1963).

Production ratios on clipping weight and *EVI*, and biomass ratios on *VP*, are not constant even for canopy individuals. Ratios of wood and bark production to twig and leaf production generally increase with tree age and size. Ratios of leaf, root, and branch mass to stem mass and *VP* decrease with age, and ratios of leaf, branch, and root production to *EVI* decrease with age in many cases (Whittaker, 1962; Whittaker and Woodwell, 1968). In a young forest, however, the ratio of branch to stem mass may increase from small to large, subordinate to dominant trees (Zavitkovski, 1971; Whittaker *et al.*, 1972). The ratios also change with environment within a species. The ratio of foliage production to *EVI* and stem-wood production increases toward less favorable environments (Whittaker, 1962; Satoo, 1966), and ratios of branch and root production and mass to stem production and mass may increase toward less favorable environments (Whittaker, 1962; Bray, 1963).

Despite these shifts with age and environment, ratios relating production or biomass to other plant measurements (of the same number of dimensional components) are much less widely variable than the plant measurements themselves. Estimates of production from ratios on *EVI*, and of biomass from ratios on *VP*, consequently can be applied to mixed-age forests in which the ranges of tree sizes make mean-tree approaches questionable. Different estimative ratios may be needed, however, for canopy and subordinate plants. For sets of canopy plants, and of subordinate plants, analyzed in the southern Appalachians, standard errors of the production ratios were in most cases between 5% and 15% of the production ratios (Table 4–2, Whittaker, 1962; Whittaker *et al.*, 1963).

Dimension analysis of forests

For forests of mixed ages the production-ratio approach may give only an approximation, and the mean tree approach is untenable. A major source of difficulty is the great span of tree sizes in a mixed forest. The masses of trees in a mature forest can extend through five orders of magnitude from canopy individuals to saplings (10,000 to 1 kg), and shrubs and seedlings extend the range by further orders of magnitude. Regression equations relating production to more easily measured dimensions of trees are needed.

These regressions must be suited to the curvilinear character of the relationships. It is a principle of engineering that substantial enlargement of a system or structure requires a redesign of its proportions; the system is unlikely to work if all its dimensions are multiplied by a constant factor. Similarly the dimensions of trees as they enlarge change in ways that maintain their functional balance, but not in ways that maintain constant ratios between the dimensions. The relationship between two dimensions, such as height and diameter, may be expressed not as $y = ax$, but as $y = ax^B$; hence $\log y = A + B \log x$. B is a slope constant expressing the manner in which the two dimensions change in relation to one another. Thus, if height y is related to diameter at breast height x as $\log y = 2.480 + 0.580 \log x$, then the slope constant 0.58 implies that

Table 4–3 Production estimation for three heath communities in the Great Smoky Mountains by estimative ratios[a,b]

	1	2	3	4	5	6	7	8	9
Sample	Community and species	Estimated volume increment (cm³/m²/year)	Clipping dry weight (g/m²)	Shoot production/ EVI ratio (g/cm³)	Shoot production/ clipping ratio (g/g)	Shoot production (g/m²/year)	Leaf-blade weight (g/m²)	Leaf-area ratio (m²/m²)	Leaf-blade chlorophyll (mg/m²)
14	Chestnut oak heath								
	Trees, all species	61.9	—	3.55	—	220	100	1.10	400
	Arborescent shrubs								
	Rhododendron maximum	24.5	—	3.87	—	94.6	164.0	1.31	578
	Kalmia latifolia	27.4	—	5.20	—	142.7	112.1	1.12	511
	Clethra acuminata	2.4	—	4.21	—	10.3	7.7	0.30	65
	Vaccinium constablaei	1.3	—	4.49	—	5.8	2.3	0.04	12
	Lyonia ligustrina	0.3	—	3.45	—	0.9	2.0	0.03	8
	Total:	55.9	—	—	—	254.3	288.1	2.80	1174
	Understory shrubs								
	Rhododendron maximum	—	8.54	—	2.45	20.9	—	—	—
	Kalmia latifolia	—	9.24	—	2.25	20.8	—	—	—
	Clethra acuminata	—	3.76	—	1.82	6.8	—	—	—
	Vaccinium constablaei	—	0.24	—	1.91	0.5	—	—	—
	Lyonia ligustrina	—	1.74	—	2	3.5	—	—	—
	Gaylussacia ursina	—	7.05	—	1.61	10.3	6.4	0.23	51
	Other	—	0.23	—	2	0.5	—	—	—
	Total:	—	30.80	—	—	63.3	—	—	—
	Total, all shrubs:	—	—	—	—	317.6	294.5	3.03	1225
	Total, herbs:	—	1.40	—	1	1.4	1.3	0.03	12
	Total, all strata:	—	—	—	—	539	395	4.2	1637

7 Mixed heath bald

Rhododendron catawbiense	—	162.4	2.31	375	416	3.12	1048
Kalmia latifolia	—	76.2	2.17	165	155	0.94	326
Viburnum cassinoides	—	9.0	2	18	8	0.19	67
Vaccinium constablaei	—	8.7	1.91	17	7	0.10	24
Gaylussacia baccata	—	5.6	1.67	9	5	0.09	24
Pyrus melanocarpa	—	4.1	2	8	3	0.09	31
Herbs	—	0.4	1	0.4	0.3	0.01	4
Total, all strata:	—	266.4	—	592	594	4.53	1524

4 Subalpine heath bald

Rhododendron catawbiense[c]	—	261	1.86	486	337	2.78	860

[a] Data from Whittaker (1962).

[b] Production ratios of trees and arborescent shrubs are based on production ratios to estimated volume increment (columns [2] and [4]), those of other shrubs on production ratios to clipping (current twig and leaf) dry weight (columns [3] and [5]). Leaf data (columns [7–9]) for understory shrubs in the oak heath are combined with those for the arborescent stratum, for species occurring in both strata.

[c] *Rhododendron catawbiense* is the only vascular plant species in the subalpine heath sample, but the stand contained thallophytes of 52 g/m² biomass and unknown productivity.

with a doubling of diameter, tree height will increase by $2.0^{0.58}$ or about 1.5 times. The constant A, in contrast, relates the scales of the two dimensions. Thus, the antilog of 2.480 is $a = 17.7 = y/x^B$. To an increase of 1 cm in (diameter)$^{0.58}$ corresponds an increase of 17.7 cm in height.

These exponential or logarithmic relationships, that characterize harmonious growth with changing proportions, are termed "allometric" (Huxley, 1931, 1932). Polynomial and other equations are often used for forestry measurements, but most investigators dealing with mixed-age forests have felt it necessary to approach them through the logarithmic regressions by which growth and dimensional relationships are best expressed. In practice a set of sample trees are cut down and subjected to intensive measurement, so that biomass, production, and other dimensions can be related (as dependent variables) to diameter (or other independent variables) in logarithmic regressions. A number of authors have used the allometric approach in production measurement (Ovington and Madgwick, 1959a; Kimura, 1963; Baskerville, 1965a; Tadaki, 1965a, b; Kimura *et al.*, 1968; Satoo, 1966, 1968a, b; Kira *et al.*, 1967; Kira and Shidei, 1967; Hozumi *et al.*, 1969a, b; Andersson, 1970, 1971; Maruyama, 1971; Kira and Ogawa, 1971; Reiners, 1972; Nihlgård, 1972; Whittaker *et al.*, 1974; Rochow, 1974). For other references on regressions, see Ogawa *et al.* (1965), Kira and Shidei (1967), Newbould (1967), Young (1971), and Whittaker and Woodwell (1971a). The most intensive and detailed approach to net productivity and related measurements is the system of "dimension analysis of woody plants" developed at Brookhaven National Laboratory (Whittaker and Woodwell, 1967, 1968, 1969, 1971a). The method is designed to use the complexity of structure of forests—through measurements on the various parts of plants including those critical marks of rates of growth that occur in most temperate forests, wood rings and bud-scale scars—to measure or estimate the productivity of the various tissues of woody plants. The method proceeds through the following steps.

1. FIELD MEASUREMENTS ON FOREST STANDS. As a separate process from the analysis of sample trees, trees are tallied by DBH and species in sample quadrats. Heights are measured and increment borings are taken to measure bark thickness, mean current wood-growth rate for the last 5 or 10 years, and age, for all large trees in the quadrat and for sets of smaller trees representing different species and size classes (hence usually 50–75 trees in 0.1-ha quadrats). Trees and shrubs reaching 1 cm or more diameter at breast height are treated as trees; tree seedlings and shrubs not reaching 1 cm at breast height and herbs are clipped in undergrowth subquadrats as described above. Coverages, light penetration, and soil characteristics are measured, and in the longer-term studies at Brookhaven, New York, and Hubbard Brook, New Hampshire, litter fall was measured (Woodwell and Marples, 1968; Gosz *et al.*, 1972). Much of the work has been based on 20 × 50 m (0.1-ha) quadrats (Whittaker, 1966; Whittaker and Woodwell, 1969). Quadrat sizes have been increased, however, for forests of large trees and decreased for small tree and shrub communities, and the 0.1-ha quadrat was replaced by scattered smaller quadrats when pro-

duction of a small watershed was to be estimated (Whittaker *et al.*, 1972; cf. Harris *et al.*, 1973). Some shrub communities have been approached as miniature forests, with diameter and increment measurements taken at 10 cm above ground level rather than at breast height (Whittaker and Niering, 1975).

2. CALCULATIONS FROM STAND MEASUREMENTS. From the preceding information the stand dimensions are calculated, for individual trees, for species, and for the quadrat as follows:

Basal area ($BA = \pi DBH^2/4$) of the stem, and of wood only (BAW) at breast height

Parabolic volume of the stem ($VP =$ one-half basal area times tree height H) for the stem wood plus bark, and for stem wood only (VPW)

Conic surface ($SC =$ one-half breast height circumference $\times$ height) for the stem, and for stem wood (SCW)

Basal area increment ($BAI =$ mean annual increase in wood area at breast height during the past 5 or 10 years)

Estimated volume increment ($EVI = BAI \cdot H/2$)

Other stand dimensions of interest are

Basal increment ratio ($\Sigma BAI/\Sigma BAW$)

Weighted mean radial increment ($\Sigma EVI/\Sigma SCW$)

Weighted mean height ($\Sigma VPW \times 2/\Sigma BAW$) or ($\Sigma VP \times 2/\Sigma BA$)

Volume-weighted mean age ($\Sigma VP \cdot AGE/\Sigma VP$)

Undergrowth clipping dry weights per square meter, by species and by strata.

3. FIELD ANALYSIS OF SAMPLE TREES. Sets of trees (and shrubs if necessary) of major species are felled, and their roots excavated if possible. Sets of 15 (or 10) individuals each of dominant tree species (or shrub species) are taken, their sizes representing the full spread of sizes in the community. It is possible, although more difficult because trees must be climbed, to obtain the necessary measurements on standing trees for a nondestructive sample (Reiners, 1972). Measurements on felled trees include

Base (10 cm) and breast-height diameter (DBH) and height of the tree

A tally of branches with distance from the top or bottom, basal diameter, age, and condition (vigorous, senescent, or dead) recorded for each branch

Sample branches (usually five per tree representing different positions and conditions) for which are recorded also branch length, number of current twigs, and fresh and dry weights of live wood and bark, dead wood, current twigs with leaves, older leaves if any, and fruits

Wood and bark diameters, and fresh and dry weights, of logs from the stem and of discs from the bases of these logs; wood and bark weights are separated for some or all discs or logs

For the discs (or ends of logs) bark, sapwood, and heartwood thickness, and mean annual wood radial increment for the most recent 5 or 10 years (and for preceding decades or pentads)

Fresh and dry weight of root crown (or shrub rhizome) and of excavated

tap and non-tap roots, and broken root ends where branch roots have
been lost

Sample roots (often five per tree), dug up as complete as possible and
measured for basal diameter, fresh and dry weights, and length

From separate twig and leaf samples, dry weights of twigs, petioles, and leaf
blades of current twigs with leaves, and dry weight, insect loss, and chlorophyll content per unit area of leaves

A set of forms was prepared on which field data can be entered and punched
onto computer cards or tape for the following calculations that are part of the
Brookhaven program. The forms may be obtained from the authors.

4. CALCULATIONS ON SAMPLE TREES. Fresh and dry weights of the major
plant fractions can be directly calculated. Weights of branches and branch fractions for the whole tree are calculated by computing for the sample branches
logarithmic regressions of these fractions on branch basal diameter. Branch
regressions have been published by Whittaker *et al.* (1963), Whittaker and
Woodwell (1968), Andersson (1970, 1971), and Whittaker and Niering (1975),
and a particularly useful set of regressions with error estimates is given by
Reiners (1972). The regressions are applied to the full number of branches
recorded by diameter in the branch tally, and the resulting estimates are totaled
for all branches of each tree. The plant-dimensional expressions listed under
step 5 are calculated. From the logs and discs are calculated actual volume and
surface of stem bark, wood, and heartwood if any, and mean annual volume
and dry weight increment of stem wood in the past decade or pentad (and
preceding decades or pentads). Mean values from the sample twigs are used to
calculate, from total weight of current twigs with leaves (and older leaves if
any) of the tree, current twig weight, leaf weight, leaf weight lost to insects,
and leaf-surface area and chlorophyll content. From measurements of base
diameter, length, and current twig number for sample branches, bark-surface area
of sample branches is estimated (Whittaker and Woodwell, 1967, 1968), and
a regression of these estimates on branch basal diameter is calculated. From
this regression, surface area estimates are computed for all branches tallied and
summed for the tree. Calculations on production are as follows:

Stem-wood growth is directly calculated log by log, by multiplying the ratio
of the mean of the annual wood-area increments at the ends of the log to
the mean cross-sectional area for the log, times the wood dry weight of
the log.

Stem bark growth has been estimated in several ways. A most direct means
applies the ratio of current growth to total weight of wood for a log to
weight of bark for that log. Corrections for bark sloughing may be needed
for older trees.

Production of branch wood and bark has been estimated from the relation
BW/A (W is dry weight of branch wood and bark, A is branch age). The
slope constant B is computed from the logarithmic regression of branch
(wood and bark) dry weight on branch age. The only checks on the

calculation so far suggest that it is reasonable but tends to overestimate branch growth (Whittaker, 1965). Overestimation may result from the effect of a higher death rate for small branches than large ones on the slope constant B; mistakes in branch ages because of missing wood rings also will lead to overestimation. Satoo (1968a) has determined branch production through piece-by-piece measurement of wood growth. Baskerville (1965a) divided weights of whorls of branches by ages of these whorls; the result W/A may be a considerable underestimate, for the factor B is in many cases 2.0 or larger (Whittaker *et al.*, 1963; Whittaker and Woodwell, 1968).

Current twig and leaf productions are obtained directly from the biomass values, but with leaf production corrected for insect loss by percent of area lost from leaves on sample twigs (Bray, 1961, Reichle *et al.*, 1973a). Estimation of growth of leaves beyond the first summer may be needed in evergreen species (Whittaker and Garfine, 1962; Whittaker, 1962; Whittaker *et al.*, 1963; Kuroiwa, 1960a, b; cf. Kimura, 1969).

Other fractions of production (flowers, fruits, stipules, bud scales) are variously estimated from sample branch regressions, collection of all fruits from sample plants, separate determinations of mean weights and ratios of these fractions to current twig and leaf production, and litter collections.

5. SUMMARY CALCULATIONS. The calculations to this point are summarized in the forms of mean-tree measurements, production ratios, and regressions for the sets of plants (Tables 4–4 and 4–5). For the final set of regressions, diameter at breast height (or 10 cm for shrubs), conic surface (defined above), parabolic volume, and estimated volume increment are used as independent variables; to them are related as dependent variables the sums, for the plants in a set, of biomass, production, volume, and surface estimates. All regressions are calculated in double logarithmic form; some of them (for which dimensional relations of the dependent variable and independent variable are closely related, such as actual volume and parabolic volume estimate) are also calculated in linear form. The logarithmic regressions have characteristics that make expected error and confidence limits difficult to express in concise form. Coefficients of correlation for dimension analysis regressions are of limited value; for the most part they take values much above 0.9, with these values strongly influenced by the range of sizes of plants sampled. In the effort to express relative tightness of the regressions more effectively, "estimates of relative error" (*e, E*) are part of the summary calculations (Table 4–4, and Whittaker and Woodwell, 1968). For a linear regression an estimate of relative error *e* is the standard error of estimate divided by the mean value of the independent variable; for a logarithmic regression *E* is the antilog of the standard error of estimate (this value is not the same as the standard error of the untransformed variables).

6. APPLICATION TO STAND DATA. The regressions are used to calculate, for each tree in the original sample quadrat, its probable biomass, production, volume, and surface dimensions. For species populations that have not been

Table 4–4 Mean dimensions of some temperate-zone trees[a]

1 Mean dimension	2 Acer spicatum	3 Quercus alba	4 Acer saccharum	5 Quercus robur	6 Pinus rigida	7 Picea rubens
Location	HB	BNL	HB	LS	BNL	HB
Number of trees in sample	15	15	14	11	15	15
Breast-height diameter (cm)	4.8	9.3	25.9	43.5	15.2	14.5
Height (m)	6.3	7.3	17.9	19.7	8.9	9.1
Age (years)	24	33	72	149	41	87
Bark thickness, breast height (mm)	1.6	5.67	6.3	16.4	12.05	2.8
Wood radial increment (mm/year)	0.53	0.64	1.13	1.59	1.08	0.72
Stem volume (dm^3)	11.7	42.1	780	1490	125.1	144
Parabolic volume estimate (dm^3)	10.5	40.7	980	1235	114.1	152
Stem-wood volume increment (dm^3/year)	0.48	1.41	12.9	24.2	3.41	4.41
Estimated volume increment (dm^3/year)	0.31	1.00	13.2	19.5	2.10	2.74
Stem surface (m^2)	0.72	1.60	10.16	—	3.17	3.29
Conic stem surface estimate (m^2)	0.59	1.26	9.14	—	2.33	2.62
Aboveground biomass (dry kg)	8.7	36.6	703	987.8	85.5	87.5
stem wood (%)	54.1	54.6	59.6	64.6	54.3	57.0
stem bark	8.0	17.1	7.5	7.3	12.3	8.0
branches	35.0	20.2	31.4	25.7	22.3	27.8
older leaves	—	—	—	—	5.9	6.3
curr. twigs and leaves	2.9	7.9	1.5	2.4	5.2	0.8

Aboveground production (dry kg/year)	0.86	4.9	30.9	42.6	9.84	3.38
(%)						
stem wood	27.2	15.4	26.3	28.5	18.6	53.0
stem bark	4.2	4.8	3.3	1.9	4.6	6.3
branches	38.0	24.4	33.6	21.2	23.0	14.2
curr. twigs and leaves	29.1	51.6	35.6	48.4	49.5	22.6
fruits	1.5	3.4	1.2	—	4.1	3.9
Biomass accumulation ratio	10.2	6.9	22.8	23.2	8.7	25.8
Aboveground production ratio to:						
Estim. vol. increment (g/cm³)	2.77	4.91	2.33	2.18	4.68	1.23
Leaf-blade area (g/m²)	175	216	177	111	155	56

[a] Based on sets of sample trees at Hubbard Brook, New Hampshire (HB, Whittaker et al. 1974), Brookhaven National Laboratory, New York (BNL, Whittaker and Woodwell 1968), and Linnebjer, Sweden (LS, Andersson 1970, 1971). The samples include two sets of small deciduous trees (columns [2] and [3]), two of small conifers ([6] and [7]), and two of medium-sized deciduous trees ([4] and [5]). Some contrasts between the Pinus rigida and Picea rubens samples reflect the growth of the former in full sunlight in and above the canopy of small oaks versus growth of the latter in the shade beneath a deciduous canopy.

Table 4–5 Interspecies allometric regressions[a,b]

1	2	3	4	5	6	7	8	9
	Brookhaven–Oak Ridge shrubs and trees		Hubbard Brook deciduous trees		Hubbard Brook *Picea rubens*		Tropical rain forest	Japanese *Abies*
Dependent variables and coefficients[b]	on DBH	on VP	on DBH	on VP	on DBH	on VP		
Stem volume (cm³)								
A	2.3269	0.7446	2.3283	0.4382	2.2203	0.7735		
B	2.3329	0.8614	2.3571	0.9127	2.3357	0.8499		
r	0.999	0.999	0.999	0.999	0.998	0.997		
E	1.317	1.229	1.165	1.152	1.141	1.170		
Stem-wood volume (cm³)								
	DBH	VP	DBH	VP	DBH	VP		
A	2.2263	0.6437	2.2342	0.3157	2.1212	0.6503		
B	2.3428	0.8636	2.3905	0.9259	2.3732	0.8638		
r	0.999	0.999	0.998	0.999	0.998	0.997		
E	1.316	1.222	1.182	1.149	1.143	1.167		
Stem-wood dry weight (g)							Stem dry weight (g) on VP	on DBH
	DBH	VP	DBH	VP	DBH	VP		
A	2.0736	0.4848	2.0011	0.0798	1.8885	0.5048	0.1111	1.45
B	2.2336	0.8423	2.3925	0.9270	2.2380	0.8138	0.9326	2.74
r	0.997	0.999	0.998	0.999	0.995	0.994		
E	1.506	1.256	1.200	1.151	1.193	1.229		
Stem-bark dry weight (g)								
	DBH	VP	DBH	VP	DBH	VP		
A	1.5487	0.0225	1.2543	−0.5300	1.3543	0.1198		
B	2.0978	0.7912	2.2292	0.8625	1.9961	0.7259		
r	0.986	0.994	0.991	0.991	0.997	0.996		
E	1.201	1.665	1.432	1.447	1.130	1.166		
Branch wood and bark dry weight (g)							VP	DBH
	DBH	VP	DBH	VP	DBH	VP		
A	1.8518	0.4209	1.0823	−1.0608	0.9115	−0.6338	−0.8588	0.44
B	2.0748	0.7646	2.7276	1.0476	2.5428	0.9188	1.027	2.97
r	0.990	0.991	0.977	0.969	0.975	0.967		
E	1.611	1.861	2.052	2.291	1.609	1.724		

Twig and leaf dry weight (g)

	DBH	VP	DBH	VP	DBH	VP	DBH
A	1.6842	0.5371	1.1669	-0.2975	1.6199	0.5765	0.43
B	1.6526	0.6021	1.8618	0.7153	1.6992	0.6163	2.86
r	0.989	0.992	0.983	0.976	0.973	0.969	
E	1.794	1.598	1.522	1.652	1.390	1.425	

Aboveground dry weight (g)

	DBH	VP	DBH	VP	DBH	VP
A	2.4667	0.8442	2.2380	0.3070	2.3151	0.9711
B	2.0980	0.8175	2.4223	0.9357	2.1830	0.7926
r	0.998	0.998	0.997	0.995	0.991	0.988
E	1.353	1.317	1.234	1.329	1.278	1.330

Root system dry weight (g)

	DBH	VP	DBH	VP	DBH	VP	VP
A	1.6483	0.4844	1.7161	-0.0537	1.7583	0.4360	0.1862
B	2.2419	0.7626	2.2230	0.8583	2.1514	0.7806	0.775
r	0.985	0.987	0.996	0.993	0.985	0.981	
E	1.850	1.745	1.272	1.364	1.359	1.411	

Root system dry weight (g)

	SBDW	SBDW
A	0.4374	-0.1826
B	0.7887	0.9037
r	0.995	0.997
E	1.445	1.213

[a] *Sample sets include:* columns [2] and [3], 42 shrubs and trees of 10 species at Brookhaven National Laboratory, New York, and Oak Ridge, Tennessee (Whittaker and Woodwell 1968); columns [4] and [5], 21 small and medium-sized trees of three species (*Acer saccharum, Betula lutea,* and *Fagus grandifolia*) at Hubbard Brook, New Hampshire (Whittaker *et al.,* 1974); columns [6] and [7], 15 small trees (maximum DBH 38 cm) of *Picea rubens* at Hubbard Brook, New Hampshire (Whittaker *et al.,* 1974); columns [8] and [9], tropical rain forest trees (Kira *et al.,* 1967), and Japanese conifers (Kimura, 1963).

[b] Regressions are in the form $\log_{10} y = A + B \log_{10} x$, where y is the dependent variable in column [1] and x is the independent variable above the constants: DBH = diameter at breast height (cm), VP = parabolic volume (cm³), $SBDW$ = stem and branch dry weight (g), EVI = estimated volume increment (cm³/m²/year), AGE = branch age (years), BBD = branch basal diameter (cm), $BRDW$ = branch wood and bark dry weight (g). E is the estimate of relative error (antilog of the standard error of estimate of a logarithmic regression), and r is the coefficient of correlation.

Table 4-5 *continued*

	2	3	4	5	6	7	8	9
1	Brookhaven–Oak Ridge shrubs and trees		Hubbard Brook deciduous trees		Hubbard Brook *Picea rubens*		Tropical rain forest	Japanese *Abies*
Dependent variables and coefficients[b]								
Stem-wood growth (g/year)	*DBH*	*EVI*	*DBH*	*EVI*	*DBH*	*EVI*		
A	0.8809	0.2975	0.6732	0.00234	0.1608	0.0795		
B	2.0828	0.8679	2.1760	0.9488	2.4391	0.9285		
r	0.996	0.996	0.988	0.996	0.953	0.983		
E	1.530	1.480	1.506	1.251	1.889	1.466		
Stem-bark growth (g/year)	*DBH*	*EVI*	*DBH*	*EVI*	*DBH*	*EVI*		
A	0.3934	−0.0999	0.0108	−0.5974	−0.4409	−0.4942		
B	1.8073	0.7473	1.9504	0.8524	2.2053	0.8324		
r	0.992	0.987	0.979	0.989	0.963	0.984		
E	1.682	1.906	1.638	1.418	1.668	1.392		
Branch growth (g/year)	*DBH*	*EVI*	*DBH*	*EVI*	*DBH*	*EVI*		
A	1.1714	0.6771	0.6369	0.0444	−0.2893	−0.3286		
B	1.8069	0.7473	2.2343	0.9461	2.3234	0.8711		
r	0.990	0.986	0.981	0.961	0.948	0.963		
E	1.726	1.976	1.701	2.140	1.905	1.726		
Current twigs and leaves (g/year)	*DBH*	*EVI*	*DBH*	*EVI*	*DBH*	*EVI*		
A	1.6842	1.2256	1.1669	0.6803	0.8703	0.8586		
B	1.6526	0.6746	1.8618	0.7863	1.6359	0.6078		
r	0.989	0.987	0.983	0.960	0.971	0.978		
E	1.794	1.806	1.522	1.894	1.391	1.338		
Aboveground growth (g/year)	*DBH*	*EVI*	*DBH*	*EVI*	*DBH*	*EVI*		
A	1.8888	1.4140	1.4008	0.8306	0.8648	0.8140		
B	1.7405	0.7197	2.0520	0.8765	2.1158	0.7985		
r	0.994	0.990	0.990	0.978	0.960	0.982		
E	1.526	1.729	1.411	1.674	1.659	1.408		

Branch wood and bark dry weight (g)

	BBD	AGE	BBD	AGE	BBD	AGE
A	1.3954	-1.0478	1.4355	-1.4375	1.5993	2.6532
B	2.9386	2.3855	2.9073	2.9257	2.8788	1.8839
r	0.889	0.705	0.992	0.803	0.975	0.685

Current twig and leaf weight (g)

	BBD		BBD	BRDW	BBD	
A	1.1866		1.0264	0.3982	1.2532	
B	1.7682		1.9437	0.6619	2.3976	
r	0.859		0.967	0.695	0.891	

a Sample sets include: columns [2] and [3], 42 shrubs and trees of 10 species at Brookhaven National Laboratory, New York, and Oak Ridge, Tennessee (Whittaker and Woodwell 1968); columns [4] and [5], 21 small and medium-sized trees of three species (*Acer saccharum*, *Betula lutea*, and *Fagus grandifolia*) at Hubbard Brook, New Hampshire (Whittaker *et al.*, 1974); columns [6] and [7], 15 small trees (maximum DBH 38 cm) of *Picea rubens* at Hubbard Brook, New Hampshire (Whittaker *et al.*, 1974); columns [8] and [9], tropical rain forest trees (Kira *et al.*, 1967), and Japanese conifers (Kimura, 1963).

b Regressions are in the form $\log_{10} y = A + B \log_{10} x$, where y is the dependent variable in column [1] and x is the independent variable above the constants: DBH = diameter at breast height (cm), VP = parabolic volume (cm³), $SBDW$ = stem and branch dry weight (g), EVI = estimated volume increment (cm³/m²/year), BBD = branch basal diameter (cm), AGE = branch age (years), $BRDW$ = branch wood and bark dry weight (g). E is the estimate of relative error (antilog of the standard error of estimate of a logarithmic regression), and r is the coefficient of correlation.

sampled for dimension analysis, the most appropriate available regressions are used. Use of a computer enables calculation of each dependent variable from regressions on two or more of the independent variables. The dependent variables computed for individual trees are summed by species and for the sample as a whole. Among the two or more sums of a given dependent variable from regressions on different independent variables, one is chosen on the basis of lower estimate of relative error for the regression, or closer dimensional correspondence of the independent and dependent variables. In most cases, for example, parabolic volume has been preferred as the independent variable for volume and biomass calculations and for current twig and leaf production, estimated volume increment has been preferred for stem wood, stem bark, and branch wood and bark production, and conic surface has been preferred for stem- and branch-surface calculation. Ratios of production and biomass to current twig and leaf dry weight, rather than regressions, have been used to estimate production and biomass of tree seedlings and smaller shrubs in the clipping subquadrats.

Results

Some results from dimension analysis may be summarized. Allometric regressions for samples for particular species have been compared by Whittaker and Woodwell (1968), Yoda (1968), Andersson (1970, 1971), Whittaker *et al.* (1972), and others. Of more interest for our present summary are the allometric patterns that link different species with one another. Japanese studies (Ogawa *et al.*, 1961, 1965; Yoda, 1968) have shown that in tropical forests the tree species may be so similar in form as to fit a single regression line; there is no evident gain in accuracy from the use of regressions for individual species. The Brookhaven study (Whittaker and Woodwell, 1968) dealt with species more widely different in form and stature; but these, too, were related by looser, interspecies trends. Figure 4–1 illustrates the trends of foliage and stem-wood production in relation to diameter for species ranging from low shrubs (*Vaccinium vacillans, Gaylussacia baccata* at Brookhaven) to medium-sized trees (*Quercus alba, Liriodendron tulipifera* at Oak Ridge, Tennessee). Differences among species and samples affect the locations of the points; and the largest individuals of a given sample often fall below the trend line (cf. Ogawa *et al.*, 1965). Yet it is striking that common dimensional trends connect woody plants as disparate as blueberry shoots and forest oaks.

Figure 4–2 combines the interspecies regression lines for the same sets of plants, and further aspects of allometric relations may be observed. (1) Slopes of the regressions increase from lineal dependent variables (e.g., height), to quadratic (surface) dimensions and surface-related production measures, to cubic variables (volume and mass). Slopes of the production relations, involving plant surfaces for gas exchange and cambial growth, are in the same range as those for the surface relations themselves. (2) Slopes for stem surface and leaf surface are nearly parallel. Mean wood radial increment thickness tends to increase with increasing plant size when many species and individuals are treated together (Table 4–4 and Whittaker, 1962). Stem-wood and bark growth conse-

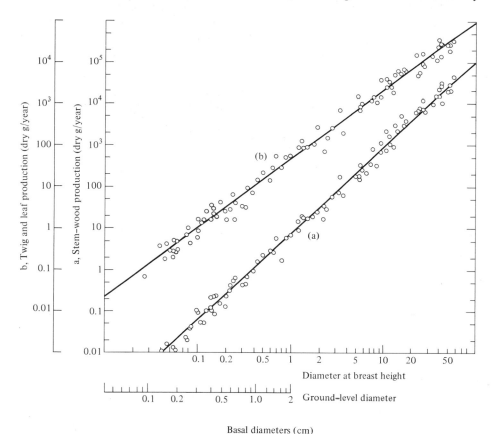

FIGURE 4–1. Interspecies regressions for production of stem wood (a) and current twigs and leaves (b), for plants ranging from small shrubs to medium-sized trees (Whittaker and Woodwell 1968).

quently increase more steeply with plant size than does the leaf surface support-ing by photosynthesis that wood and bark growth. (3) Although stem-wood weight increases more steeply than branch weight, the dissected forms of branches imply that their surfaces increase much more steeply than does stem surface. Branch wood and bark production increase less steeply than stem wood production, but more steeply than leaf production and surface. This fact, along with (2), implies that the larger the woody plant, the larger the surface and mass for growth and respiration in fractions other than leaves, which the photo-synthesis of a unit leaf surface must support (see also Fig. 4–6). (4) The point at which the ratio of foliage surface to nonphotosynthetic tissue supported by that surface becomes unfavorable (in relation to a given level of light and other resources) is quite variable within and between species, but it probably has much to do with the limits on sizes of woody plants.

Table 4–5 gives some of the most useful of the interspecies regressions. The Brookhaven regression is appropriate for shrubs and smaller, open-growth

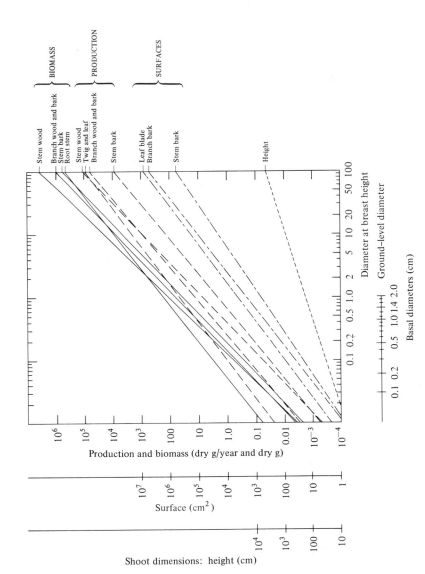

FIGURE 4–2. Interspecies regression lines for various dimensions of plants ranging from small shrubs to medium-sized trees, plotted against diameter at breast height or (for small shrubs) ground-level diameter (Whittaker and Woodwell 1968). Stem-bark biomass and stem-wood production are steeper lines in pairs of convergent lines.

deciduous trees; the Hubbard Brook regression is for small- to medium-sized mesic deciduous trees, the larger of which may have had a history of growth in partially open conditions following cutting. These regressions should be useful for other temperate deciduous forests; but neither is appropriate for climax forests. No production regressions for large coniferous trees have been published. Some of the Japanese interspecies biomass regressions have been converted into units corresponding to American practice, and are given in Table 4–5.

Table 4–6 illustrates the results of dimension analysis applied to seven forest and woodland communities. The Brookhaven and Hubbard Brook values are based on intensive studies; the Santa Catalina samples used aboveground dimension analyses of certain major species; the Smokies samples are based on estimative ratios and some regressions from the literature. The crucial results are the aboveground net productivity values; for four of the forests (columns 4–6 and 8) these are 860–1050 $g/m^2/year$. Aboveground net productivities of many temperate forests of favorable environments converge in the range 1000–1200 $g/m^2/year$; the corresponding range including root production is 1200–1500 $g/m^2/year$ (Whittaker, 1966). The forests of columns 5, 6, and 8 are in this range; the Brookhaven forest of column 4 is, with allowance for its high root production, on the lower border of the range (1195 $g/m^2/year$, above- and belowground). The range applies both to climax forests, and to many young forests averaged through their growth to maturity. Some forests of especially favorable environments (e.g., coast redwood forests, floodplain forests) may have productivities considerably above this range, as do some fast-growing young forests such as the stand of *Liriodendron tulipifera* in Table 4–6, column 7. Forests of less favorable environments have productivities below the range given and generally in the range 600–1000 aboveground, 800–1200 above- and belowground.

Columns 2 and 3 are woodlands—communities of small trees in open growth, not forming a dense canopy, and generally with well-developed undergrowth. The productivities of temperate woodlands and shrublands (excluding deserts) appear to be mostly between 250 and 800 $g/m^2/year$ (Whittaker and Niering, 1975). Columns 2 and 3 of Table 4–6, and the oak heath and heath balds of Table 4–3, are all in this range, with aboveground productivities from 285 to 592 $g/m^2/year$. Many grasslands have productivities in this range also; these three types of communities (woodland, shrubland, dry grassland) occur in environments less favorable than those of closed forests, more favorable than those of deserts, with parallel, intermediate, ranges of productivities.

As Table 4–6 also illustrates, the productivity of forests is strongly concentrated (98%, 99%, or more) in the tree stratum itself. Many woodlands have an appreciable fraction of their productivity in the undergrowth; and in some (e.g., Table 4–6, column 2) undergrowth production exceeds that of the very open tree stratum. The largest shares of forest production are in the stem wood and in the current twigs and leaves; each of these makes up 30–40% of aboveground production in many forests. Branch production is 20–30% of aboveground production in many forests, but lower in dense stands, particularly of conifers Madgwick, 1970; Satoo, 1971), and higher in some young and open-growth stands. Estimates of fruit production are often between 1% and 5%.

Table 4–6   Summary descriptions of seven temperate-zone forests and woodlands[a]

1	2	3	4	5	6	7	8
Forest stand measurements	Pygmy conifer oak scrub, mature SCM-52	Pine–oak woodland, mature SCM-51	Oak–pine forest, young BNL-60	Mixed deciduous forest, young HB-71	Cove forest, mature GSM-23	Tulip poplar forest, young GSM-22	Spruce–fir forest, mature GSM-29
Stems (> 1 cm/0.1 ha)	57	278	185	129	145	182	84
Canopy height (m)	3	10	9	20	36	27	25
Weighted mean tree height (m)	2.7	7.5	7.6	16.9	34.0	22.4	21.3
Weighted mean tree age (years)	65	46	43.3	124	222	29	161
Stem basal area, m²/ha	4.3	26.0	15.6	26.3	54.2	34.2	55.6
Mean wood radial increment (mm/year)	0.28	0.39	0.86	1.12	0.73	2.28	0.96
Basal area increment (m²/ha/year)	0.034	0.238	0.356	0.464	0.445	1.325	0.54
Stem volume, m³/ha	9.5	99.4	75.4	194	720	310	650
Stem-wood volume (m³/ha)	7.1	76.1	59.4	176	650	275	590
Parabolic volume estimate (m³/ha)	6.6	97.8	70	204	851	346	547
Estimated volume increment (cm³/m²/year)	4.6	66.2	159	379	547	1444	534
Biomass accumulation ratio (g/g)	10.1	25.5	7.7	14.4	45.6	9.1	34.7
Stem-surface area (m²/m²)	0.03	0.27	0.30	0.41	0.6	0.6	0.6
Conic stem-wood surface	0.02	0.17	0.21	0.34	0.50	0.51	0.52
Branch-surface estimate (m²/m²)	0.26	1.7	1.2	1.98	1.6	2.2	—
Leaf-area ratio (m²/m²)	2.0	3.7	3.8	6.1	6.2	7.4	14.8
Chlorophyll in leaves (g/m²)	1.0	1.8	1.9	2.4	2.2	2.1	3.0
Light penetration (%)							
Through trees	91.2	42.7	13.0	—	0.9	0.9	3.8
shrubs	12.7	34.2	6.0	—	0.8	0.7	2.6
herbs	12.7	34.0	5.9	—	0.2	0.7	1.4

Aboveground biomass (dry g/m²)							
Trees	1530	11,350	6403	16,085	50,000	22,000	34,000
Shrubs	341	17	158	15	7	2	96
Herbs	3.1	3.4	2.2	4	38	1.5	22
Thallophytes	4.4	0.2	tr.	tr.	20	4.9	40
Tree percentage:							
stem wood	46.5	54.1	54.8	65.0	77.4	73.5	76.3
stem bark	12.7	15.5	12.7	6.8	7.0	6.9	7.8
branches	31.4	24.6	24.6	26.3	14.8	17.7	11.9
leaves and twigs	9.4	5.8	7.9	1.9	0.8	1.9	4.0
Aboveground net productivity (dry g/m²/year)							
Trees	65	435	796	898	1050	2400	980
Shrubs	117	6.7	61	4.3	1.5	7	22
Tree percentage:							
stem wood	10.1	17.3	18.7	28.5	38.1	42.0	38.0
stem bark	2.6	4.6	3.3	3.0	4.5	5.1	4.4
branches	19.3	20.7	24.3	29.7	19.7	26.2	18.1
leaves and twigs	61.8	54.0	50.9	35.7	35.5	22.1	36.9
fruits	6.2	3.4	2.8	3.1	2.2	4.6	2.6

[a] Santa Catalina Mountains, Arizona (SCM: Whittaker and Niering 1975, and Whittaker and Woodwell 1969), Brookhaven National Laboratory, New York (BNL, Whittaker and Woodwell, 1969), Hubbard Brook, New Hampshire (HB, Whittaker et al. 1974, low-elevation belt), and Great Smoky Mountains, Tennessee (GSM, Whittaker 1966).

Flower and bud-scale productions are smaller (0.2% and 0.8%, respectively, at Hubbard Brook, Gosz *et al.*, 1972; cf. Ovington, 1963; Hytteborn 1975). The fraction of forest production directly harvested by herbivorous animals is surprisingly small. Leaf consumption by insects, as a major part of this harvest, seems to be mostly 1–8% of leaf production and less than 3% of aboveground net production (Bray, 1961, 1964; Whittaker and Woodwell, 1969; Andersson, 1970; and Reichle *et al.*, 1973a, b).

Toward the less favorable environments of woodlands, rates of wood growth decrease (as indicated by the mean radial increments and basal area increments of Table 4–6). Correspondingly, the distribution of production among tissues shifts toward less favorable environments, with the fraction in stem wood decreasing to 10–20%, that in twigs and leaves increasing to 50–60%. The fraction in branch wood and bark may also be higher in woodlands in most cases. (As will be discussed, root productivity is probably 15–20% of the total in forests, probably higher in many woodlands.) Young forests (columns 4, 5, and 7) compared with mature forests (columns 6 and 8) have lower biomasses and biomass accumulation ratios and, in many cases, larger fractions of productivity in branch wood and bark. Biomass of forests is even more strongly concentrated in the tree stratum, and in the stem wood of the trees, than is the productivity. The general relationship of forest biomass to productivity, finally, is indicated in Fig. 4–3. The oblique band includes the climax and near-climax samples; for these a trend of increasing biomass with increasing productivity is evident. For immature forests, in contrast, the relation of biomass to productivity is highly variable and age-dependent. The samples in the oblique band below about 15 kg/m² are woodlands. The intersection of the band with the horizontal axis marks a range of productivity below which vegetation dominated by trees is not supported, and shrublands and grasslands occur. The level of productivity at which this replacement occurs can be very different with difference in climate and effects of fire.

Reliability

In two studies dispersions were measured for productivity estimates in sets of five standard 0.1-ha samples. In a set of spruce–fir samples in the Great Smoky Mountains (Whittaker, 1966) the coefficient of variation for the most critical measurement, estimated volume increment, was 4.0%, versus higher values for basal area (15.8%) and volume (23.0%). The spruce–fir samples varied in density and volume because of the reproductive cycle—death of greater numbers of old and heavier reproduction of young trees in some samples than in others. The lower wood radial increments in the denser stands suggested, however, that volume increments in these sets of stands were convergent despite the differences in volume. The Brookhaven oak–pine samples (Whittaker and Woodwell, 1969) differed in volume growth since the last fire, and a wider dispersion of estimated volume increments ($CV = 11.8\%$) probably includes effects of place-to-place difference in habitat.

Although the allometric approach to mixed-age forests is almost inescapable, statistical aspects of the treatment are subject to questions which cannot be resolved here. In dealing with forest trees the dispersions of points about the

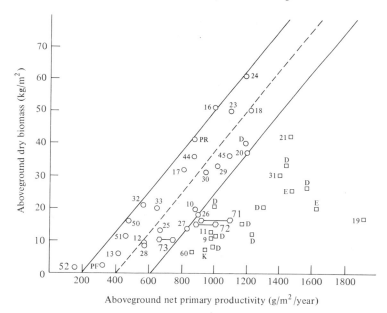

FIGURE 4–3. Forest and woodland biomass in relation to net primary production, both aboveground. Circles represent climax and near-climax stands, squares represent immature stands; trend line, fitted to circles only, is Mass = 0.0625 Prod − 25. Data are from forest production samples of first author (numbers): Whittaker (1963, 1966), Olson (1971), Whittaker and Woodwell (1969), Whittaker *et al.* (1974), Whittaker and Niering (1975); and from Duvigneaud *et al.* (1971, D), Kestemont (1971, K), and Arvisto (1970, E).

regression lines, and consequently the probable errors of estimates for individual trees, are large. The logarithmic regressions of current twig and leaf production on *DBH* in *Pinus rigida* and *Quercus alba* at Brookhaven give coefficients of correlation of 0.98 and 0.96, but the estimates of relative error are 1.27 and 1.45 (Whittaker and Woodwell, 1968; cf. Bunce, 1968). The latter values imply expected departures of points from the line of the order of 21–27% in the first case, 31–45% in the second. (For a logarithmic regression, the estimate of relative error of 1.27 implies a range from 1.27 y to y/1.27, hence from +27% to −21%, in the central part of the size distribution of trees.) As a partial estimate of sample error Andersson (1970) has summed the deviations from regressions for the species in his forest quadrats and estimated the 95% confidence limits as 3.5–6.5% for biomass values, 6.5–9.7% for production values.

   Although regressions in the form log $y = A + B$log x seem biologically and mathematically the most generally appropriate means of relating dimensions of trees, the logarithmic calculations can cause systematic error (Zar, 1968; Madgwick, 1970; Crow, 1971; Baskerville, 1972, Beauchamp and Olson, 1973). (Alternative formulas also have their problems.) For each range of the independent variable x the log transformation reduces the effect of high values of the dependent variable y relative to that of low values on the calculation of the regression. The effect is as if the regression line were fitted to the geometric means of y for different ranges at x, rather than to arithmetic means of y. The geometric mean is smaller than the arithmetic mean; and the regression estimate of y, for a given value of x, is smaller than the arithmetic mean of a set of actual measurements of y for that value of x. When the regressions are applied to the trees in forest samples, biomass underestimates of 10–20% can result (Baskerville 1972). Means of correcting for this error are discussed by Baskerville (1972) and Beauchamp and Olson (1973).

A further property of allometric relations should be observed. For a given regression and value of x, the values of y are apparently lognormally distributed (our data, *unpublished*) with a dispersion that is proportional to the mean value of y. Errors (E) of y are consequently in the form $y = EAx^B$, not of $y = Ax^B + E$ (Baskerville, 1972); and the errors of estimation for the largest trees in a forest sample have far greater effect on the biomass estimate than the errors for the smaller trees. Tests of the effect of log transformation have been made by calculating logarithmic regressions for sets of sample trees, predicting biomass and production values for these trees from the regressions, and comparing the sums of these predictions with the sums of the actual values (Whittaker *et al.*, 1974). The estimates were not consistently high, but variously high or low largely depending on the error in estimating the value for the largest trees in the sets. The correction for log transformation should give improved biomass estimates in some cases; but there are probably others in which the improvement is negligible in relation to other errors (Beauchamp and Olson, 1973). Productivity estimates involve two other errors—tendencies to overestimate production of branch wood and bark (Whittaker, 1965) and of large and senescent trees (Ogawa *et al.*, 1965; Whittaker and Woodwell, 1968); these errors are opposite in direction to that from the log transformation. Correction for log transformation probably will not improve most estimates of forest productivity significantly.

Studies testing regression estimates against known biomass values based on clear-cutting quadrats are few and do not treat mixed-age stands; and there are as yet no studies thus testing production estimates. Satoo (1966), in such a study of biomass estimation, found that logarithmic regressions tended to overestimate foliage weight by 1–9%. Satoo (1966, 1968b) and Ovington *et al.* (1968) obtained mean-tree weight estimates for stems, branches, and foliage that were within 7% of actual values, when based on the mean of several trees of average basal area. Madgwick (1971) found that the mean of replicate regression estimates of stem weight was within 2% of the actual value regardless of choice of independent variable; standard deviations of these replicate estimates were 4.5% when based on 20 trees, 5–7% when based on five trees. Foliage was consistently overestimated 6–10% with different independent variables. Estimates of branch weight were more highly variable with standard deviations 11–16% of mean values when based on 20 sample trees, 16–27% when based on five sample trees. It appeared that the mean tree and regression estimates did not differ greatly for this old-field pine stand. Crow's (1971) weight estimates based on trees of mean basal area departed from regression estimates by 1.6% for stem wood, 4.7% for foliage, and 7.0% for branches. Ribe (1973) found that dimension analysis using logarithmic regressions for a brush of small deciduous trees overestimated total biomass by 6%, leaves by 11%, branches by 13%, and stems by 3%. These studies offer a degree of reassurance on estimation techniques for dry weights in single-age stands, but much more limited encouragement for the more complex problems of production in mixed-age forests.

Experience suggests certain cautions. First, there is a tendency to select vigorously growing trees of good form for the dimension analysis, unless this tendency is consciously counteracted. The preference for "good" sample trees implies overestimation of productivity when regressions from these trees are applied to field quadrat data. Second, the largest errors result from applying regressions to the largest trees in the samples (Ogawa *et al.*, 1965; Whittaker and Woodwell, 1968). If, from the population of large trees in the stand, many of them senescent or with partly broken crowns, a particularly "good" individual has been chosen, the slope of the regression as it extends to larger tree sizes is biased by this individual. The production estimates for the few large trees in the sample quadrat will be overestimates for most of these trees. It is therefore important that errors of estimation for large trees be controlled by some means: (1) selection for dimension analysis of large trees of as typical condition for their size range as possible, (2) correction of the production estimates for senescent large trees in the sample quadrat by the growth rates or conditions of individual trees, or (3) use of a hyperbolic equation (Ogawa *et al.*, 1965; Yoda, 1968). The latter (Fig. 4–4) is not needed for most production estimates,

FIGURE 4–4. Hyperbolic fit for leaf dry weight against stem dry weight for individual trees of tropical rainforest in Thailand (Ogawa *et al.*, 1965). Fitted line is $1/y = 13.75/x + 0.025$, with x and y in kilograms.

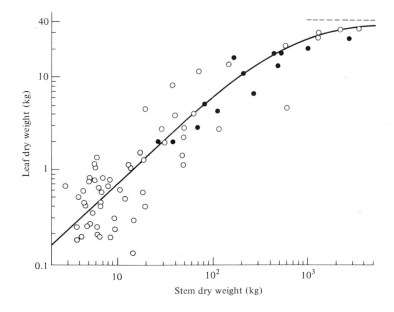

but for some properties (particularly foliage production) of large trees in some stands provides a better fit than the usual allometric equation.

Third, some sources of error may be reduced by using dimensions other than *DBH* alone as independent variables (Whittaker and Woodwell, 1968; Madgwick, 1971). The relative errors of estimate do not show consistent advantage for regressions using dimensionally related variables (e.g., leaf surface on conic surface, branch weight on stem weight or volume) over regressions using *DBH* as independent variable. Other independent variables may have some advantage, however, for their better expression of individual differences in trees. Stem and branch weight of a tree with a broken top may be less widely overestimated if the independent variable is parabolic volume rather than *DBH*. Stem-wood and branch production of a suppressed or senescent tree with narrow wood rings may be better calculated using estimated volume increment as an independent variable, than by using *DBH*. When regressions based on sample trees from one forest stand are applied to another stand, the latter stand may differ in the relation of height to diameter and in mean wood radial increment for trees of a given size. The calculations for the second stand may be less in error if the biomass estimates are based on an independent variable expressing tree height (parabolic volume), and the production estimates on an independent variable expressing wood-growth rate (estimated volume increment).

We may, finally, comment on results from parallel calculations of biomass and productivity of a given quadrat sample from different regressions. It is possible to construct and use with fair success regressions for biomass on *DBH* or parabolic volume that combine data from a number of species (Figs. 1 and 2 in Ogawa *et al.*, 1965; Kira and Shidei, 1967; Whittaker and Woodwell, 1968; Bunce, 1968; Andersson, 1970, 1971). Differences in slope of regressions have been shown for a given species in different environments (Satoo, 1962; Bunce, 1968; Whittaker *et al.*, 1974), but it may be reasonable to estimate biomass of a quadrat sample from regressions for other tree populations of similar growth form. Estimates of productivity, in contrast, may show wide differences when based on either (1) different independent variables for regressions calculated from the same dimension analysis sample, or (2) regressions using the same independent variable, calculated from dimension analysis samples for different, but apparently similar, tree populations. Results of parallel calculations of productivity from different regressions are not reassuring. Productivity estimates for a forest should be based on dimension analysis samples and regressions that are as directly appropriate as possible, with due care regarding possible sources of error in the use of these regressions.

Further research into the method is clearly needed. Apart from the evident need for study of root and branch production estimation, there is need for research into the sources of error and the limits of confidence of the method. Results of this research may contribute to another objective—understanding of where the techniques may be shortened. A production measurement that includes dimension analyses of sets of trees is laborious. There are diminishing returns from dimension analyses of additional plant populations, but further work is needed on the extent to which production estimates—as distinguished from biomass estimates—can be based on interspecies regressions or on dimension analyses of populations different from those to which regressions are applied. Most of the regressions available are based on small- to medium-sized trees, and these regressions cannot be extrapolated with confidence to large trees. Dimension analyses of large forest trees are needed, and in the future such analyses should include the whole tree and should obtain measurements and regressions for production as well as biomass. Both foresters and ecologists may gain by such work, by which both merchantable timber and total forest productivity may be measured more accurately along with a wide range of supporting information on forest dimensions.

Application to tropical forests

Dimension analysis as developed at Brookhaven is wholly dependent on those marks of age and growth rate—wood rings and bud-scale scars—that occur in most temperate-zone woody plants for its estimation of productivity. It cannot be used in this form to answer a principal question for the tropics— the net productivity of old or climax, mixed-age forests in climates without seasonal contrasts that result in wood rings. Extension of the method to tropical forests (cf. Müller and Nielsen, 1965; Kira *et al.*, 1967; Kira and Ogawa, 1971; Jordan, 1971) may involve: (1) sample quadrats in which the growths and deaths of trees are followed through a year or a longer period, with measurements of the increase in diameter (and if possible, height) of the individual trees during this period; (2) dimension analysis of trees for biomass relations to diameter (and height) at least to obtain regressions; (3) calculation, using these regressions, of the stem and branch biomass of the trees in the sample quadrat at the beginning and end of the study period, subtraction of the values to obtain stem production (partitioned if possible betweeen wood and bark) and increase in branch (wood and bark) mass. Addition to the latter value of lost branch wood and bark, collected as litter; (4) estimation of root production either from the root/shoot biomass ratio times shoot production or (in young forests) from the root mass increase during the study period with some correction for root loss; (5) independent estimation of foliage production from litter collections corrected for loss, or from foliage biomasses and leaf turnover times for the species.

Application of these and related methods to tropical forests permits only preliminary conclusions. The rate of growth of some tropical successional forests is legendary; it has been thought that tropical forest productivity much exceeded that of temperate-zone communities (Becking, 1962). It appears now, however, that the same range of 1000–3000 $g/m^2/year$ includes most temperate and tropical forests (Whittaker, 1966; Brunig, 1974); but that high values are more common in the tropics. From them result higher means, as discussed in Chapter 11 by Murphy in this volume.

Root and Shoot Relationships

Excavation

The study of productivity aboveground is affected by complexity of structure, but study of productivity belowground is affected also by the inaccessibility of roots. Knowledge of root production is consequently more primitive than that of shoot production; and much of what we know is limited to, or based on, a first, crude datum—the mass of roots present. Even this mass is not so simply obtained. The two major approaches to its determination are based on the roots in volumes of soil, and the roots of individual plants.

Bray *et al.* (1959) (see also Ovington *et al.*, 1963; Wein and Bliss, 1974) used a cylindrical corer driven into the soil with a sledgehammer to obtain samples from which the roots could be separated, and the mass of roots per unit surface area determined. The place-to-place variability of such samples (like that of other soil characteristics) is high, and a considerable number of samples are necessary to give a reliable mean value. Lieth (1968) (see also Schuster, 1964; Jeník, 1971; White *et al.*, 1971) obtained the root mass of maize (*Zea mays*) by digging pits somewhat more than 1 m deep. One side of the trench was a carefully flattened vertical surface, and from this soil cubes, 20 cm on a side, were removed to represent the different depths in the soil. Given either soil cores, or soil cubes, the root mass may be obtained by first crumbling the soil samples and removing the larger roots by hand, and then washing the remaining soil in a sieve that retains the fine roots. It is sometimes possible to distinguish the roots of different species, and to separate them to obtain biomass values for species from the root samples.

Root cores and trenches are feasible in grasslands, but for forests the more common practice is excavation of the roots of individual plants. Such excavations are, however, laborious. As observed by Lieth (1968), obtaining one figure for root mass for a sample may require three to five times more labor than all the other tissues together. The difficulty of obtaining root data increases exponentially with the size of the plants, and for trees some labor-saving means of excavation become necessary if useful data are to be obtained.

In some cases the mass washing of roots from the soil is feasible, if the community under study is accessible to a tank truck with a power hose. With or without prior excavation, the jet of water from the hose may wash soil from the root systems while leaving them largely intact. Some loss of fine roots is inescapable, and enough larger roots may be broken that correction by the procedure that follows is needed. If washing is feasible, however, the investigator may count himself fortunate for both quality and ease of collecting of root data.

The student of mountain vegetation distant from a road does not share in this fortune. In various studies of mountain forests and shrublands, root systems have had to be excavated without the aid of water and with the acceptance of substantial loss of roots by breakage. In the authors' work with shrubs and successional trees (Whittaker, 1962; Marks, 1971, 1974) roots were dug out by hand using spades, trowels, and patience to obtain root systems that were reasonably complete. For larger shrubs and trees the demands on time and patience to obtain complete root systems are excessive. It may be feasible, however, to dig out the major share of the roots attached to a plant, together with loose pieces of its own and other plants' roots. Those pieces that have all their branch roots can be matched by the diameters at their bases, to the diameters of the broken ends of roots attached to the plant, in order to obtain an estimated fresh weight of roots lost in excavation. This weight is later converted, along with that of the rest of the root system, to a dry weight on the basis of root systems or samples weighed both fresh and dry.

Finally, roots of trees have been excavated with the assistance of dynamite (Whittaker and Woodwell, 1968; Whittaker *et al.*, 1972). Dynamite sticks, increasing in number with size of the root system, are placed around and under the root crown; these can be placed so that they both lift the root systems from the soil and break larger root crowns into manageable pieces. The crown pieces are assembled, and larger roots remaining in the crater are hand excavated as far as possible and matched by their bases to broken ends on the crown. The roots attached to the crown are cut off and combined with the excavated roots, crown pieces and roots are weighed, and the diameters of distal broken ends of roots are recorded. During the excavations for a set of trees of a given species, additional sample roots of a wide range of sizes are dug up by hand or washing, as complete as possible. Regressions are computed for this set of sample roots, relating root dry weight to root basal diameter. The regressions are used to estimate, from the broken ends recorded, the dry weight of roots lost in excavation. An interspecies regression for roots of deciduous forest trees in the Hubbard Brook forest (Whittaker *et al.*, 1974) is, $\log_{10} RDW = -2.1604 + 2.0705 \log_{10} RBD$, $r = 0.947$ for 190 roots, RDW is root dry weight in grams, RBD root basal diameter in centimeters. The regression for 64 roots of *Picea rubens* at Hubbard Brook was, $\log RDW = -2.1427 + 2.0442 \log RBD$; a regression for shrub roots of three species (*Quercus ilicifolia, Gaylussacia baccata, Vaccinium vacillans*) at Brookhaven National Laboratory was $\log RDW = -1.1208 + 2.2085 \log RBD$ (Whittaker and Woodwell, 1968). Roots differ, however, in taper between some species and habitats as well as between upper and lower roots on the same root system of some species. Some species have clearly distinguished tap roots that taper rapidly with depth, and horizontal roots of slow taper. Regressions for these two root types in *Pinus rigida* at Brookhaven (Whittaker and Woodwell, 1968) were tap roots $\log RDW = -1.4309 + 2.2907 \log RBD$, horizontal roots $\log RDW = -0.8303 + 2.1325 \log RBD$.

Root/shoot ratios

Extraction of roots from soil samples can give a direct value of root mass per unit ground-surface area. More commonly, however, it is a ratio of root to shoot dry weight that is sought through excavation of roots of individual plants. A mean root/shoot ratio for these, applied to a measure of aboveground biomass for the community, gives the desired belowground biomass. (For plants with belowground stems the below- /aboveground mass ratio is not really, of course, a root/shoot ratio although we shall refer to it as such.)

Root/shoot ratios are not consistent and even within a single community divergence in these ratios and in root patterns may be part of the niche differentiation among species—differentiation toward different use of space and resources in the community. Table 4–7 compiles some root/shoot ratios; as column 6 shows, these ratios differ widely in different plants and communities (Bray, 1963; Monk, 1966; Rodin and Bazilevič, 1967, 1968). Some annual herbs are most economical in use of their production for belowground tissues, and these may have root/shoot ratios below 0.1. Some perennial herbs in con-

90

Part 2: Methods of Productivity Measurement

trast transport the greatest share of their production belowground, to root systems three to five times more massive than their shoots. Their belowground structures (that may include rhizomes or other underground stems) give the plant a protected base of survival through the unfavorable season, as well as underground storage of food to support the next summer's early shoot growth. Some shrubs with rhizome systems (Table 4–7, *Vaccinium, Gaylussacia*, etc.) also have high "root"/shoot ratios, as do some shrubs with heavy root crowns in fire-adapted communities (*Quercus ilicifolia*). Other shrubs (*Clethra acuminata, Viburnum alnifolium*) more nearly resemble small trees in their root/shoot ratios.

Two of the trees in Table 4–7 (*Quercus alba* and *Q. coccinea* at Brookhaven) have high root/shoot ratios because in this forest, as in a fire-adapted shrubland, root crowns have survived past fires and are old and heavy in comparison with the shoots they now support. For many trees, the ratios range downward from somewhat over 0.4 for seedlings, to 0.2–0.3 for young trees, to below 0.2 for large trees (Ovington, 1962; Art and Marks, 1971). In a given species the root/shoot ratios decrease with age (note *Rhododendron maximum* and *Acer saccharum*), and increase toward drier environments (Bray, 1963; Whittaker, 1962; Harris *et al.*, 1973). A ratio of 0.2 has been used as an approximate intermediate value for forest trees since the early work of Möller (1945, 1947), and a recent intensive study (with 81 root systems excavated at Hubbard Brook, Whittaker *et al.*, 1974) gave mean values for elevation belts in deciduous forest quite close to this (0.18 to 0.21). The mean value for the fire-adapted Brookhaven forest is much higher, 0.59. The over-all relation of root/shoot ratios to plant size, for several species of shrubs and small trees, was expressed by Whittaker and Woodwell (1968) as a regression: log (root/shoot ratio) $= -0.0473 - 0.414$ log (shoot basal diameter, cm). Regressions relating root-system dry weight (*RSDW*) to woody shoot system dry weight (*SSDW*) are: log $RSDW = 0.4374 + 0.7887$ log $SSDW$ (shrubs and small trees only, Brookhaven and Great Smoky Mountains), and log $RSDW = -0.1826 - 0.9037$ log $SSDW$ (small and medium-sized trees, Hubbard Brook). Figure 4–5 gives root versus shoot weights for individual trees, from the authors' and others' data. In this plot, the smallest individuals are *Acer saccharum* and young successional *Prunus* and *Populus*. These, in contrast to most shrubs and many small-tree species, have low root/shoot ratios. With shrub species with heavier roots excluded from Figure 4–5, a single allometric trend connects the wide range of seedlings, saplings, and trees represented. The slope B of this trend is different from those for the Brookhaven and Hubbard Brook regressions, and close to 1.0.

Root production

Root production is in most cases impossible to measure directly in the field. In principle, the same approaches through wood rings can be applied to root crowns and rhizomes as to stems, and the same approach through weight–age relationships can be applied to roots as to branches. The editors of this volume have experimented with such applications (Lieth, 1968; Whittaker, 1962), but they have had no systematic use on communities. The fact that wood rings, and

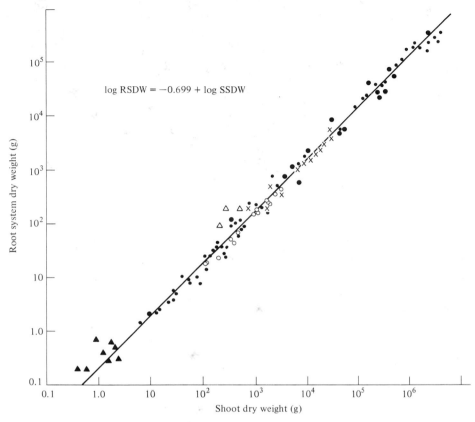

FIGURE 4–5. Root system dry weight against shoot system dry weight (including foliage) for individual trees of various species, from data of authors and others. (L. K. Forcier supplied data for *Acer saccharum* seedlings.) Smallest individuals are seedlings of *Prunus pensylvanica* and *Acer saccharum*. Intermediate size individuals are *Populus tremuloides* and *Prunus pensylvanica*; the largest are *Acer saccharum*, *Betula allegheniensis* and *Fagus grandifolia*. All trees were sampled in New Hampshire.

consequently ages, seem even more uncertain in roots than in branches makes such applications questionable. In some plants increase in root mass can be measured late in the season, as profit from the summer's photosynthesis is transferred underground. Therefore, in a grassland root mass at its minimum in late winter can be subtracted from the maximum in late summer as an indication of belowground growth (Dahlman and Kucera, 1965; Evans and Wiegert, 1966; Singh and Yadava, 1974). This difference is at best a low estimate of root production, and for many plants the change in root mass is not easily related to aboveground and total net primary productivity. If the leaves are exposed to tracer CO_2, then after a suitable time the distribution of the tracer above- and belowground may express the ratio of net primary productivity above and belowground. The reliability of such measurement may be affected by the rates of transfer and respiration of the tagged organic matter in the plant. Extensive

Table 4-7 Root/shoot relationships of terrestrial plants[a]

Terrestrial plant	Shoot		Root system		Root/shoot ratio	Reference
	Foliage and twigs	Stem and branches	Roots	Crown* or rhizome†		
Annual herbs						
Solanum lycopersicum (tomato)	92		8		0.09	Lieth (1972)
Cassia fasciculata (prairie senna)	88		12		0.133	Monk (1966)
Helianthus annuus (sunflower)	91		9		0.10	Lieth (1972)
Zea mays (maize)	66		34		0.05	Lieth (1972)
Secale cereale (rye)	82		18		0.22	Lieth (1962)
Galium aparine (bedstraw)	86		14		0.09	Struik (1965)
Ambrosia artemisifolia (ragweed)	92		8		0.09	Struik (1965)
Perennial herbs						
Carex lacustris (sedge, in marsh)	67		33		0.50	J. M. Bernard (*personal communication*)
Andropogon scoparius (grass, prairie)	82		18		0.22	Lieth (1968)
Panicum aciculare (grass, old-field)	74		26		0.354	Monk (1966)
Sorghastrum nutans (grass, old-field)	17		83		4.8	Bray *et al.* (1959)

Festuca ovina (grass, alpine)	69		31		0.44	Scott and Billings (1964)
Arid steppe	15		85		5.7	Rodin and Bazilevič (1967)
Trifolium parryi (alpine clover)	23		77		3.4	Scott and Billings (1969)
Lespedeza cuneata	69		31		0.45	Monk (1966)
Typha latifolia (cattail, Texas)	34		66		1.98	McNaughton (1966)
Beta sp. (beets)	25		75		3.0	Lieth (1962, 1972)
Solanum tuberosum (potato)	42		58		1.38	Lieth (1962, 1972)
Phlox caespitosa (alpine phlox)	62		38		0.6	Scott and Billings (1964)
Circaea quadrisulcata (enchanter's nightshade)	67		33		0.5	Struik (1965)
Shrubs						
Vaccinium vacillans (blueberry)	30.9	12.0	7.6	49.5†	2.52	Whittaker and Woodwell (1968)
Gaylussacia baccata (huckleberry)	17.6	25.8	6.4	50.2†	2.20	Whittaker and Woodwell (1968)
Quercus ilicifolia (shrubby oak)	15.4	11.7	38.8	34.1*	6.23	Whittaker and Woodwell (1968)
Kalmia latifolia (mountain laurel)	10.0	31.0	55.0		1.78	Whittaker (1962)
Rhododendron maximum (small)	51.2	18.6	30.2		1.62	Whittaker (1962)
Rhododendron maximum (large)	13.1	49.7	37.2		0.75	Whittaker (1962)
Calluna vulgaris (heather)	19.4	28.4	52.2		1.09	Forrest (1971)
Clethra acuminata (white alder)	9.7	54.1	36.2		0.67	Whittaker (1962)
Viburnum alnifolium (hobble bush)	8.1	56.7	35.2		0.62	Whittaker (1962)
Larrea divaricata (creosote bush)	6.6	67.3	13.8	12.3*	0.39	Chew and Chew (1965)

[a] Root/shoot ratios for woody plants compare the root system with the woody shoot (stem and branch wood and bark).

Table 4–7 *continued*

1	2	3	4	5	6	7
	Shoot		Root system			
			Dry-weight per cents in:			
Terrestrial plant	Foliage and twigs	Stem and branches	Roots	Crown* or rhizome†	Root/ shoot ratio	Reference
Trees						
Phyllostachys bambusoides (bamboo)	16.9	45.0	16.5	21.6†	0.85	Numata (1965)
Prunus pensylvanica (pin cherry, successional)	10.0	76.0	14.0		0.19	Marks (1971)
Quercus alba (white oak, fire affected)	7.3	48.6	24.6	19.5*	0.91	Whittaker and Woodwell (1968)
Quercus coccinea (scarlet oak, fire affected)	7.9	62.8	16.4	12.9*	0.47	Whittaker and Woodwell (1968)
Acer spicatum (mountain maple, small)	2.8	74.4	9.1	13.7*	0.305	Whittaker *et al.* (1974)
Corylus avellana (hazel, small tree)	5.7	74.8	19.5		0.26	Andersson (1970)
Acer saccharum (sugar maple, mature)	1.5	83.2	9.8	5.5*	0.187	Whittaker *et al.* (1974)
Acer saccharum (sugar maple, small)	4.4	71.0	13.6	10.9*	0.345	Whittaker *et al.* (1974)
Quercus robur (oak, mature)	2.4	82.4	15.2		0.185	Andersson (1970)
Pinus rigida (pitch pine)	9.8	70.4	10.0	9.8*	0.28	Whittaker and Woodwell (1968)
Picea rubens (red spruce)	7.4	66.9	16.0	9.7*	0.384	Whittaker *et al.* (1974)
Tropical deciduous forest		91.3	8.7		0.095	Greenland and Kowal (1960)

[a] Root/shoot ratios for woody plants compare the root system with the woody shoot (stem and branch wood and bark).

transactions beneath the soil surface, involving multidirectional movements of material between roots, mycorrhiza and other microorganisms, and soil, may be not merely unmeasured but largely unknown.

Most estimates of root production for trees are based on the assumption that the ratio of production to mass must be similar for the root system and the shoot system (Whittaker, 1962; Newbould, 1968; Kira and Ogawa, 1968; Andersson, 1970). The amounts of annual loss of root hairs and roots from plants in the field are almost unknown. The ratio of wood and bark production to mass, aboveground, times root system mass is likely to give an underestimate of root production because it omits root loss from consideration. The ratio of total aboveground production (including leaves) to mass, aboveground, times root system mass probably gives an overestimate. True root production may well lie between these two values (Whittaker, 1962). Some improvement of the estimate may be possible by summing: (1) the ratio of stem-wood and bark growth to stem mass times root crown mass, and (2) the ratio of branch wood and bark growth plus current twigs to branch mass, times root mass without crowns (Whittaker *et al.*, 1974). In the *Liriodendron* forest at Oak Ridge, Tennessee, change of root biomass with season was followed by extracting roots from soil cores and pits (Reichle *et al.*, 1973). The measurements indicated an increase of 750 g/m^2/year in lateral root mass, this increase being about 80% of the amount fixed in aboveground and root crown tissues. These results, which suggest that the conventional forest root/shoot production ratio of 0.2 may be a serious underestimate, have not yet been tested in other forests. No independent measure of forest root production, by which these divergent estimates might be checked, is available. For other discussions of root production see Bray (1963), Lieth (1962, 1968), Newbould (1967, 1968), and Ghilarov *et al.* (1968).

Gas-Exchange Approaches

Cuvettes

Gas-exchange approaches to measuring productivity have varied widely with investigators' purposes and objects of study. Infrared analysis of CO_2 content of air has made possible extensive application of gas-exchange measurements to terrestrial communities. Recent papers (Eckardt, 1968; Larcher, 1969; Woodwell and Botkin, 1970; Lange and Schulze, 1971; Schulze and Koch, 1971; Mooney, 1972; Tranquillini and Caldwell, 1972) cover details of the most commonly used leaf and plant cuvettes or chambers and equipment for CO_2 analysis.

The largest number of studies deal with photosynthesis of individual leaves or twigs in cuvettes or transparent cylinders, and measure the CO_2 content of air as it enters and leaves the cuvette. The possibility that CO_2 dissolved in the transpiration stream may be fixed photosynthetically in the leaves has not been investigated to our knowledge. The preferred measurement in gas-exchange work would be gross photosynthesis per unit leaf area. Because of various difficulties, notably those of measuring photorespiration (Zelitch, 1964; Botkin *et al.*, 1970; Black, 1971), net photosynthesis or net assimilation rate is usually measured (Schulze and Koch, 1971). Any single value, whether gross or net, must fail to express differences between plant species in light and dark respiration and assimilate use, to say nothing of community-level differences in respiration and photosynthetic efficiency in relation to leaf area and arrangement (see Watson, 1958). The many advantages, primarily in measurement, of working with CO_2 flux in a single, attached leaf in a small leaf chamber under conditions that can be reasonably measured and controlled have been summarized by Wallace *et al.* (1972).

Monitoring of net assimilation of CO_2 in individual, attached leaves is probably most useful for study of difference in assimilation rates in different species and in different environments; it is of questionable value for study of community productivity. There are problems enough in monitoring CO_2 flux for an individual leaf in a growth chamber (minimizing chamber effects, selecting a leaf of standard age, as age affects net assimilation rate, taking into consideration leaf position on the plant, water status of the entire plant and of the leaf in the chamber, and so on). For communities, these problems are joined by others involving respiratory losses of branches, stems and roots, and different and complexly changing exposure of leaves and plants to light and other environmental factors (Botkin *et al.*, 1970; Woodwell and Botkin, 1970). Bark photosynthesis may supplement leaf photosynthesis and may need to be allowed for in calculating branch respiration and total production (e.g., Pearson and Lawrence, 1958; Strain and Johnson, 1963; Perry, 1971). The further labor of a dimension analysis of the forest may be needed to convert photosynthesis measurements on individual leaves or twigs and respiration measurements on particular bark surfaces to estimates for the full foliage and bark surface of the community. In principle the full range of measurements are possible; extensive labor and technological support made possible both dimension analysis and the necessary gas-exchange measurements in numerous cuvettes for two young forests at Brookhaven National Laboratory (Woodwell and Whittaker, 1968; Woodwell and Botkin, 1970; Botkin *et al.*, 1970) and Oak Ridge National Laboratory (Reichle *et al.*, 1973b). In practice for many research projects, the effort necessary to close the gap between gas exchange for leaves or twigs in cuvettes, and community productivity, is prohibitive. Gas-exchange measurement of terrestrial productivity is not lightly to be undertaken.

Micrometeorologic approach

The difficulties of integrating measurements in cuvettes make attractive an alternative—study of whole-community gas exchange, with measurement of daytime depletion and nighttime accumulation of CO_2 in different strata of the community (Baumgartner, 1969). Such techniques have been applied to agricultural communities (Lemon, 1967, 1968; Monteith, 1968; and Inoue, 1968), grasslands (Totsuka *et al.*, 1968), tundra (Johnson and Kelley, 1970), and forests (Baumgartner, 1968, 1969, Woodwell and Dykeman, 1966; Lemon *et al.*, 1970; Allen *et al.*, 1972). The approach has advantage over use of cuvettes in that natural conditions are maintained during the course of measurement. Community characteristics of obvious importance to production such as leaf arrangement and canopy architecture are undisturbed; measurements of these become part of the basis of the production estimate. Study of such community-level characteristics should bring closer together gas-exchange measurements and actual dry matter production. Lemon (1969) reports good agreement between light saturation curves for corn determined by his own work monitoring diurnal course of CO_2 flux intensities at different heights in the community, and curves determined by Musgrave using individual, attached corn leaves in plastic chambers.

H. T. Odum and Jordan (1970) sought to measure community gas exchange in a giant plastic cylinder (60 ft across) enclosing a piece of Puerto Rican rain forest. Ordway (1969) gives a critical evaluation of both giant cylinder and micrometeorologic approaches.

The Brookhaven inversion approach

A variant of the micrometeorologic approach was used at Brookhaven National Laboratory, New York (Woodwell and Dykeman, 1966). Local temperature inversions, which served as a barrier to CO_2 escape, were used to measure nocturnal accumulation of CO_2. As would be expected, the accumulations of CO_2 were temperature dependent. The nighttime accumulation of CO_2, which is most pronounced near the soil surface, represents total plant and consumer respiration. Woodwell and Dykeman measured this nocturnal buildup of CO_2 under temperature inversions throughout the year as meteorologic conditions permitted, and then plotted rate of CO_2 production on a daily basis against mean temperatures. Relations between CO_2 production and temperature differed with season; cold adaptation of the organisms and community appeared in lower rates of CO_2 release at a given temperature in the dormant, than in the growing, season. Using these CO_2–temperature relationships total ecosystem respiration for the year was estimated from local temperature records. As in other field studies, light and dark respiration were treated as the same (lacking means of correcting for this recognized error). We might expect that, in a fully mature, climax community total respiration (of plant, animals, and saprobes) should approximate gross primary productivity. In the young Brookhaven forest it did not, but total respiration could be related to gross and net primary productivity and net ecosystem production on the basis of other studies of the forest (Woodwell and Whittaker, 1968; Whittaker and Woodwell, 1969).

Soil respiration

Soil respiration is a useful index of overall biologic activity in the soil, and has been suggested as an index of primary productivity (Waksman and Starkey, 1924; see also Voigt, 1962). However, the evolution of CO_2 from the soil, which is what is actually measured, is not necessarily equivalent to soil respiration because of losses to deep percolating water, and anerobic respiration (Lieth and Ouellette, 1962; Woodwell and Botkin, 1970; Kucera and Kirkham, 1971). Smirnov (1955; see Voigt, 1962) found a good relationship between CO_2 evolution from the soil and net productivity in various forest stands. The relationship should be expected, as temperature and moisture conditions favorable for soil heterotroph activity and root respiration should also be favorable for primary production, most of which feeds into the soil heterotroph system in mature forests.

Lieth and Ouellette (1962) and H. T. Odum *et al.* (1970) have discussed usefulness and problems of method. Chief among the latter is the disruption of natural air circulation when any closed chamber or funnel-like device is placed on the soil surface. At one extreme, when a closed chamber is used so that CO_2 from the soil diffuses into still air, measurements of CO_2 flux are probably low (H. T. Odum *et al.*, 1970). At the other extreme, when air is pumped through an open chamber into a gas analyzer, CO_2 can actually be pulled out of the soil to produce an overestimation the degree of which depends on flow rate (Kucera and Kirkham, 1971). Reiners (1968) and Kucera and Kirkham have devised sampling systems that strike a compromise between these extremes.

The sources of the CO_2 liberated from soils include decomposition of many components of aboveground litter, organic compounds in stemflow, throughfall, and root exudates, and sloughed root tissue and dead roots, along with respiration of roots and of animals. Carbon evolution from soils is consequently greater than the carbon contribution to the soil in litter (Reiners, 1968; H. T. Odum *et al.*, 1970; Kucera and Kirkham, 1971). Much as we should like to separate the components of soil CO_2 release, the problems seem insurmountable for the present, at least, in forests. For tall-grass prairie, Kucera and Kirkham (1971) offer as a tentative breakdown 60% of total CO_2 release from decomposer respiration and the balance from root metabolism.

Gross primary productivity

We trust a sense of the difficulty of gas-exchange measurements may be communicated. Further details of carbon cycling in the plant that affect the interpretation of gross productivity and its relationship to net productivity are beyond the scope of this chapter. Despite ecologists' interest in total energy flow through the community, practicalities of method have pressed ecologists' concerns away from gross and toward net primary productivity. The standard measurement by which productivity of land communities is to be expressed and compared is consequently net primary productivity in dry matter, $g/m^2/year$, the basic datum from which concern may variously proceed to biomass accumulation and turnover, to gross productivity by way of respiration measurement or estimate, to energy flow by way of caloric equivalents, to nutrient cycling by way of elemental contents of tissues, to animal productivities by way of consumption, and to productivity as carbon for comparison with aquatic productivities thus expressed. Although research feasibility has influenced the emphasis on net productivity, it can be argued that this is as fundamental a community characteristic as gross productivity; for net productivity is the basis for biomass accumulation and community structure, and for the function of all trophic levels above the plants. A few generalizations on gross primary productivity can be offered.

The fraction of gross primary productivity expended in plant respiration is variable over a range of probably 20–80%. In an early study of maize, Transeau

(1926) reported respiration to be 23% of gross productivity; values for other annual plants are 20–40% (Müller, 1962). For a young ash woods Möller *et al.* (1954a, b) estimated 29%; estimates for other temperate successional forests are 40–60% (Möller *et al.*, 1954a, b; Ogawa *et al.*, 1961; Müller, 1962; Yoda *et al.*, 1965; Woodwell and Whittaker, 1968). Higher respiration rates have been obtained in tropical forests, as observed above (Müller and Nielsen, 1965; Hozumi *et al.*, 1969a; Kira *et al.*, 1964, 1967; Kira, 1968). Some of these forests, with high net productivity and high plant respiration have gross primary productivities of 10,000–12,000 g/m²/year, as a probable maximum for terrestrial natural communities.

Two major correlations for respiration rates suggest themselves. For plant communities of comparable structures, the respiration rate increases with temperature. For communities at comparable temperatures, the respiration rate increases with massiveness of community structure. The latter may be expressed as biomass or, perhaps more appropriately, as the biomass accumulation ratio (biomass/annual production), since the latter more directly represents the "load" of respiring tissue to be supported per unit of productivity. The relationship of community respiration to biomass accumulation ratio in temperate and tropical communities is shown in Fig. 4–6.

FIGURE 4–6. Plant respiration rate (as percentage of gross primary productivity) against biomass accumulation ratio (biomass/net annual production). Data for temperate communities (circles) from H. T. Odum (1971), Ovington (1962), Whittaker and Woodwell (1969), Ogawa *et al.* (1961), Möller *et al.* (1954b), and Maruyama (1971). Data for tropical communities (squares) from Müller and Nielsen (1965), H. T. Odum (1971), Ogawa *et al.* (1961), and Kira *et al.* (1967). Hand-fitted trend line for temperate communities is % Ra = 35 + 20 log BAR.

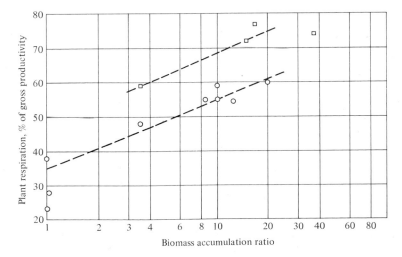

For some communities data obtained by different techniques permit calculation of production balances. For a fully mature climax community on land, total community respiration should equal approximately gross primary productivity; and the production balance becomes rather simply, as illustrated with values for a tropical forest (Kira and Ogawa, 1971, assuming the forest to be climax):

$$\begin{array}{ccccccccc} \text{GPP} & = & \text{NPP} & + & R_a & = & R_h & + & R_a \\ 12{,}300 & = & 3000 & + & 9300 & = & 3000 & + & 9300 \end{array} \text{ g/m}^2\text{/year}$$

(R_a and R_h are autotroph and heterotroph respiration, respectively.)
For a typical cereal crop

$$\begin{array}{ccccccccccc} \text{GPP} & = & \text{NPP} & + & R_a & = & R_h & + & R_a & + & \text{Yield} \\ 800 & = & 650 & + & 150 & = & 450 & + & 150 & + & 200 \end{array} \text{ g/m}^2\text{/year}$$

if about 30% of the NPP is harvested as grain and the remainder is left in the field as mulch until decomposed. For forests the difference between GPP and total respiration is "yield" only if harvested. The difference otherwise appears in the accumulation of wood and bark and soil organic matter as net ecosystem production. For the Brookhaven forest, various techniques of dimension analysis, leaf harvest measurement, and gas exchange in cuvettes and beneath inversions were brought to bear on the determination of the production balance (Woodwell and Whittaker, 1968; Whittaker and Woodwell, 1969). A comparable analysis has been carried out by Reichle (1973b) at Oak Ridge National Laboratory, Tennessee, in a young *Liriodendron tulipifera* (tulip poplar) forest similar to column 7, Table 4–6, but less productive. Only incomplete measurements of plant respiration and no measurements of animal consumption belowground are available. With inferences regarding these, however, the production balances for these two forests become

$$\text{GPP} = \text{NPP} + R_a = R_h + R_a + \text{NEP}$$

Brookhaven

$$2646 = 1195 + 1451 = 653 + 1451 + 542 \text{ g/m}^2\text{/year}$$

Oak Ridge

$$3280 = 1380 + 1900 = 1060 + 1900 + 320 \text{ g/m}^2\text{/year}$$

The net ecosystem production is in these cases wood and bark accumulating as net community growth. These young forests are characterized by ratios of NEP to NPP of 0.45, and 0.23, and of total respiration to GPP of 0.80 and 0.90. As the forests mature to climax stature these ratios should approach 0.0 and 1.0, respectively. ("Approach" rather than equality is indicated because in climax there may be some net import, or export, of leaf litter or soil organic solutes. Also, in some climax communities there is slow accumulation of net ecosystem production as peat.)

Leaves, Chlorophyll, and Light

It is natural to seek short-cuts to estimation of productivity through indices that relate to photosynthesis. Among the indices that suggest themselves are: the dry weight of current twigs and leaves (previously discussed in connection with estimative ratios), annual fall of leaves in litter, leaf-area index, chlorophyll content per unit area, and light extinction by the foliage.

Of these, the use of the dry weight of current twigs and leaves as an approach to production has been discussed. "Clipping dry weight" can indeed serve as a basis of production estimate, but this use requires knowledge of the ratios of clipping weight to total production. Much the same is true of annual litter fall. The amount of leaves collected in litter baskets tends seriously to understate actual leaf production (Bray and Gorham, 1964). Furthermore, similar amounts of leaf litter can be obtained from forests differing significantly in productivity of woody tissues. The more productive the forest is in growth of woody tissues, the less effectively litter collections express its productivity. Many temperate-zone forests are convergent in the amounts of their leaf productions (300–400 g/m²/year) while differing widely in rate of wood and bark growth.

Leaf-area index (mean number of square meters of leaf surface above a square meter of ground surface) and chlorophyll content are more directly expressive of the photosynthetic apparatus of the community. For a given species or kind of community, these may be strongly related to productivity (Fig. 4–7); for different plants and communities their relation to productivity is weak (Figs. 4–8 and 4–9). As expressions of forest productivity leaf area index and chlorophyll content are subject to the same limitations as leaf mass—they are convergent in forests of quite different growth rates. They are at the same time divergent in evergreen, as compared with deciduous forests of the same productivities (see Fig. 4–9 and Chapter 5). Light extinction, from the upper surface of the community to the ground surface, is also correlated with foliage mass, leaf area and chlorophyll, and productivity; but the correlation with productivity again is loose. These measurements are surely of interest in the study of productivity; but their bearing on the amount of productivity is, in general, suggestive rather than effective (Medina and Lieth, 1963, 1964; Whittaker, 1966; Whittaker and Woodwell, 1971).

Some of their limitations might well be escaped by a more detailed analysis of community structure and photosynthetic function. Models of community function based on structure and light relationships, or these plus gas exchange, as determinants of photosynthesis have been developed by a number of authors (Monsi and Saeki, 1953; Saeki, 1963; Monsi, 1968; Kuriowa, 1968; Duncan *et al.*, 1967; Maruyama, 1971; Lemon, 1967; Lemon *et al.*, 1970; see also the section on gas exchange). Such models are approximate when they are simple, and of formidable complexity when they are detailed enough to be accurate.

FIGURE 4–7. Relationship of productivity (as indicated by the seasonal maximum biomass) in meadows to total chlorophyll content at the same time. From Medina and Lieth (1964).

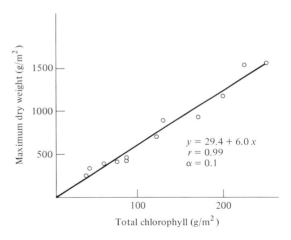

$y = 29.4 + 6.0\,x$
$r = 0.99$
$\alpha = 0.1$

FIGURE 4–8. Correlation between chlorophyll content and net primary production in various communities, data of Bray (1960, 1962) and Medina and Lieth (1963, 1964) from (Lieth 1972, Fig. 2).

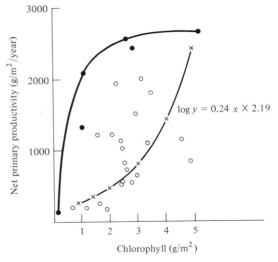

$\log y = 0.24\,x \times 2.19$

They may be better regarded as directions of research toward understanding of productivity than as bases of measuring productivity.

Conclusions

It would indeed be a welcome circumstance if one simple measurement were a sufficient index of relative productivity of land communities. There may be no such measurement. Figure 4–10 shows relationships of forest production to three other accessible measurements: mean tree height, basal area, and estimated volume increment.

Tree height at a given age has long been used in practical forestry as an index of site quality and relative productivity of plantations. In natural forests of mixed ages production is related to height, but less simply. Figure 4–10

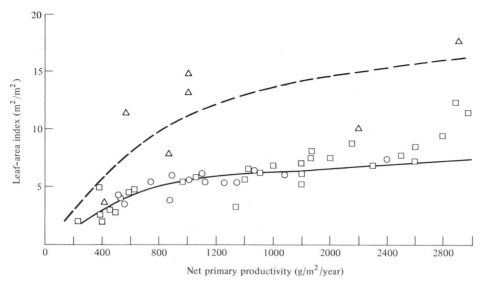

FIGURE 4–9. Leaf-area index (square meters of leaf blade surface per square meter of ground surface) in forest and shrub communities, in relation to aboveground net primary productivity. (○) Deciduous broadleaf species; (□) evergreen broadleaf species, and triangles evergreen needleleaf species. Surfaces are based on one side only of broad leaves, but full perimeter of needles. Visual trend lines are for evergreen needle leaf and deciduous broadleaf species. Data are from Art and Marks (1971), Kira *et al.* (1967), and work of the authors.

shows that for climax forests in the Great Smoky Mountains the relationship is significant (coefficient of correlation $r = 0.85$, Whittaker, 1966). It is not, however, tight enough to use as an index of climax productivity; and it does not apply to young forests. Because the trees of a climax forest are of mixed heights, a best number for canopy height is not easily determined. Figure 4–10 is based not on canopy but on weighted mean tree height (2 × parabolic volume/basal area). Basal area (square meters of stem cross-sectional area at breast height per hectare), a common measure of forest structure, is poorly related to productivity (Fig. 4–10). Estimated volume increment, which includes radial wood growth at breast height as an index of growth rate, is a useful first index of forest production (Fig. 4–10). The variation of radial increments at different heights on a tree stem implies that estimated volume increment is an expression, not a measure, of actual stem wood growth. Ratios of these two in sets of woody plants that have been analyzed are 1.17–1.57 in arborescent shrubs, 1.35–1.62 in small and 0.94–1.18 in larger trees (Whittaker, 1962; Whittaker and Woodwell, 1968; Whittaker *et al.*, 1974). Ratios of leaf production and branch production to stem-wood production also are variable. Although no single index of productivity seems adequate, a combination of indices—such as estimated volume increment with a correction for tree branching form and an independent

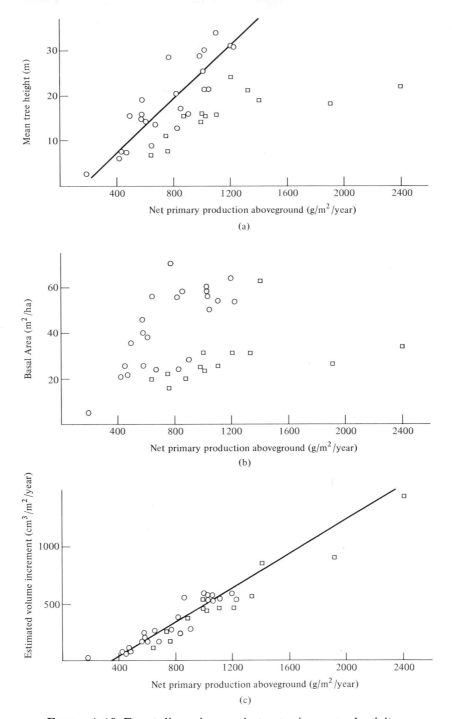

FIGURE 4–10. Forest dimensions against net primary productivity, aboveground. (○) Climax and near-climax forests and woodlands; (□) young stands, both from production samples of the first author (R. H. W.).

estimate of foliage from litter collections or regression on parabolic volume—might prove serviceable. A variety of indirect methods useful for the assessment of NPP were discussed in Lieth (1962) and Lieth (1965). The use of phytosociologic information cannot be included here.

Prediction of productivity from environmental variables may also be considered. For terrestrial communities principal variables are moisture availability and temperature; additional ones are sunlight intensity, nutrient availability, and seasonal change in climatic factors. A number of people have established correlations of productivity with these variables, or combinations of them. Walter (1939, 1964) showed that in grasslands of fairly dry climates aboveground production increased with precipitation in a nearly linear manner, at 1 $g/m^2/year$ per millimeter of precipitation. Particularly favorable circumstances (in consistency of method and character of the communities) may be necessary to give data with so tight a fit. Paterson (1961) has employed formulas using several climatic variables (mean temperature of the warmest month, range between warmest and coldest months, precipitation amount, length of growing season, and insolation). Rosenzweig (1968) has shown an effective, logarithmic relationship between net primary production of climax vegetation and actual evapotranspiration; the relation is further discussed in Chapters 7 and 12 in this volume and by Whittaker and Niering (1975). Lieth and Box have made extensive use of models predicting primary productivity from environmental parameters. Those assessments are described in section 4 of this volume.

Russian work (Drozdov, 1971; Bazilevich *et al.*, 1971a and b) has related productivity to the ratio of radiation intensity and the amount of heat needed to evaporate the annual precipitation. Productivity has been correlated with elevation by Filzer (1951), Whittaker (1966) and Maruyama (1971). Considering climax forests only, aboveground net annual production decreased at a mean rate of 356 g/m^2 and aboveground biomass at a mean rate of 230 t/ha per 1000 m gain in elevation in the Great Smoky Mountains (Whittaker, 1966). Maruyama (1971) illustrates highly dispersed relations of biomass and production to elevation in Japanese beech forests, with a trend of 2000 $g/m^2/year$ decrease in gross primary productivity per 1000 m; and Kira and Shidei (1967) and Yoda (1968) illustrate complex, curvilinear relations of biomass to elevation (Fig. 4–11). Figure 4–12 indicates, for climax forests in the Great Smoky Mountains, some of the relationships underlying the decrease in mean production with elevation. Both deciduous and coniferous forests of moist sites have aboveground net productivity in the range of 1000–1300 $g/m^2/year$ below a 1500-m elevation. Above that elevation production of deciduous forests decreases rapidly, but that of coniferous stands, apparently better adapted to subalpine climates, decreases less rapidly. Production of pine forests of dry sites is lower throughout the elevation range sampled and decreases more rapidly with elevation. Both Whittaker (1966) and Maruyama (1971) found prediction of production from multiple correlation with elevation and indices of topographic moisture conditions feasible, but such correlations are not easily applied to other areas. It is not hard to establish correlations of productivity with environmental factors for a limited set of climax communities; but for wider ranges of communities

Part 2: Methods of Productivity Measurement

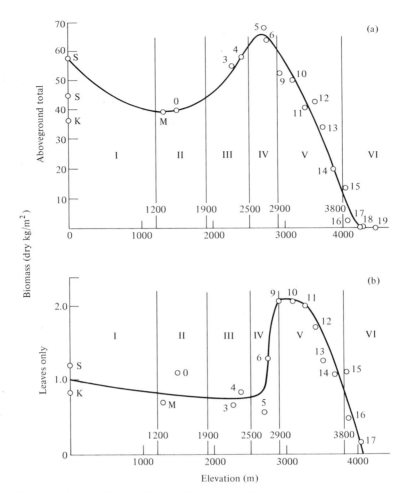

FIGURE 4–11. Biomass in relationship to elevation for a series of climax forest and shrub communities of the eastern Nepalese Himalayas (Yoda *et al.*, 1968). Arabic numerals and letters represent sample plots; roman numerals are elevation zones: I tropical and subtropical, II warm-temperate, III cool-temperate, IV alpine.

affected by additional factors and of different ages, the data scatter widely. The extent of the scatter, and an approach to summarizing trends in relation to climate, are illustrated in Chapter 12.

Measurements of rates in the complex function of living systems are not easy, whether the systems in question are cells, organisms, or communities. This review may indicate some of the uncertainties and directions in which research is needed, in the measurement and prediction of terrestrial primary production. The growth in knowledge of production amounts and factors affecting these, since the pioneer work of Boysen Jensen (1932), Burger (1929, 1953), Möller

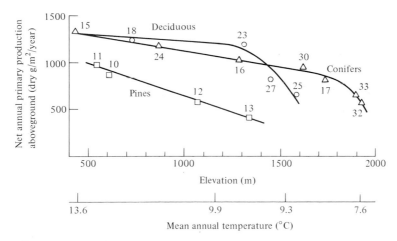

FIGURE 4–12. Aboveground net primary production of climax forests against elevation in Great Smoky Mountains, Tennessee. Different patterns of response are shown by conifers (abietine forests of mesic environments, dominated by *Abies fraseri, Picea rubens,* and *Tsuga canadensis*), deciduous forests of mesic environments, and pine forests and pine heaths of xeric environments. Numbers at points are those of the samples of Whittaker (1966); temperature data are from Shanks (1954).

(1945, Filzer (1951), Satoo *et al.* (1955), and Ovington (1956), nonetheless seems impressive. Enough is known about productivity to permit some generalizations for certain kinds of land communities and for the total land surface of the earth in subsequent chapters.

Acknowledgments

This chapter is an outgrowth of the authors' research on forest and shrubland productivity in projects supported by the National Science Foundation, and work by the first author and G. M. Woodwell at Brookhaven National Laboratory under the auspices of the U. S. Atomic Energy Commission.

References

Allen, L. H. Jr., E. Lemon, and L. Müller. 1972. Environment of a Costa Rican forest. *Ecology* 53:102–111.

Andersson, F. 1970. Ecological studies in a Scanian woodland and meadow area, southern Sweden. II. Plant biomass, primary production and turnover of organic matter. *Bot. Notiser* 123:8–51.

———. 1971. Methods and preliminary results of estimation of biomass and primary production in a south Swedish mixed deciduous woodland. (French summ.) In *Productivity of Forest Ecosystems: Proc. Brussels Symp., 1969*, P. Duvigneaud, ed. *Ecology and Conservation*, Vol. 4, 281–288. Paris: UNESCO.

Art, H. W., and P. L. Marks. 1971. A summary table of biomass and net annual primary production in forest ecosystems of the world. In *Forest Biomass Studies*, H. E. Young, ed., pp. 3–32. Orono, Maine: Univ. of Maine, Life Sciences and Agriculture Experiment Station.

Arvisto, E. 1970. Content, supply, and net primary production of biochemical constituents in the phytomass of spruce stands on brown forest soils. *Estonian Contributions to the International Biological Programme*, T. Frey, L. Laasimer, and L. Reintam, eds., Vol. 1, 49–70. Tartu: Acad. Sci. Estonian SSR.

Attiwill, P. M. 1966. A method for estimating crown weight in Eucalyptus and some implications of relationships between crown weight and stem diameter. *Ecology* 47: 795–804.

———, and J. D. Ovington. 1968. Determination of forest biomass. *Forest Sci.* 14: 13–15.

Baskerville, G. L. 1965a. Dry matter production in immature balsam fir stands. (French summ.) *Forest Sci. Monogr.* 9:1–42.

———. 1965b. Estimation of dry weight of tree components and total standing crop in conifer stands. *Ecology* 46:867–869.

———. 1972. Use of logarithmic regression in the estimation of plant biomass. (French summ.) *Can. J. For. Res.* 2:49–53.

Baumgartner, A. 1968. Ecological significance of the vertical energy distribution in plant stands. (French summ.) In *Functioning of Terrestrial Ecosystems at the Primary Production Level: Proc. Copenhagen Symp. 1965*, F. E. Eckardt, ed. *Natural Resources Research*, Vol. 5, 367–374. Paris: UNESCO.

———. 1969. Meteorological approach to the exchange of CO_2 between the atmosphere and vegetation, particularly forest stands. *Photosynthetica* 3:127–149.

Bazilevich, N. I., A. V. Drozdov, and L. E. Rodin. 1971a. World forest productivity, its basic regularities and relationships with climatic factors. (French summ.) In *Productivity of Forest Ecosystems: Proc. Brussels Symp. 1969*, P. Duvigneaud, ed. *Ecology and Conservation*, Vol. 4, 345–353. Paris: UNESCO.

———, L. Ye. Rodin, and N. N. Rozov. 1971b. Geographical aspects of biological productivity. *Sov. Geogr. Rev. Transl.* 12:293–317.

Beauchamp, J. J., and J. S. Olson. 1973. Corrections for bias in regression estimates after logarithmic transformation. *Ecology* 54:1403–1407.

Becking, J. H. 1962. Ein Vergleich der Holzproduktion im gemässigten und in tropischen Klima. In *Die Stoffproduktion der Pflanzendecke*, H. Lieth, ed., pp. 128–133. Stuttgart: Fischer.

Black, C. C. 1971. Ecological implications of dividing plants into groups with distinct photosynthetic production capacities. *Advan. Ecol. Res.* 7:87–114.

Bliss, L. C. 1966. Plant productivity in alpine microenvironments on Mt. Washington, New Hampshire. *Ecol. Monogr.* 36:125–155.

———. 1970. Primary production within arctic tundra ecosystems. In *Proc. Conf. Productivity and Conservation in Northern Circumpolar Lands; Edmonton, Alberta, 1969*, pp. 77–85. Morges, Switzerland: Int. Union Conserv. Natl. Res.

Botkin, D. B., G. M. Woodwell, and N. Tempel. 1970. Forest productivity estimated from carbon dioxide uptake. *Ecology* 51:1057–1060.

Boyd, C. E. 1970. Production, mineral accumulation and pigment concentrations in *Typha latifolia* and *Scirpus americanus*. *Ecology* 51:285–290.

Boysen Jensen, P. 1932. *Die Stoffproduktion der Pflanzen*, 108 pp. Jena: Fischer.

Bray, J. R. 1960. The chlorophyll content of some native and managed plant communities in central Minnesota. *Can. J. Bot.* 38:313–333.

———. 1961. Measurement of leaf utilization as an index of minimum level of primary consumption. *Oikos* 12:70–74.

———. 1962. The primary productivity of vegetation in central Minnesota, U.S.A. and its relationship to chlorophyll content and albedo. In *Die Stoffproduktion der Pflanzendecke*, H. Lieth, ed., pp. 102–109. Stuttgart: Fischer.

———. 1963. Root production and the estimation of net productivity. *Can. J. Bot.* 41:65–72.

———. 1964. Primary consumption in three forest canopies. *Ecology* 45:165–167.

———, and L. A. Dudkiewicz. 1963. The composition, biomass and productivity of two *Populus* forests. *Bull. Torrey Bot. Club*, 90:298–308.

———, and F. Gorham. 1964. Litter production in forests of the world. *Advan. Ecol. Res.* 2:101–157.

———, D. B. Lawrence, and L. C. Pearson. 1959. Primary production in some Minnesota terrestrial communities for 1957. *Oikos* 10:38–49.

Brünig, E. F. 1974. Ökosysteme in den Tropen. *Umschau* 74:405–410.

Bunce, R. G. H. 1968. Biomass and production of trees in a mixed deciduous woodland. I. Girth and height as parameters for the estimation of tree dry weight. *J. Ecol.* 56:759–775.

Burger, H. 1929. Holz, Blattmenge und Zuwachs. I. Die Weymouthsföhre. (French summ.) *Mitt. Schweiz. Anst. Forstl. Versuchsw. Zürich* 15(2):243–292.

———. 1940. Holz, Blattmenge und Zuwachs. IV. Ein 80-jähriger Buchenbestand. (French summ.) *Mitt. Schweiz. Anst. Forstl. Versuchsw. Zürich* 21(2):307–348.

———. 1953. Holz, Blattmenge und Zuwachs. XIII. Fichten im gleichalterigen Hochwald. (French summ.) *Mitt. Schweiz. Anst. Forstl. Versuchsw. Zürich* 29(1)38–130.

Chew, R. M., and A. E. Chew. 1965. The primary productivity of a desert shrub (*Larrea tridentata*) community. *Ecol. Monogr.* 35:355–375.

Clymo, R. S. 1970. The growth of *Sphagnum*: Methods of measurement. *J. Ecol.* 58:13–49.

Crow, T. R. 1971. Estimation of biomass in an even-aged stand—regression and "mean tree" techniques. In *Forest Biomass Studies*, H. E. Young, ed., pp. 35–48. Orono, Maine: Univ. of Maine, Life Sciences and Agriculture Experiment Station.

Dahlman, R. C., and C. L. Kucera. 1965. Root productivity and turnover in native prairie. *Ecology* 46:84–89.

Drozdov, A. V. 1971. The productivity of zonal terrestrial plant communities and the moisture and heat parameters of an area. *Sov. Geogr.: Rev. Transl.* 12:54–60.

Duncan, W. G., R. S. Loomis, W. A. Williams, and R. Hanau. 1967. A model for simulating photosynthesis in plant communities. *Hilgardia* 38:181–205.

Duvigneaud, P. 1971. Concepts sur la productivité primaire des écosystèmes forestiers. (Engl. summ.) In *Productivity of Forest Ecosystems: Proc. Brussels Symp. 1969*, P. Duvigneaud, ed. *Ecology and Conservation*, Vol. 4, 111–140. Paris: UNESCO.

———, P. Kestemont, and P. Ambroes. 1971. Productivité primaire des forêts tem-

110
Part 2: Methods of Productivity Measurement

pérées d'essences feuillues caducifoliées en Europe occidentale. (Engl. summ.) In *Productivity of Forest Ecosystems: Proc. Brussels Symp. 1969*, P. Duvigneaud, ed. *Ecology and Conservation*, Vol. 4, 259–270. Paris: UNESCO.

Eckardt, F. E. 1968. Techniques de mesure de la photosynthèse sur le terrain basées sur l'emploi d'enceintes climatisées. (Engl. summ.) In *Functioning of Terrestrial Ecosystems at the Primary Production Level: Proc. Copenhagen Symp., 1965*. F. E. Eckardt, ed. *Natural Resources Research*, Vol. 5, 289–319. Paris: UNESCO.

Filzer, P. 1951. *Die natürlichen Grundlcgen des Pflanzenertrages in Mitteleuoropa*, 198 pp. Stuttgart: Sweizerbart.

Ford, E. D., and P. J. Newbould. 1970. Stand structure and dry weight production through the sweet chestnut (*Castanea sativa* Mill.) coppice cycle. *J. Ecol.* 58: 275–296.

Forrest, G. I. 1971. Structure and production of North Pennine blanket bog vegetation. *J. Ecol.* 59:453–479.

Ghilarov, M. S., V. A. Kovda, L. N. Novichkova-Ivanova, L. E. Rodin, and V. M. Sveshnikova. (eds.) 1968. *Methods of Productivity Studies in Root Systems and Rhizosphere Organisms*, 240 pp. Leningrad: U.S.S.R. Academy of Sciences.

Golley, F. B. 1965. Structure and function of an old-field broomsedge community. *Ecol. Monogr.* 35:113–137.

Gosz, J. R., G. E. Likens, and F. H. Bormann. 1972. Nutrient content of litter fall on the Hubbard Brook Experimental Forest, New Hampshire. *Ecology* 53: 769–784.

Greenland, D. J., and J. M. L. Kowal. 1960. Nutrient content of the moist tropical forest of Ghana. *Plant Soil* 12:154–174.

Harris, W. F., R. A. Goldstein, and G. S. Henderson. 1973. Analysis of forest biomass pools, annual primary production and turnover of biomass for a mixed deciduous forest watershed. In *IUFRO Biomass Studies: International Union of Forest Research Organizations Papers*, H. E. Young, ed., pp. 41–64. Orono, Maine: Univ. of Maine, College of Life Sci. and Agr.

Hozumi, K., K. Yoda, and T. Kira. 1969a. Production ecology of tropical rain forests in southwestern Cambodia. II. Photosynthetic production in an evergreen seasonal forest. *Nature Life SE Asia* 6:57–81.

———, K. Yoka, S. Kokawa, and T. Kira. 1969b. Production ecology of tropical rain forests in southwestern Cambodia. I. Plant biomass. *Nature Life SE Asia* 6:1–51.

Huxley, J. S. 1931. Notes on differential growth. *Amer. Nat.* 65:289–315.

———. 1932. *Problems of Relative Growth*, 276 pp. New York: Dial.

Hytteborn, H. 1975. Deciduous woodland at Andersby, Eastern Sweden. Aboveground tree and shrub production. *Acta Phytogeogr. Suecica* 61:1–96.

Inoue, E. 1968. The CO_2-concentration profile within crop canopies and its significance for the productivity of plant communities. (French summ.) In *Functioning of Terrestrial Ecosystems at the Primary Production Level: Proc. Copenhagen Symp. 1965*, F. E. Eckardt, ed. *Natural Resources Research*, Vol. 5, 359–366. Paris: UNESCO.

Jeník, J. 1971. Root structure and underground biomass in equatorial forests. (French summ.) In *Productivity of Forest Ecosystems: Proc. Brussels Symp., 1969*. P. Duvigneaud, ed. *Ecology and Conservation*, Vol. 4, 323–331. Paris: UNESCO.

Johnson, P. L., and J. J. Kelley. 1970. Dynamics of carbon dioxide and productivity in an arctic biosphere. *Ecology* 51:73–80.

Jordan, C. F. 1971. Productivity of a tropical forest and its relation to a world pattern of energy storage. *J. Ecol.* 59:127–142.

Kestemont, P. 1971. Productivité primaire des taillis simples et concept de nécromasse. (Engl. summ.) In *Productivity of Forest Ecosystems: Proc. Brussels Symp. 1969*, P. Duvigneaud, ed *Ecology and Conservation*, Vol. 4, 271–279. Paris: UNESCO.

Kimura, M. 1963. Dynamics of vegetation in relation to soil development in northern Yatsugatake Mountains. *Jap. J. Bot.* 18:255–287.

———. 1969. Ecological and physiological studies on the vegetation of Mt. Shimagare. VII. Analysis of production processes of young *Abies* stand based on the carbohydrate economy. *Bot. Mag. Tokyo* 82:6–19.

———, I. Mototani, and K. Hogetsu. 1968. Ecological and physiological studies on the vegetation of Mt. Shimagare. VI. Growth and dry matter production of young *Abies* stand. (Jap. summ.) *Bot. Mag. Tokyo*, 81:287–296.

Kira, T. 1968. A rational method for estimating total respiration of trees and forest stands. (French summ.) In *Functioning of Terrestrial Ecosystems at the Primary Production Level: Proc. Copenhagen Symp. 1965*. F. E. Eckardt, ed. *Natural Resources Research*, Vol. 5, 399–409. Paris: UNESCO.

———, and H. Ogawa. 1968. Indirect estimation of root biomass increment in trees. In *Methods of Productivity Studies in Root Systems and Rhizosphere Organisms*, M. S. Ghilarov, V. A. Kovda, L. N. Novichkova-Ivanova, L. E. Rodin, and V. M. Sveshnikova, eds., pp. 96–101. Leningrad: U.S.S.R. Academy of Sciences.

———, and H. Ogawa. 1971. Assessment of primary production in tropical and equatorial forests. (French summ.) In *Productivity of Forest Ecoystems: Proc. Brussels Symp. 1969*, P. Duvigneaud, ed. *Ecology and Conservation*, Vol. 4, 309–321. Paris: UNESCO.

———, H. Ogawa, K. Yoda, and K. Ogino. 1964. Primary production by a tropical rain forest of southern Thailand. *Bot. Mag. Tokyo* 77:428–429.

———, H. Ogawa, K. Yoda, and K. Ogino. 1967. Comparative ecological studies on three main types of forest vegetation in Thailand. IV. Dry matter production, with special reference to the Khao Chong rain forest. *Nature Life SE Asia* 5:149–174.

———, and T. Shidei. 1967. Primary production and turnover of organic matter in different forest ecosystems of the Western Pacific. *Jap. J. Ecol.* 17:70–87.

Kucera, C. L., and D. R. Kirkham. 1971. Soil respiration studies in tall grass prairie in Missouri. *Ecology* 52:912–915.

Kuroiwa, S. 1960a. Ecological and physiological studies on the vegetation of Mt. Shimagare. IV. Some physiological functions concerning matter production in young *Abies trees*. (Jap. summ.) *Bot. Mag. Tokyo* 73:133–141.

———. 1960b. Ecological and physiological studies on the vegetation of Mt. Shimagare. V. Intraspecific competition and productivity difference among tree classes in the *Abies stand*. (Jap. summ.) *Bot. Mag. Tokyo* 73:165–174.

———. 1968. A new calculation method for total photosynthesis of a plant community under illumination consisting of direct and diffused light. (French summ.) In *Functioning of Terrestrial Ecosystems at the Primary Production Level: Proc. Copenhagen Symp., 1965*, F. E. Eckardt. ed. *Natural Resources Research*, Vol. 5, 391–398. Paris: UNESCO.

Lange, O. L., and E. D. Schulze. 1971. Measurement of CO_2 gas-exchange and transpiration in the beech (*Fagus silvatica* L.). In *Integrated Experimental*

Ecology, H. Ellenberg, ed., *Ecological Studies* 2:16–28. New York: Springer-Verlag.

Larcher, W. 1969. Physiological approaches to the measurement of photosynthesis in relation to dry matter production by trees. *Photosynthetica* 3:150–166.

Lemon, E. R. 1967. Aerodynamic studies of CO_2 exchange between the atmosphere and the plant. In *Harvesting the Sun: Photosynthesis in Plant Life*, A. San Pietro, F. A. Greer, and T. J. Army, eds., pp. 263–290. New York: Academic Press.

————. 1968. The measurement of height distribution of plant community activity using the energy and momentum balance approaches. (French summ.) In *Functioning of Terrestrial Ecosystems at the Primary Production Level: Proc. Copenhagen Symp., 1965*, F. E. Eckardt, ed. *Natural Resources Research*, Vol. 5, 381–389. Paris: UNESCO.

————. 1969. Gaseous exchange in crop stands. In *Physiological Aspects of Crop Yield: Proc. Symp. Univ. Nebraska, Lincoln*, J. D. Eastin, ed., pp. 117–137. Madison: Amer. Soc. Agron.

————, L. H. Allen, Jr., and L. Müller. 1970. Carbon dioxide exchange of a tropical rain forest. II. *BioScience* 20:1054–1059.

Lieth, H., ed. 1962. Die Stoffproduktion der Pflanzendecke. *Vorträge und Diskussionen des Symposiums in Hohenheim im Mai 1960*, 156 pp. Stuttgart: Fischer.

————. 1962. Stoffproduktionsdaten. In *Die Stoffproduktion der Pflanzendecke*, H. Lieth, ed., pp. 117–127. Stuttgart: Fischer.

————. 1965. Indirect methods of measurement of dry matter production. In *Methodology of Plant Eco-physiology: Proc. Montpellier Symp., 1962*, F. E. Eckardt, ed., pp. 513–518. Paris: UNESCO.

————. 1968. The determination of plant dry-matter production with special emphasis on the underground parts. (French summ.) In *Functioning of Terrestrial Ecosystems at the Primary Production Level: Proc. Copenhagen Symp., 1965*, F. E. Eckardt, ed. *Natural Resources Research*, Vol. 5, 179–186. Paris: UNESCO.

————. 1972. Über die Primärproduktion der Pflanzendecke der Erde. *Zeit. Angew. Bot.* 46:1–27.

————, and R. Ouellette. 1962. Studies on the vegetation of the Gaspé peninsula. II. The soil respiration of some plant communities. *Can. J. Bot.* 40:127–140.

————, D. Osswald, and H. Martens. 1965. Stoffproduktion, Spross-Wurzel-Verhältnis, Chlorophyllgehalt und Blattfläche von Jungpappeln. *Mitt. Ver. Forstl. Standortsk. Forstpflanz.*, Vol. 15, pp. 70–74. Stuttgart: Fischer.

————, and E. Box. 1972. Evapotranspiration and primary productivity; C. W. Thornthwaite Memorial Model. *Publications in Climatology*, Vol. 25(2), pp. 37–46. Centerton/Elmer, New Jersey: C. W. Thornthwaite Assoc.

Lomnicki, A., E. Bandola, and K. Jankowska. 1968. Modification of the Wiegert-Evans method for estimation of net primary production. *Ecology* 49:147–149.

McNaughton, S. J. 1966. Ecotype function in the *Typha* community-type. *Ecol. Monogr.* 36:297–325.

Madgwick, H. A. I. 1970. Biomass and productivity models of forest canopies. In *Analysis of Temperate Forest Ecosystems*, D. E. Reichle, ed., *Ecological Studies* 1:47–54. New York: Springer-Verlag.

————. 1971. The accuracy and precision of estimates of the dry matter in stems, branches and foliage in an old-field *Pinus virginiana stand*. In *Forest Biomass*

Studies, H. E. Young, ed., pp. 105–112. Orono, Maine: Univ. of Maine, Life Sciences and Agriculture Experiment Station.

Marks, P. L. 1971. The role of *Prunus pensylvanica L.* in the rapid revegetation of disturbed sites. Ph.D. thesis, 119 pp. New Haven: Yale Univ.

———. 1974. The role of pin cherry (*Prunus pensylvanica L.*) in the maintenance of stability in northern hardwood ecosystems. *Ecol. Monogr.* 44:73–88.

———, and F. H. Bormann. 1972. Revegetation following forest cutting: Mechanisms for return to steady-state nutrient cycling. *Science* 176:914–915.

Maruyama, K. 1971. Effect of altitude on dry matter production of primeval Japanese beech forest communities in Naeba Mountains. *Mem. Fac. Agr. Niigata Univ.* 9:85–171.

Medina, E., and H. Lieth. 1963. Contenido de clorofila de algunas associaciones vegetales de Europa Central y su relacion con la productividad. *Qual. Plant. Mat. Veg.* IX:217–229.

———, and ———. 1964. Die Beziehungen zwischen Chlorophyllgehalt, assimilierender Fläche und Trockensubstanzproduktion in einigen Pflanzengemeinschaften. *Beitr. Biol. Pflanzen* 40:451–494.

Milner, C., and R. E. Hughes. 1968. Methods for the Measurement of the Primary Production of Grassland. In *International Biological Program Handbook,* Vol. 6, pp. 1–82. Philadelphia, Pennsylvania: Davis.

Möller, C. Mar.: 1945. Untersuchungen über Laubmenge, Stoffverlust und Stoffproduktion des Waldes. (Engl. and Danish summs.) *Forstl. Forsøgsv. Danm.* 17: 1–287.

———. 1947. The effect of thinning, age, and site on foliage, increment, and loss of dry matter. *J. Forestry* 45:393–404.

———, D. Müller, and J. Nielsen. 1954a. Loss of branches in European beech. *Forstl. Forsø gsv. Danm.* 21:253–271.

———, D. Müller, and J. Nielsen. 1954b. Graphic presentation of dry matter production of European beech. *Forstl. Forsøgsv. Danm.* 21:327–335.

Monk, C. 1966. Ecological importance of root/shoot ratios. *Bull. Torrey Bot. Club* 93:402–406.

Monsi, M. 1968. Mathematical models of plant communities. In *Functioning of Terrestrial Ecosystems at the Primary Production Level: Proc. Copenhagen Symp., 1965,* F. E. Eckardt, ed., *Natural Resources Research,* Vol. 5, 131–149. Paris: UNESCO.

———, and T. Saeki. 1953. Über den Lichtfaktor in den Pflanzengesellschaften und seine Bedeutung für die Stoffproduktion. *Jap. J. Bot.* 14:22–52.

Monteith, J. L. 1968. Analysis of the photosynthesis and respiration of field crops from vertical fluxes of carbon dioxide. (French summ.) In *Functioning of Terrestrial Ecosystems at the Primary Production Level: Proc. Copenhagen Symp., 1965,* F. E. Eckardt, ed., *Natural Resources Research,* Vol. 5, 349–358. Paris: UNESCO.

Mooney, H. A. 1972. Carbon dioxide exchange of plants in natural environments. *Bot. Rev.* 38:455–469.

Müller, D. 1962. Wie gross ist der prozentuale Anteil der Nettoproduktion von der Bruttoproduktion? In *Die Stoffproduktion der Pflanzendecke,* H. Lieth, ed., pp. 26–28. Stuttgart Fischer.

———, and J. Nielsen. 1965. Production brute, pertes par respiration et production nette dans la forêt ombrophile tropicale. (Danish summ.) *Forstl. Forsøgsv. Danm.* 29:69–160.

Nemeth, J. C. 1973. Dry matter production in young loblolly (*Pinus taeda* L.) and slash pine (*Pinus elliottii* Engelm.) plantations. *Ecol. Monogr.* 43:21–41.

Newbould, P. J. 1967. Methods for estimating the primary production of forests. *International Biological Programme, Handbook 2*, 62 pp. Oxford and Edinburgh: Blackwell.

————. 1968. Methods of estimating root production. (French summ.) In *Functioning of Terrestrial Ecosystems at the Primary Production Level: Proc. Copenhagen Symp., 1965*, F. E. Eckardt, ed. *Natural Resources Research*, Vol. 5, 187–190. Paris: UNESCO.

Nihlgård, B. 1972. Plant biomass, primary production and distribution of chemical elements in a beech and a planted spruce forest in South Sweden (Russ. summ.) *Oikos* 23:69–81.

Numata, M. 1965. Ecology of bamboo forests in Japan. *Advan. Frontiers Plant Sci.* 10:89–120.

Odum, E. P. 1960. Organic production and turnover in old field succession. *Ecology* 41:34–49.

Odum, H. T., and C. F. Jordan. 1970. Metabolism and evapotranspiration of the lower forest in a giant plastic cylinder. In *A Tropical Rain Forest: A Study of Irradiation and Ecology at El Verde, Puerto Rico*, H. T. Odum, ed., Part I, pp. 165–189. Washington, D.C.: U.S. Atomic Energy Commission.

————, A. Lugo, G. Cintron, and C. F. Jordan. 1970. Metabolism and evapotranspiration of some rain forest plants and soil. In *A Tropical Rain Forest: A Study of Irradiation and Ecology at El Verde, Puerto Rico*, H. T. Odum, ed., Part I, pp. 103–164. Washington, D.C.: U.S. Atomic Energy Commission.

Ogawa, H., K. Yoda, and T. Kira. 1961. A preliminary survey on the vegetation of Thailand. *Nature Life SE Asia* 1:21–157.

————, K. Yoda, K. Ogino, and T. Kira. 1965. Comparative ecological studies on three main types of forest vegetation in Thailand. II. Plant biomass. *Nature Life SE Asia* 4:49–80.

Olson, J. S. 1971. Primary productivity: temperate forests, especially American deciduous types. (French summ.) In *Productivity of Forest Ecosystems: Proc. Brussels Symp. 1969*, P. Duvigneaud, ed. *Ecology and Conservation*, Vol. 4, 235–258. Paris: UNESCO.

Ordway, D. E. 1969. An aerodynamicist's analysis of the Odum cylinder approach to net CO_2 exchange. *Photosynthetica* 3:199–209.

Ovington, J. D. 1956. The form, weights and productivity of tree species grown in close stands. *New Phytol.* 55:289–304.

————. 1957. Dry-matter production by *Pinus silvestris* L. *Ann. Bot. N.S.* 21:287–314.

————. 1962. Quantitative ecology and the woodland ecosystem concept. *Advan. Ecol. Res.* 1:103–192.

————. 1963. Flower and seed production. A source of error in estimating woodland production, energy flow and mineral cycling. *Oikos* 14:148–153.

————. 1965. Organic production, turnover and mineral cycling in woodlands. *Biol. Rev.* 40: 295–336.

————, and H. A. I. Madgwick. 1959a. Distribution of organic matter and plant nutrients in a plantation of Scots pine. *Forest Sci.* 5:344–355.

————, and H. A. I. Madgwick. 1959b. The growth and composition of natural stands of birch. I. Dry-matter production. *Plant Soil* 10:271–283.

————, and W. H. Pearsall. 1956. Production ecology. II. Estimates of average production by trees. *Oikos* 7:202–205.

————, W. G. Forrest, and J. S. Armstrong. 1968. Tree biomass estimation. In *Symp. Primary Productivity and Mineral Cycling in Natural Ecosystems*, H. E. Young, ed., pp. 4–31. Orono, Maine: Univ. of Maine.

————, D. Heitkamp, and D. Lawrence. 1963. Plant biomass and productivity of prairie, savanna, oakwood, and maize field ecosystems in central Minnesota. *Ecology* 44:52–63.

Paterson, S. S. 1961. Introduction to phyochorology of Norden. (Swedish summ.) *Medd. Stat. Skogsforsk. Inst.* 50(5):1–145.

Pearson, L. C., and D. B. Lawrence. 1958. Photosynthesis in aspen bark. *Amer. J. Bot.* 45:383–387.

Perry, T. O. 1971. Winter-season photosynthesis and respiration by twigs and seedlings of deciduous and evergreen trees. *Forest Sci.* 17:41–43.

Peterken, G. F., and P. J. Newbould. 1966. Dry-matter production by *Ilex aquifolium* L. in the New Forest. *J. Ecol.* 54:143–150.

Reader, R. J., and J. M. Stewart. 1972. The relationship between net primary production and accumulation for a peatland in southeastern Manitoba. *Ecology* 53:1024–1037.

Reichle, D. E., R. A. Goldstein, R. I. Van Hook, Jr., and G. J. Dodson. 1973a. Analysis of insect consumption in a forest canopy. *Ecology* 54:1076–1084.

————, B. E. Dinger, N. T. Edwards, W. F. Harris, and P. Sollins. 1973b. Carbon flow and storage in a forest ecosystem. In *Carbon and the biosphere*, G. M. Woodwell and E. V. Pecan, eds. *Brookhaven Symp. Biol.* 25:345–365. Springfield, Virginia: Natl. Tech. Inform. Serv. (CONF-720510).

Reiners, W. A. 1968. Carbon dioxide evolution from the floor of three Minnesota forests. *Ecology* 49:471–483.

————. 1972. Structure and energetics of three Minnesota forests. *Ecol. Monogr.* 42:71–94.

Ribe, J. H. 1973. A study of multi-stage and dimensional analysis sampling of puckerbrush stands. *IUFRO Biomass Studies: International Union of Forest Research Organizations Papers*, H. E. Young, ed., pp. 119–130. Orono, Maine: Univ. of Maine, College of Life Sci. and Agr.

Rochow, J. J. 1974. Estimates of above-ground biomass and primary productivity in a Missouri forest. *J. Ecol.* 62:567–577.

Rodin, L. E., and N. I. Bazilevich. 1966. The biological productivity of the main vegetation types in the northern hemisphere of the Old World. *Forestry Abstr.* 27:369–372. [*Dokl. Akad. Nauk SSSR* 157:215–218 (1964).]

————, and N. I. Bazilevich. 1967. *Production and Mineral Cycling in Terrestrial Vegetation*, 288 pp. Edinburgh: Oliver and Boyd.

————, and N. I. Bazilevich. 1968. World distribution of plant biomass. (French summ.) In *Functioning of Terrestrial Ecosystems at the Primary Production Level: Proc. Copenhagen Symp., 1965*, F. E. Eckardt, ed. *Ecology and Conservation*, Vol. 4:45–52. Paris: UNESCO.

Rosenzweig, M. L. 1968. Net primary productivity of terrestrial communities: prediction from climatological data. *Amer. Nat.* 102:67–74.

Rothacher, J. S., F. W. Blow, and S. M. Potts. 1954. Estimating the quantity of tree foliage in oak stands in the Tennessee Valley. *J. Forestry*, 52:169–173.

Saeki, T. 1963. Light relations in communities. In *Environmental Control of Plant Growth*, L. T. Evans, ed., pp. 79–92. New York and London: Academic Press.

Satoo, T. 1962. Notes on Kittredge's method of estimation of amount of leaves of forest stand. (Jap. summ.) *J. Jap. Forestry Soc.* 44:267–272.

———. 1966. Production and distribution of dry matter in forest ecosystems. *Misc. Inform. Tokyo Univ. Forests* 16:1–15.

———. 1968a. Materials for the study of growth in stands. 7. Primary production and distribution of produced dry matter in a plantation of Cinnamomum camphora. (Jap. summ.) *Bull. Tokyo Univ. Forests* 64:241–275.

———. 1968b. Primary production relations in woodlands of *Pinus densiflora*. In *Symp. Primary Productivity and Mineral Cycling in Natural Ecosystems*, H. E. Young, ed., pp. 52–80. Orono, Maine: Univ. of Maine.

———. 1970. A synthesis of studies by the harvest method: Primary production relations in the temperate deciduous forests of Japan. In *Analysis of Temperate Forest Ecosystems*, D. E. Reichle, ed. *Ecological Studies* 1:55–72. New York: Springer-Verlag.

———. 1971. Primary production relations of coniferous forests in Japan. (French summ.) In *Productivity of Forest Ecosystems: Proc. Brussels Symp. 1969*, P. Duvigneaud, ed. *Ecology and Conservation*, Vol. 4:191–205. Paris: UNESCO.

———, and M. Senda. 1966. Materials for the studies of growth in stands. VI. Biomass, dry-matter production, and efficiency of leaves in a young *Cryptomeria* plantation. (In Jap. with Engl. summ.) *Bull. Tokyo Univ. Forests* 62:117–146.

———, K. Nakamura, and M. Senda. 1955. Materials for the studies of growth in stands. I. Young stands of Japanese red pine of various density. *Bull. Tokyo Univ. Forests* 48:65–90.

Schultz, G. 1962. Blattfläche und Assimilationsleistung in Beziehung zur Stoffproduktion. Untersuchungen an Zuckerrüben. *Ber. Deut. Bot. Ges.* 75:261–267.

Schulze, E. D., and W. Koch. 1971. Measurement of primary production with cuvettes. (French summ.) In *Productivity of Forest Ecosystems: Proc. Brussels Symp. 1969*, P. Duvigneaud, ed., pp. 141–157. Paris: UNESCO.

Schuster, J. L. 1964. Root development of native plants under three grazing intensities. *Ecology* 45:63–70.

Scott, D., and W. D. Billings. 1964. Effects of environmental factors on standing crop and productivity of an alpine tundra. *Ecol. Monogr.* 34:243–270.

Shanks, R. E. 1954. Climates of the Great Smoky Mountains. *Ecology* 35:345–361.

Singh, J. S., and P. S. Yadava. 1974. Seasonal variation in composition, plant biomass, and net primary productivity of a tropical grassland at Kurukshetra, India. *Ecol. Monogr.* 44:351–376.

Stephens, G. R., and P. E. Waggoner. 1970. Carbon dioxide exchange of a tropical rain forest. I. *BioScience* 20:1050–1053.

Strain, B. R., and P. L. Johnson. 1963. Corticular photosynthesis and growth in *Populus tremuloides*. *Ecology* 44:581–584.

Struik, G. J. 1965. Growth patterns of some native annual and perennial herbs in southern Wisconsin. *Ecology* 46:401–420.

Tadaki, Y. 1965a. Studies on production structure of forests (VII). The primary production of a young stand of *Castanopsis cuspidata*. (Jap. summ.) *Jap. J. Ecol.* 15:142–147.

———. 1965b. Studies on production structure of forest (VIII). Productivity of an *Acacia mollissima* stand in higher stand density. (In Jap. with English summ.) *J. Jap. Forest Soc.* 47:384–391.

Totsuka, T., N. Nomoto, T. Oikawa, T. Saeki, Y. Ino, and M. Monsi. 1968. The energy balance method and the half-leaf method as applied to the photosynthetic production of a *Miscanthus sacchariflorus* community. In *Photosynthesis and Utilization of Solar Energy, Level III Experiments*, pp. 17–19. Tokyo: JIBP/PP —Photosynthesis Level III Group, Japanese National Subcommittee for PP (JPP).

Tranquillini, W., and M. M. Caldwell. 1972. Integrated calibrations of plant gas exchange systems. *Ecology* 53:974–976.

Transeau, E. N. 1926. The accumulation of energy by plants. *Ohio J. Sci.* 26:1–10.

Voigt, G. K. 1962. The role of carbon dioxide in soil. In *Tree Growth*, T. T. Kozlowski, ed., pp. 205–220. New York: Ronald.

Waksman, S. A., and R. L. Starkey. 1924. Microbiological analysis of soil as an index of soil fertility. VII. Carbon dioxide evolution. *Soil Sci.* 17:141–161.

Wallace, D. H., J. L. Ozbun, and H. M. Munger. 1972. Physiological genetics of crop yield. *Advan. Agron.* 24:97–146.

Walter, H. 1939. Grasland, Savanne und Busch der ariden Teile Afrikas in ihrer ökologischen Bedingtheit. *Jahrb. Wiss. Bot.* 87:850–860.

———. 1962. *Die Vegetation der Erde in ökologischer Betrachtung*, Vol. 1: *Die tropischen und subtropischen Zonen*. 538 pp. Jena: Fischer.

Watson, D. J. 1958. The dependence of net assimilation rate on leaf area index. *Ann. Bot. Lond., N.S.* 22:37–54.

Wein, R. W., and L. C. Bliss. 1974. Primary production in arctic cottongrass tussock tundra communities. *Arctic Alpine Res.* 6:261–274.

White, E. H., W. L. Pritchett, and W. K. Robertson. 1971. Slash pine root biomass and nutrient concentrations. In *Forest Biomass Studies*, H. E. Young, ed., pp. 165–176. Orono, Maine: Univ. of Maine, Life Sciences and Agriculture Experiment Station.

Whittaker, R. H. 1961. Estimation of net primary production of forest and shrub communities. *Ecology* 42:177–180.

———. 1962. Net production relations of shrubs in the Great Smoky Mountains. *Ecology* 43:357–377.

———. 1963. Net production of heath balds and forest heaths in the Great Smoky Mountains. *Ecology* 44:176–182.

———. 1965. Branch dimensions and estimation of branch production. *Ecology* 46:365–370.

———. 1966. Forest dimensions and production in the Great Smoky Mountains. *Ecology* 47:103–121.

———, and V. Garfine. 1962. Leaf characteristics and chyorophyll in relation to exposure and production in *Rhododendron maximum*. *Ecology* 43:120–125.

———, and G. M. Woodwell. 1967. Surface area relations of woody plants and forest communities. *Amer. Bot.* 54:931–939.

———, and G. M. Woodwell. 1968. Dimension and production relations of trees and shrubs in the Brookhaven forest, New York. *J. Ecol.* 56:1–25.

————, and G. M. Woodwell. 1969. Structure, production and diversity of the oak-pine forest at Brookhaven, New York. *J. Ecol.* 57: 157–174.

————, and G. M. Woodwell. 1971a. Measurement of net primary production of forests. (French summ.) In *Productivity of Forest Ecosystems: Proc. Brussels Symp. 1969*, P. Duvigneaud, ed. *Ecology and Conservation*, Vol. 5:159–175. Paris: UNESCO.

————, and G. M. Woodwell. 1971b. Evolution of natural communities. In *Ecosystem Structure and Function*, J. A. Wiens, ed., Corvallis: *Oregon State Univ. Ann. Biol. Colloq.* 31:137–159.

————, and W. A. Niering. 1975. Vegetation of the Santa Catalina Mountains, Arizona. (V) Biomass, production, and diversity along the elevation gradient. *Ecology* (In press.)

————, N. Cohen, and J. S. Olson. 1963. Net production relations of three tree species at Oak Ridge, Tennessee. *Ecology* 44:806–810.

————, F. H. Bormann, G. E. Likens, and T. G. Siccama. 1974. The Hubbard Brook ecosystem study: Forest biomass and production. *Ecol. Monogr.* 44:233–254.

Wiegert, R. G., and F. C. Evans. 1964. Primary production and the disappearance of dead vegetation in an old field in southeastern Michigan. *Ecology* 45:49–63.

Woodwell, G. M., and D. B. Botkin. 1970. Metabolism of terrestrial ecosystems by gas exchange techniques: The Brookhaven approach. In *Analysis of Temperate Forest Ecosystems*, D. E. Reichle, ed., *Ecological Studies* 1:73–85. New York: Springer-Verlag.

————, and W. R. Dykeman. 1966. Respiration of a forest measured by carbon dioxide accumulation during temperature inversions. *Science* 154:1031–1034.

————, and T. G. Marples. 1968. Production and decay of litter and humus in an oak–pine forest and the influence of chronic gamma irradiation. *Ecology* 49: 456–465.

————, and R. H. Whittaker. 1968. Primary production in terrestrial communities. *Amer. Zool.* 8:19–30.

Yoda, K. 1968. A preliminary survey of the forest vegetation of eastern Nepal. III. Plant biomass in the sample plots chosen from different vegetation zones. *J. College Arts Sci. Chiba Univ. Nat. Sci. Ser.* 5:277–302.

————, K. Shinozaki, H. Ogawa, K. Hozumi, and T. Kira. 1965. Estimation of the total amount of respiration in woody organs of trees and forest communities. *J. Biol. Osaka City Univ.* 16:15–26.

Young, H. E., ed. 1971. *Forest Biomass Studies: Symp. Int. Union Forest Res. Orgs., Gainesville, Fla. 1971.* 205 pp. Orono, Maine: Univ. Maine, Life Sciences and Agricultural Experiment Station.

————, L. Strand, and R. Altenberger. 1964. Preliminary fresh and dry weight tables for seven tree species in Maine. *Tech. Bull. Maine Agr. Exp. Sta., Orono,* 12: 1–76.

Zar, J. H. 1968. Calculation and miscalculation of the allometric equation as a model in biological data. *BioScience* 18:1118–1120.

Zavitkovski, J. 1971. Dry weight and leaf area of aspen trees in northern Wisconsin. In *Forest Biomass Studies*, H. E. Young, ed., pp. 193–205. Orono, Maine: Univ. of Maine, Life Sciences and Agricultural Experiment Station.

————, and R. D. Stevens. 1972. Primary productivity of red alder ecosystems. *Ecology* 53:235–242.

Zelitch, I. 1964. Organic acid and respiration in photosynthetic tissue. *Ann. Rev. Plant Physiol.* 15:121–142.

5

Measurement of
Caloric Values

Helmut Lieth

Most models and essays that consider the basic processes of photosynthesis and primary production start with solar energy or CO_2 as input and end with dry-matter weight produced as output. The importance of interpreting productivity in terms of energy also has long been realized but the large amount of extra work necessary to convert dry-matter values into caloric values has in many cases discouraged further investigation of energy of productivity. In this volume the data that allow this conversion on a world scale are evaluated. The original table (Lieth, 1972, 1973) that presented such a conversion was based primarily on energy measurements that were made using the method described in this chapter. Similar evaluations were attempted by Golley (1972) and Jordan (1971), many of whose data were from the paper by Cummins and Wuycheck (1971), which was available in prepublished form.

Although the method of energy determination described in this chapter is still the one used most extensively in primary productivity work, a variety of other devices are available that are based on the same or similar principles but are specially designed for small samples or other special situations. Most of the available methods were summarized by Paine (1971).

The data presented in this book are evaluated either directly with the bomb calorimeter (see Fig. 5-1), or are calculated with conversion tables using known chemical compositions of plant material and known caloric values of the chemical compounds. Compilations useful for this purpose may be found in Morowitz (1968) and Runge (1973). The tables included in this chapter are intended as a guide for future attempts to arrive at gross energy calculations in ecosystems similar to those presented in this book. The description of the com-

KEYWORDS: Caloric values; plant material; methods.

Primary Productivity of the Biosphere, edited by
Helmut Lieth and Robert Whittaker.
© 1975 by Springer-Verlag New York Inc.

bustion-value measuring procedure is based on an earlier work (Lieth and Pflanz, 1968).

Preparation of Samples

Definition of the measured values

Energy measurements of biologic material require determination of the thermo-chemical caloric content, which is defined in the German Standard leaflet DIN 51708 as follows. The calorific value of a fuel is the amount of calories liberated while one unit of fuel is completely burned, provided that

1. The fuel at the time of ignition and the resulting combustion products are at a temperature of 20°C.
2. The water originally present in the fuel and that formed during the burning process are in the liquid phase.
3. The combustion products of carbon and sulfur are present only as carbon dioxide and sulfur dioxide gases.
4. No oxidation of nitrogen takes place.

For the calculation of ecologic efficiency, caloric values should be based upon ash-containing matter, whereas for studies of translocation and growth analysis the values should be based upon ash-free dry matter. The method of calculation is described for both quantities in the following sections.

Collection and preparation of material in the field

Sample collection for energy studies often calls for greater care in separating the total yield than is normally necessary for dry-matter measurements. Therefore each component of a stand of vegetation that differs from the other components of the harvested yield or that cannot be milled to a homogeneous powder must be handled separately.

Suppose that a large annual forb is sampled. One would separate this at first into the main groups: roots, stems, leaves, flowers, and fruits. Such grouping usually appears as shown in Tables 5–1 and 5–2. But for the actual measurements, additional separations of the material are often necessary, such as leaves of different age classes or roots or stems of different diameters. Sometimes the need for further separation appears only while the sample is being milled after drying. As mentioned above, the main purpose of subdivisions within each group is to permit subsequent homogenization of the material to be so thoroughly carried out that the necessary three replicates show minimal variations of caloric values. The deviations should not exceed 25 cal.

The harvested and sorted fresh material is weighed in its entirety. If the quantity is too large, subsamples of each group must be separated, and each subsample must be properly labeled and packed into a plastic bag. The crop should be brought to the laboratory as soon as possible. The weight of one subsample should not exceed 250–500 g fresh weight, as the further processes of drying and milling become increasingly difficult with larger samples.

Drying

Once in the laboratory, all samples must be unpacked, transferred into paper bags, weighed, and put loosely into a drying oven. Using forced ventilation at 80°C, it takes less than 24 hr until the weight of the samples is constant. The dried samples are weighed again, and the difference between fresh and dry weight allows the conversion of the total fresh weights into dry weights.

It should be noted that great care must be taken in determining the dry-matter production. The variability of the energy estimates depends almost entirely on the accuracy of this dry-matter determination, as its variability is about 20 to 100 times larger than that of the calorie measurements.

Milling

The dried material must be homogenized before smaller samples can be taken for energy determination. The easiest way to do this is by milling and mixing the total sample into a uniform powder. For this purpose we use a disk-type swing mill, the containers of which (100- or 250-g capacity) are completely closed, so that no dust can be lost or separated. The milling process should not exceed 5 min during which time the normal material is milled to a fine powder. Longer processing is useless and may even overheat or partially burn the material.

Some material is difficult or impossible to mill (e.g., stems with strong fibers covered by soft parenchyma or material rich in liquid compounds and resins such as seeds or young buds). Such material has to be prepared and homogenized as well as possible by hand, and usually more replicates must be burned in the calorimeter. The milled powder is transferred carefully from the milling container into plastic bags or glass containers, which should be closed carefully and labeled with the sample number. In this form the samples can be accumulated and stored easily in a dry place until they can be analyzed.

Preparation of tablets or other combustion units

To measure the calorific values, compact units should be formed from the powder as it is troublesome to weigh and process loose powder in a crucible. Units are prepared either by packing the powder into small combustion capsules, by melting it into waxes or paraffin, or by compressing it into tablet form. The method chosen depends on local conditions and the nature and quantity of the powder. Detailed descriptions for making tablets are provided below.

For making tablets, the powder should have a moisture content of $\sim 5\%$. This can be attained by leaving the powder overnight in open dishes in a room with high humidity. On the following day, part of the powder can be used for pressing tablets and the rest can be used for the determination of moisture content and ash content. Pressing the tablets requires a pressing set, which is available from the factory. The tablets should weigh ~ 1 g. The weight should be less when the caloric content is expected to be close to 10 kcal/g, and more when the energy content is expected to be < 4000 cal/g.

The amount of pressure required varies according to the condition of the material. For example benzoic acid requires > 10 atm, whereas some powdered wood materials require ~ 100 atm to ensure smooth surfaces and sufficient compactness, so that nothing is lost during the operations described below. Every manipulation of the tablet, from compressing to putting it into the bomb, should receive the greatest care to ensure accuracy of the caloric determination. Three combustion units should be prepared from each sample: two will be burned and one is kept in reserve to double-check errant reading. Delicate material may be weighed and handled in the crucible that will hold it within the bomb later on. The wire (e.g., iron) needed to ignite the tablet in the bomb can be obtained from the calorimeter factory with a known caloric value per centimeter.

Preparation of the Oxygen Bomb

The preweighed tablet should be placed in the oxygen bomb, a heavy stainless steel container with a capacity of ~ 0.3 liter. It has a screw cap that contains all the necessary devices, such as inlet and outlet valves, terminals for the electric ignition, a holding device for a small stainless steel crucible or quartz cup, a shield to protect the upper part of the bomb against sparks, and a rubber washer. The construction of the calorimeter bomb and the position of the tablet are shown in Figure 5–1. The combustion unit is placed in the quartz cup, and

FIGURE 5–1. Oxygen bomb, ready for measuring.

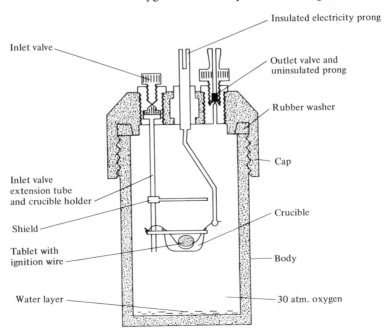

the two ends of the ignition wire are carefully attached to the electrodes. About 5 ml of water should be poured into the bottom of the bomb and the bomb body slowly screwed to the cap. The closed bomb then can be filled with 30 atm of oxygen through the inlet valve. This takes 1 min. At the beginning of the filling we open the outlet valve for a few seconds to replace most of the nitrogen-containing air with oxygen. The filled bomb is now ready to be placed in the water bath, where we have to check whether the bomb is gas tight.

Preparation of the Water Bath

The water bath of an adiabatic calorimeter is kept in a thin-walled kettle, which can be removed from the instrument. We remove this kettle and put the bomb into the holding device inside the kettle. The kettle then is filled with clean water until the bomb cap is covered and only the two ignition prongs are above the water level. Calorie measurements should always start at the same temperature. Therefore we adjust the temperature of the water, which is normally cooler than desired, with an immersion heater–electric stirrer system. The common reference temperature is 22°C; in hotter climates, in laboratories without air-conditioning, one should start at 35°C.

The warmed bath is taken quickly to a balance and the exact weight, calibrated from the very beginning for each pair of bomb and kettle, is adjusted with a pipette. Immediately after weighing, the bath should be placed in the calorimeter, the temperature probe inserted, and the necessary electrical connection made. The cover of the water jacket can be closed and the Beckmann thermometer dipped through the cover into the bath. About 5 min will bring the entire system to an even temperature.

The Measurement

When a stable temperature is obtained, as shown by the behavior of the various pilot devices of the calorimeter, the Beckmann thermometer is read. The ignition button is pressed, and a little later the Beckmann thermometer shows that the temperature is increasing. After about 10 min, when no further increase of temperature can be obtained, the highest constant temperature is read. The difference between the two readings, corrected by means of the calibration table for the Beckmann thermometer, is used for calculation of the calorific value.

Additional Procedures after the Measurement

After the final temperature reading we remove the bomb from the bath, carefully unscrew the body from the cover, and check whether the sample was totally burned. If not, the experiment must be repeated. If the tablet has burned correctly, we can go on to check the amount of nitric and sulfuric acids formed by first collecting all the liquid into an Erlenmeyer flask. We then titrate first against $\frac{1}{10}$ mol $Ba(OH)_2$ until phenolphthalein changes to pink, then add

Table 5–1 Calorie equivalents for sulfuric and nitric acid in different plant materials

Species	Plant part	Correction in calories[a]
Zea mays	Roots	1.3
	Leaves	2.7
	Male flowers	3.5
	Young female flowers	3.1
	Grains	4.4
Helianthus annuus	Fruits	5.1

[a] Should be subtracted from the calorific value of 1 g dry matter.

5 or 10 ml of a $\frac{1}{10}$-mol Na_2CO_3 solution and heat for 20 min. The liquid is cooled and filtered, a few drops of methyl orange are added, and the unused part of the carbonate is backtitrated. The first titration indicates the total amount of acid formed; the second titration indicates the amount of sulfuric acid alone. For each milliliter of N/10–HNO_3 we should consider a surplus of 1.5 cal and for each milliliter of N/10–H_2SO_4, 3.6 cal.

The titration involves considerable work, but normal material contains only a small amount of sulfur and nitrogen. Table 5–1 shows the values we have found for different materials; the normal error caused by the formation of acid is $> 0.1\%$ and is therefore within the accuracy limits of the energy determination itself, which may show deviations of about 11 cal with standardized material. For our own experiments, we have accepted a deviation of 20 cal between two replicates of our material.

Calculation of Caloric Values

The "water value" of the instrument

To calculate the caloric value of any substance the calorimeter should be calibrated first as a whole. This is done with a small sample of benzoic acid (NBS, 6323 cal/g) or succinic acid (Merck, 3022 cal/g). These samples undergo the treatment that we described for the ordinary samples.

The calibration is made to ensure that subsequent calculations arrive at the number of calories necessary to raise the temperature of the water bath by 1 degree centigrade. This is the so-called water value (W) of the system. We need the following information for the calculation of this value: the caloric value per gram sample (V), the sample dry weight (G), the corrected temperature-difference reading at the Beckmann thermometer before and after burning (Δt), the correction values for the acid formed and for the ignition wire (Σc). Among these quantities the following relationship exists:

$$W = (VG + \Sigma c)/\Delta t \qquad (5\text{-}1)$$

This water value must be estimated for each bomb–water bath pair used in connection with any calorimeter. The values may vary from time to time with changing climatic conditions.

Calculation of the Caloric Values of the Samples

If we have all the above-mentioned information, including the water value, we can transcribe formula (5-1) into the form

$$V = [W (\Delta t - \Sigma c)]/G \qquad (5\text{-}2)$$

and calculate in this way the caloric value of any sample. If all the known values are substituted in this formula, the values for V can be calculated as calories per gram dry weight, including ash content. These values can be used to calculate the stored energy from the dry-matter determinations.

Calculation of the Stored Energy

We can calculate the total amount of stored energy if we have determined all the components of one harvest. Table 5–2 shows the procedure for two different crops, sunflower (*Helianthus annuus*) and maize (*Zea mays*). The measurements in this table allow a good comparison between the accuracy of the dry-matter harvest and the energy determinations. The values for the dry-matter production are averages from four replicates for maize and six replicates for sunflowers. The average variation for maize was calculated to be 12.3% of the total harvest; for sunflowers, the variability is $> 10\%$. The determinations of the caloric values show variations from zero to 6% (for the positions marked

Table 5–2   Energy content of two different annual crops[a]

Plant parts	*Helianthus annuus* (local breed) Growing time: 19 April–10 Sept. 1963			*Zea mays* (INRA 258) Growing time: 20 April–13 Sept. 1963		
	Dry-matter mean[b] (g)	cal/g	10^6 cal/m²	Dry-matter mean[c] (g)	cal/g	10^6 cal/m²
Roots	284.2	4611	1.31	89.5	3192	0.29
Stems	1203.5	4014	4.83	325.0	4155*	1.38
Leaves	566.0	3404	1.93	331.4	4045	1.31
Male flowers	—	—	—	13.2	4197*	0.05
Fruits	1158.9	5014*	5.81	1176.0	4291	5.05
Total:	3212.6		13.87	1935.1		8.08

[a] Replicates showing more than 20-cal deviation are marked with an asterisk.
[b] Average of six replicates, 12% variation.
[c] Average of four replicates, 10% variation.

Part 2: Methods of Productivity Measurement

with an asterisk). The average variation for the normal samples is $\sim 0.4\%$. This shows that the accuracy of the calculation of total energy content, $\sim 15\%$ variability, depends almost entirely on the dry-matter determinations.

Calculation of energy from chemical analyses

Chemical analysis of biologic material is so important for agriculture, forestry, and technology, that a vast body of literature exists for this purpose (e.g., Watt and Merrill, 1963). Major categories such as N-free extract, crude protein, crude fiber, resin and fat and ash are analyzed on a routine basis everywhere and these data can be converted with sufficient accuracy into average caloric values for geoecologic comparison.

The procedure is explained in Table 5–3 and 5–4. Table 5–3 contains a compilation of the caloric content of chemical compound or matter classes. The accounting procedure for the total combustion value of material of interest is demonstrated in Table 5–4 for some woody and herbaceous material as well. (For aquatic plants see also Table 3–1.)

Such determinations are for gross comparative purposes as useful as the direct determinations. As we have indicated, the caloric value calculated per unit area is in most cases more dependent on the accuracy of the dry-matter determination than on the calorific value conversion. Data such as those in Table 5–4 not only permit comparisons of energy content between different communities, but have further interest in future productivity research, as discussed in Chapter 14.

Table 5–3   Caloric content of chemical compounds important for ecologic calculations[a]

Compound or matter class	kcal/g
Starch	4.18
Cellulose	4.2
Saccharose	3.95
Glucose	3.7
Raw fiber	4.2
N-free extract	4.1
Glycine	3.1
Leucine	6.5
Raw protein	5.5
Oxalic acid	0.67
Ethanol	7.1
Tripalmitin	9.3
Palmitinic acid	9.4
Isoprene	11.2
Lignin	6.3
Fat	9.3

[a] Compiled from Pflanz (1964), Morowitz (1968), and Runge (1973).

Table 5–4 Examples for calculating caloric values from standard chemical analyses[a]

Biologic material	Average caloric value	Beech wood		Spruce wood		Grass		Legumes	
		Wt (%)	kcal contributed to total	Wt (%)	kcal contributed to total	Wt (%)	kcal contributed to total	Wt (%)	kcal contributed to total
Crude fiber	4.2	—	—	—	—	30.6	1.28	20.1	0.84
Lignin	6.3	22.7	1.43	28.0	1.76	5.0	0.32	5.0	0.32
Cellulose	4.2	45.4	1.91	41.5	1.74	—	—	—	—
N-free extract	4.1	—	—	—	—	45.1	1.85	40.9	1.68
Woodpolyoses	4.1	22.2	0.91	24.3	1.00	—	—	—	—
Resins and fats	8.8	0.7	0.06	1.8	0.16	—	—	—	—
Crude protein	5.5	—	—	—	—	3.0	0.28	6.2	0.58
Rest fraction	5.5	7.4	0.41	3.2	0.18	9.4	0.52	19.9	1.09
Ash	—	1.6	—	1.2	—	6.9	—	7.9	—
Caloric content of material (kcal/g)			4.72		4.84		4.25		4.51
Caloric content for ash-free material							4.56		4.89

[a] After Runge (1973).

References

Anon. 1956. *Prüfung fester Brennstoffe. Bestimmung der Verbrennungswärme und des Heizwertes.* DIN 51708 Beuth Vertrieb GmbH, Berlin W15 and Cologne.

————. 1958. *National Bureau of Standards Certificate, standard sample 39 h benzoic acid.* Washington, D.C.: U.S. Government Printing Office 464336.

————. 1961. *American Oil Chemist's Official and Tentative Methods*, 2nd ed. Chicago, Illinois: American Oil Chemical Society.

Begg, J. E. 1965. High photosynthetic efficiency in a low latitude environment. *Nature (London)* 203:1025–1926.

Berger-Landefeldt, U. 1964. Über den Strahlungshaushalt verschiedener Pflanzenbestände. *Ber. Deutsch. Bot. Ges.* 77:27–48.

Bliss, L. C. 1962. Caloric and lipid content in alpine tundra plants. *Ecology* 43:753–757.

Cummins, K. W., and J. C. Wuycheck. 1971. Caloric equivalents for investigations in ecological energetics. *Mitt. Int. Ver. Limnol.* 18:1–158.

Gates, D. M. 1962. *Energy Exchange in the Biosphere*, 151 pp. New York: Harper and Row.

————. 1968. Energy exchange in the biosphere. In *Functioning of Terrestrial Ecosystems at the Primary Production Level: Proc. Copenhagen Symp. 1965*, F. E. Eckardt, ed., *Natural Resources Res.* 5:33–43. Paris: UNESCO.

Golley, F. B. 1960. Energy dynamics of a food chain of an old field community. *Ecol. Monogr.* 30:187–206.

Golley, F. B. Energy values of ecological material. *Ecology* 42:581–584.

————. 1972. Energy flux in ecosystems. In *Ecosystem Structure and Function*, J. A. Wiens, ed. Corvallis, Oregon: Oregon State Univ. *Ann. Biol. Colloq.* 31:69–90.

Jordan, C. F. 1971. A world pattern of plant energetics. *Scient. Amer.* 59:425–433.

Kaishio, Y., T. Hashizume, and H. Morimoto. 1961. Energy values of wild grass hay and rice straw for maintenance. In *2nd Symp. Energy Metabolism, Wageningen, 1961*. (EAPP Publ.) 10:165–176.

Kamel, M. S. 1958. Efficiency of solar energy conversion as related with growth in barley. *Mededel. Landbouw. Hogesch. Wageningen* 58: 1–19.

Kreh, R. 1965. *Produktivitätsstudien an Sonnenblumen*, 71 p. Stuttgart: Staatsexamensarbeit der Technischen Hochschule.

Lieth, H. 1964. Versuch einer kartographischen Darstellung der Produktivität der Pflanzendecke auf der Erde. *Geogr. Taschenbuch* 1964/1965, 72–80. Wiesbaden: Max Steiner Verlag.

————. 1965. Ökologische Fragen bei der Untersuchung der biologischen Stoffproduktion. 1. *Qual. Plant. Mater. Veg.* 11:241–261.

————, and B. Pflanz. 1968. The measurement of calorific values of biological material and determination of ecological efficiency. In *Functioning of Terrestrial Ecoystems at the Primary Production Level: Proc. Copenhagen Symp. 1965*, F. E. Eckardt, ed., *Natural Resources Research* 5:233–242. Paris: UNESCO.

Lindeman, R. 1942. The trophic–dynamic aspect of ecology. *Ecology* 23:399–418.

Long, F. L. 1934. Application of calorimetric methods to ecological research. *Plant Physiol.* 9:323–338.

Medina, E., and H. Lieth. 1964. Die Beziehungen zwischen Chlorophyllgehalt, assimilierender Fläche und Trockensubstanzproduktion in einigen Pflanzengemeinschaften. *Beitr. Biol. Pflanz.* 40: 451–494.

Mörikofer, W. 1940. Meteorologische Strahlungsmessmethoden für biologische und ökologische Untersuchungen. *Ber. Geobot. Forsch. Inst. Rübel* 12:13–75.

Morowitz, H. J. 1968. *Energy Flow in Biology,* 179 pp. New York: Academic Press.

Müller, D. 1960. Ökologische Energetik der Photosynthese. *Handbuch Pflanzenphysiol.* 5(pt. 2):255–268.

———, and J. Nielsen. 1965. Production brute, pertes par respiration et production nette dans la forêt ombrophile tropicale. *Forstlige Forsøgsvaesen Danmark* 29:69–160.

Nijkamp, H. J. 1961. Some remarks about the use of the bomb calorimeter. In *2nd Symp. Energy Metabolism, Wageningen, 1961.* (EAAP Publ.) 10:86–93.

Odum, E. P., and H. T. Odum. 1959. *Fundamentals of Ecology,* 2nd ed., 546 pp. Philadelphia, Pennsylvania: Saunders.

Ovington, J. D. 1961. Some aspects of energy flow in plantations of *Pinus silvestris* L. *Ann. Bot. London* 25:12–21.

Paine, R. 1971. The measurement and application of the calorie to ecological problems. *Annu. Rev. Ecol. Syst.* 2:145–164.

Pflanz, B. 1964. *Der Energiegehalt und die ökologische Energieausbeute verschiedener Pflanzen und Pflanzenbestände,* 42 pp. Stuttgart: Staatsexamensarbeit der Technischen Hochschule.

Rossini, F. E. (ed.) 1956. *Experimental Thermochemistry.* New York: Wiley (Interscience).

Roth, W. A., and F. Becker. 1956. Kalorimetrische Methoden. *Verfahrens- und Messkunde der Naturwissenschaft.* Braunschweig: Viehweg.

Runge, M. 1973. Energieumsätze in den Biozönosèn terrestrischer Ökosysteme. *Scripta Geobotanica,* Vol. 4, 77 pp. Göttingen: Goltze Verlag.

Sauberer, F., and O. Härtel. 1959. *Pflanze und Strahlung.* Leipzig.

Schneider, W. 1962. Zur Brennwertbestimmung von organischen Substanzen mit Hilfe eines adiabatischen Kalorimeters. *Janke & Kunkel Nachr.* 7:2–4.

Slobodkin, L. B. 1962. Energy in animal ecology. *Advan. Ecol. Res.* 1:69–99.

Tanner, C. B., G. T. Thurtell, and J. B. Swan. 1963. Integration systems using a commercial coulometer. *Proc. Soil Sci. Soc. Amer.* 27:478–481.

Watt, B. K., and A. L. Merrill. 1963. Composition of foods. *U. S. Dept. Agr., Handbook* 8.

6

Assessment of
Regional Productivity
in North Carolina

Douglas D. Sharp, Helmut Lieth,
and Dennis Whigham

One goal of productivity research is to present and analyze patterns of actual
and potential primary productivity of landscapes. The data available for such
determinations have greatly increased since initiation of the International Biological Program (IBP); and models developed from these studies will be used
eventually to predict production for almost any ecosystem. At present, however,
productivity data from intensive site studies are of limited value for estimating
landscape productivity patterns. No matter how careful the production measurements, the values obtained are in the strict sense valid only for the particular
sites and time periods of investigation. For a proper assessment of landscape
production patterns we need numerous measurements that can be related to the
pattern of the landscape itself. Various agricultural and forestry statistics are
readily available and can be used to demonstrate landscape production patterns.
Most of these data express primary production for the commercially usable portion of each land-use category (e.g., seed production for crops and mercantile
lumber for forests). The data can be converted, however, to estimates of total
primary production through the use of appropriate conversion factors—the
ratios of total primary production to the commercial yield. Similar analyses can
be performed on any land-use category if the production–yield ratios are known.
This chapter summarizes 2 years of research in North Carolina, which included
the utilization of U. S. Forest Service and state agricultural and land-use data
for estimating rates of net primary productivity for all land-use categories in the
state's 100 counties. Additional estimates were made of total net primary production for each county, an estimated net primary production rate for the entire
state, and an estimate of the state's total net primary production. More complete documentation is available in Whigham *et al.* (1971) and Sharp (1973).

KEYWORDS: Primary productivity; plants, agriculture;
North Carolina; forestry.

Primary Productivity of the Biosphere, edited by
Helmut Lieth and Robert H. Whittaker.
© 1975 by Springer-Verlag New York Inc.

Table 6–1   Sources of data used to determine county coverage and
commercial yield statistics for each land-use category[a]

I. Forests
    a. *North Carolina's Timber.* 1966. U.S. Forest Service Bulletin SE-5.
    b. *Forest Statistics for the Southern Coastal Plain of North Carolina.* 1952. U.S. Forest Service, Forest Survey Release No. 41.
    c. *Forest Statistics for the Mountain Regions of North Carolina.* 1955. U.S. Forest Service, Forest Survey Release No. 46.
    d. *Forest Statistics for the Northern Coastal Plain of North Carolina.* 1955. U.S. Forest Service, Forest Survey Release No. 45.
    e. *Forest Statistics for the Piedmont of North Carolina.* 1956. U.S. Forest Service, Forest Survey Release No. 48.
    f. *Preliminary Forest Survey Statistics for the Southern Coastal Plain of North Carolina.* 1962. U.S. Forest Service publication by the Division of Forest Economics Research.
    g. *Preliminary Forest Survey Statistics for the Northern Coastal Plain of North Carolina.* 1963. U.S. Forest Service publication by the Division of Forest Economics Research.
    h. *Preliminary Forest Survey Statistics for the Piedmont of North Carolina.* 1964. U.S. Forest Service publication by the Division of Forest Economic Research.
    i. *Preliminary Forest Survey Statistics for the Mountain Regions of North Carolina.* 1964. U. S. Forest Service publication by the Division of Forest Economic Research.

 II. Land-use categories
    U.S. Department of Agriculture, Soil Conservation Service:
        Form S-1: *State Land Use Summary (Acres)*
        Form S-2a: *Summary—Land by Land Capability Classes*

III. Crop statistics
    North Carolina Agricultural Statistics. 1967–1973. Available through the Federal Crop Reporting Service. Raleigh, North Carolina.

IV. Water acreages
    Profile, North Carolina Counties. 1970. Statistical Services Section, Budget Division, Department of Administration, Raleigh, North Carolina.

---

[a] See Table 6–2.

History

Filzer (1951) was the first to evaluate production patterns for a large region. With detailed statistics available from pre-World War I Germany he used agricultural yield as an indicator. This treatment did not include forestry statistics nor did Filzer attempt to calculate total primary production. Further attempts to utilize statistical data for mapping production patterns were made by Weck (1955) for forest yield in Germany, and by Paterson (1956) for forests of the world. Their data were presented as yield in lumber and not as total primary production. Because lumber production and total production occur in predictable ratios to one another, it was possible for Lieth (1964), to use Paterson's data to construct his first world primary productivity map. Whittaker (1961, 1966), Lieth (1964), Monsi (1968), Whittaker and Woodwell (1968, 1969), Kira

6. *Assessment of Regional Productivity in North Carolina*

State __North Carolina__   __Orange__ County Name
                                  __68__ County Number
Investigator __Douglas Sharp__   __36.0 N   79.1 W__ Geographic Coordinates
                                        of County Center

Data Sources   1: Yield information   __Timber in N.C.   N.C. Farm__ Summary
               2: Acerage data   __Timber in N.C.   N.C. Farm__ Summary, USDA S.C.S.
               3: Other

	1	2	3 [a]	4	5 [b]	6 [c]	7 [d]	
Land-use category	Hectares	Comm. yield (t/ha)	Conver. factor	Water content	Corr. conver. factor	Adjusted product. rate	Total prod.	
Softwood Forests	62526	.63	2.0	0	2.0	1.26	78783	e 12.90
Hardwood Forests		.81	2.0	0	2.0	1.62	101292	f 80658
Corn	2711	3.32	2.62	.12	2.31	7.67	20793.37	
Soy Bean	627	1.28	4.52	.12	3.98	5.09	3191.43	
Tobacco	991	2.16	2.03	.12	1.79	3.87	3835.17	
Wheat	1153	3.03	3.69	.12	3.25	9.85	11357.05	
Oats (Winter)	910	1.69	5.30	.12	4.66	7.88	7170.80	
Peanut	0	0	2.00	.12	1.76	0	0	
Cotton	0	0	2.08	0	2.08	0	0	
Irish Potato	8	11.77	2.47	.75	.62	7.30	58.40	
Sweet Potato	4	17.93	2.47	.75	.62	11.70	44.68	
Hay	2286	3.47	1.30	.14	1.12	3.89	8892.54	
Urban areas	5972	.50	.50	0	.50	.25	1495	
Water	81	5.00	1.00	0	1.00	5.00	405	
Pasture—Range	10164	3.47	.60	0	.60	2.08	21141.12	
Orchards—Vinyards	176	.81	2.00	0	2.00	1.62	285.12	
Open land	1753	3.47	.60	0	.60	2.08	3646.24	
Tillage rotation	6944	3.47	.60	0	.60	2.08	14443.52	
Other								
Oats (Spring)	0	0	5.22	.12	4.59	0	0	
Total Land Area	96306						276834	g 90334

Actual = 103077

Total County Production

[a] Total prod./yield
[b] Column 3 times (1.00 – Column 4)
[c] Column 5 times Column 2
[d] Column 6 times Column 1
[e] Potential forest prod. rate
[f] Potential total forest prod. (t/ha/year)
[g] Potential total county prod. (t/ha/year)

Weighted county production rate: 2.87
(Total production/Total area)
(t/ha/year)

Potential weighted county prod. rate: 9.38
(t/ha/year)

FIGURE 6–1. Tally sheet for calculating net primary production for counties by land-use categories. See text for explanation.

et al. (1969), and Satoo (1970) all have shown the feasibility of calculating total productivity figures from partial production values. These authors worked independently during the same period; from their work and that of others has come the essential knowledge of production ratios by which agricultural and forestry statistics can be used for productivity mapping.

Table 6–2 Summary of state net primary production data by region (in grams and tons dry matter)

North Carolina productivity profile (year)	Region	Area (10³ ha)	% total	Rate of net productivity (g/m²/year)	Area net production (10³ t/year)	% total
1971	Mountains	2,175	16.0	251	5,454	13.9
	Piedmont	4,018	29.5	259	10,411	26.5
	N.E. Coastal Plain	4,067	29.9	321	13,076	33.2
	S.E. Coastal Plain	3,350	24.6	310	10,384	26.4
	Totals:	13,609	100.0		39,325	100.0
1972 Without adjustments to forest production data	Mountains	2,160	16.6	260	5,623	13.7
	Piedmont	3,988	30.7	277	11,029	27.0
	N.E. Coastal Plain	3,446	26.6	391	13,483	32.9
	S.E. Coastal Plain	3,391	26.1	318	10,793	26.4
	Totals:	12,984	100.0		40,928	100.0
With adjustments to forest production data	Mountains	2,160	16.64	628	13,564	13.0
	Piedmont	3,988	30.17	592	23,593	22.6
	N.E. Coastal Plain	3,446	26.54	927	31,949	30.6
	S.E. Coastal Plain	3,391	26.11	1046	35,453	33.9
	Totals:	12,984	100.0	$805 = \overline{X}$	104,559	100.0

Table 6–3   Comparison of crop conversion factors
used in 1971 and 1972

Land-use category	Conversion factor[a]	
	1971	1972
Corn	2.03	2.31
Soybeans	3.92	3.92
Tobacco	2.68	1.79[b]
Wheat	2.15	5.24[b]; 3.25[c]
Oats	2.64	
Winter	—	5.49[b]; 4.66[c]
Spring	—	5.41[b]; 4.59[c]
Peanuts	2.64	2.64
Cotton	2.08	2.08
Irish potato	0.60	0.60
Sweet potato	0.60	0.60
Hay	1.12	1.12

[a] Dry weight productivity/wet weight yield.
[b] Experimentally derived conversion factors.
[c] These conversion factors were utilized in the 1972 report.

Methods

Various sources (Table 6–1 and references cited therein) were used to determine county coverage and commercial yield statistics for each land-use category shown in Figure 6–1. Using Figure 6–1 as a model, computational procedures were as follows. To estimate primary productivity rates for each land-use category, commercial yield data (column 2) were multiplied by appropriate conversion factors (column 3). For each land-use category, the conversion factor represents the ratio of estimated total primary production to commercial yield.

Conversion factors used in 1971 were determined from a literature review and through the cooperation of Dr. Ray Noggle and Dr. Douglas Gross of North Carolina State University (Whigham *et al.*, 1971). For several crops conversion factors were verified by actual sampling at several North Carolina agricultural experiment stations. Conversion factors used in 1972 were somewhat different and are discussed subsequently in more detail.

Commercial yield statistics for crop types were based upon wet-weight figures, and it was necessary to adjust the conversion factors to their dry-weight equivalents (column 5) using an estimated water content (column 4). Commercial yield statistics taken from Dorman *et al.* (1970) (column 2) then were multiplied by the corrected conversion factors (column 5). To determine total county production for each land-use category, the adjusted productivity rates (column 6) were multiplied by the coverage data (column 1). Total primary production estimates for each land-use category were summed, and a weighted county productivity rate (total production/total area) was determined. Computer maps

Table 6–4 Computation of conversion ratio used in estimating net forest production from merchantable timber data in North Carolina

Measurements	Mountains	Piedmont	Coastal Plain
Merchantable timber growth, U.S. Forest Service survey data, g/m²/year	Mean for N. C. mountain Counties	Alamance County 265 Orange County 287 Durham County 247	Beaufort County 250 Washington County 307 Pitt County 319
Means:	266	266	292
Forest net production, Plantation survey data, g/m²/year	Mean of the production-rate of Tennessee mountain counties (DeSelm et al., 1971)	Duke Forest[a] Pinus echinata (18 stands) 564 Pinus taeda (47 stands) 998	Beaufort County Plantations[b] (Pinus taeda, 28 stands) Trees aged 10 and 11 years 2230 Trees aged 12 years 1840 Trees aged 4 years 430
Means:	765	781	1500
Conversion ratio (forest net production/merchantable timber growth)	2.88[c]	2.94	5.14
Mean conversion ratio:		3.65 (used in all regions)	

[a] Ralston, C. W., and C. F. Korstian (1957–1958), *Piedmont Plantation Data* (*unpublished*).
[b] Nemeth, J. C. (1971). *Dry-Matter Production in Young Loblolly* (*Pinus taeda*) *and Slash Pine* (*Pinus elliottii*) *Plantations*, Doctoral dissertation. Raleigh, North Carolina: North Carolina State Univ.
[c] Based on the premise that green volume growth rates of trees in the mountains are similar to those rates for the Piedmont or stated in De Selm *et al.,* (1971) *Tennessee Productivity Profiles.* (Mountain counties only.)

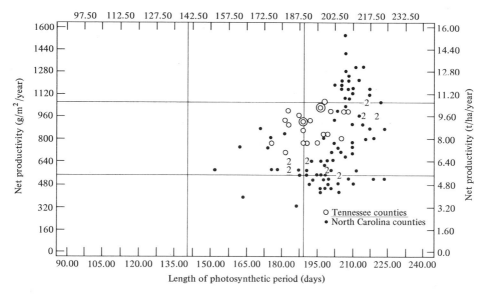

FIGURE 6–2. Net primary productivity rates of North Carolina and Tennessee counties compared to growing season length. Abscissa: photosynthetic period in days; ordinate: net primary productivity ($100g/m^2 = 1^t/_{ha}$). The number 2 is inserted in places where two North Carolina points occupy the same position; double circles indicate the same for Tennessee.

(Reader, 1972) were used to compare graphically weighted county primary productivity rates and rates for each land-use category.

Results

Table 6–2 summarizes the 1971 primary production estimates for the state's four regions. The average regional production rates were lower than might be predicted for a humid temperate climate (Art and Marks, 1971; Whittaker, 1970; Bray and Dudkiewicz,1963; Duvigneaud and Denaeyer-DeSmet, 1967; Satoo, 1967; Madgwick, 1968; Post, 1970; Woodwell and Whittaker, 1970; Odum, 1971). Based on the assumption that the low 1971 estimates were due to inaccuracies in the original set of conversion factors, 1972 efforts focused on a reassessment of the latter. Five crops (Table 6–3) were intensively sampled at agricultural experiment stations throughout the state and, to some degree, all of the 1971 crop conversion factors were changed (Sharp, 1973). When the 1972 conversion factors were used for crops, estimated production rates for the four regions, were increased but still lower than might be anticipated (Table 6–2). Because changes in the crop conversion factors did not significantly alter the estimates of county primary productivity rates, forest-yield conversion factors were examined and changed significantly in 1972 (Table 6–4).

The changes were based upon comparison between Forest Service statistics

138

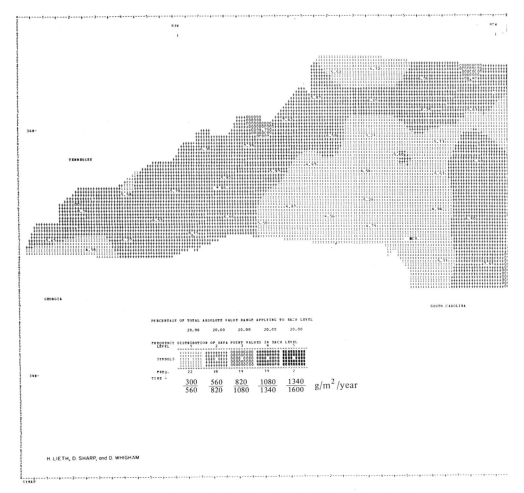

FIGURE 6–3. Map shows distribution of estimated county net primary productivity rates in North Carolina in 1972. Datum points on map represent production rates for each county. Values are based on three distinct mean conversion ratios determined for each region of state. $1^t/_{ha} = 100g/m^2$.

and results of two previous productivity studies. Selecting three counties in the coastal plain region, Forest Service statistics showed a mean production rate for merchantable lumber of 292 g/m²/year (2.92 t/ha/year). Nemeth (1971) demonstrated a mean production rate of 1500 g/m²/year for mixed stands of *Pinus taeda* and *P. elliotii* in one of those counties (Beaufort). An earlier study by Ralston and Korstian (*unpublished*) in the Piedmont permitted further comparisons with Forest Service data. For Alamance, Orange, and Durham counties, the Forest Service estimates indicate an average productivity of 266 g/m²/year (Table 6–4). For stands of *Pinus taeda* and *P. echinata* in the same counties, Ralston and Korstian estimates yield a forest productivity rate of 781

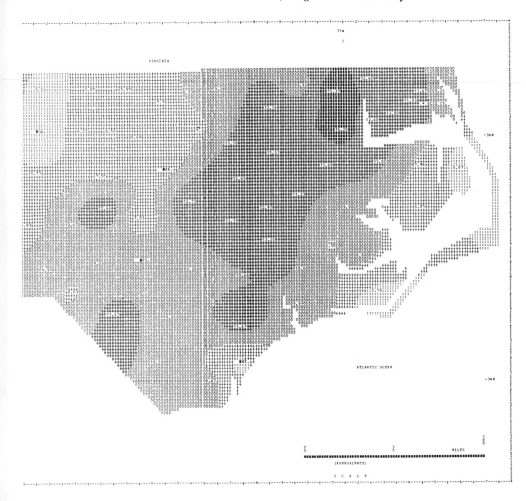

g/m²/year. The higher production values were determined for managed plantations, and it was assumed that most unmanaged forests had somewhat lower
production rates. On that assumption, mean conversion ratios (Snedecor and
Cochran, 1967; Lieth, 1972) were determined for each region (Table 6–4).
Based on DeSelm *et al.* (1971) and on the premise that growth rates of trees
in the mountains are similar to those in the Piedmont, similar mean conversion
ratios were used for those two regions of the state. Distinct, mean conversion
ratios computed for each physiographic province and the average mean conversion ratio of 3.65 for the entire state (Table 6–4) were then used to estimate
forest production for each county (county forest production based on Forest
Survey data times the mean conversion ratio for each province or the average
mean conversion ratio for the state). Using the adjusted forestry conversion
factors, Table 6–2 shows that the estimated production rates for the four regions
were greatly increased. Figure 6–2 shows that these county estimates agree with
county primary production estimates made in Tennessee (DeSelm *et al.* 1971).

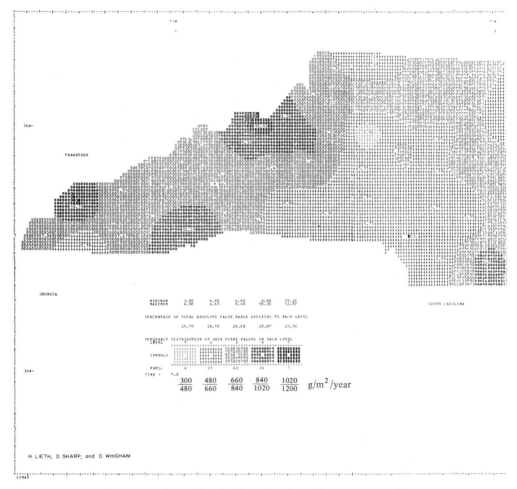

FIGURE 6–4. Map shows distribution of estimated county net primary productivity rates in North Carolina in 1972. Datum point values were determined through use of average mean conversion ratio applied throughout all regions of state. $1^{t}/_{ha} = 100g/m^2$.

Patterns of Productivity in North Carolina

The range of estimated net primary productivity was 400 g/m²/year (Alexander County, Piedmont) to 1538 g/m²/year (Hertford County, Coastal Plain) utilizing distinct, mean conversion ratios for each province of North Carolina. Most counties were estimated to have productivity rates between 600 and 1200 g/m²/year (Fig. 6–3). The average rate of primary production for the state was 805 g/m²/year. The average mean conversion ratio, tabulated for use throughout the entire state, produced the image presented in Figure 6–4. For most crops (Fig. 6–5 is an example), productivity was highest in the eastern counties and lower in the Piedmont and Mountain counties. In the 1971 study,

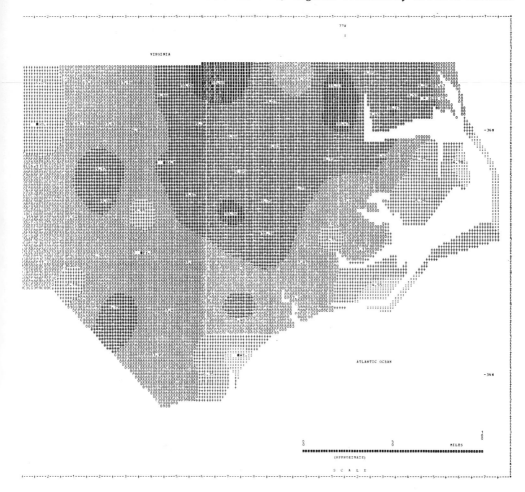

a distinct east–west productivity gradient occurred when all land-use categories were considered (Lieth, 1972). When the 1972 conversion factors were used, the east–west pattern was still present but was less distinct (cf. Figures 6–3 and 6–4). This was caused by doubling to tripling the adjusted productivity rates for counties in the mountain region. This result was expected because of the extremely high percentage of forested lands in the mountain counties. It might also be concluded that the higher productivity estimates for mountain counties are the result of more favorable edaphic, climatic, and topographic factors.

Control of Productivity Rates by Environmental Factors

Productivity rates of natural and man-influenced vegetation units are controlled by a complex of edaphic, climatic, topographic, and time-related factors. Man's utilization also influences the range of primary productivity values for an area. It has been assumed that for most land-use categories in a humid climate the total net primary productivity is most highly correlated to the onset and

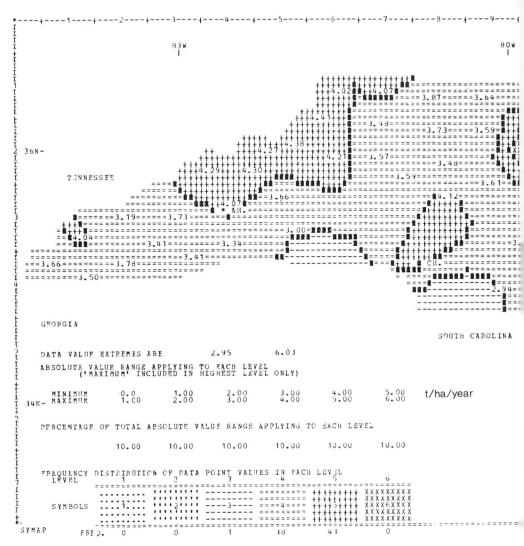

FIGURE 6–5. Map shows distribution of county productivity rates for tobacco in 1972. $1^t/_{ha} = 100g/m^2$.

length of favorable growing conditions. In North Carolina, phenologic studies have shown that the growing season begins earlier and lasts longer in the eastern counties than in the western counties (Radford, 1971). One might expect that there is a relationship between the length of the growing season and rates of primary production throughout the state. Figure 6–6 shows that the correlation between the two factors is limited. Several Mountain counties (Mitchell, Transylvania, Avery, and Graham) have high estimated primary production rates even though the length of the growing season is short. Furthermore, the Coastal Plain counties of Dare, Carteret, and New Hanover exhibit low rates of production under long growing-season conditions. In essence, a clear correlation

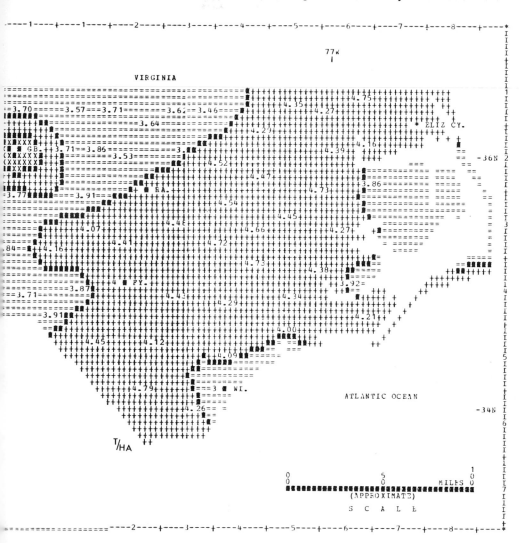

between the productivity rate and the length of the growing season cannot be demonstrated by the relationship shown for an area the size of North Carolina. Reader (1973) has shown, however, that the controlling effects of climate over the primary productivity of a large area (biome) can clearly be delineated, and that there exists a correlation between the length of the growing period and rate of net primary production. This correlation is evaluated in Chapter 12.

Conclusions

Since the initial investigation, the techniques for using extant data sources to predict the patterns of landscape primary productivity have much improved. The most critical part of such an analysis is the determination of conversion factors to translate agricultural and forestry statistics into total net primary production

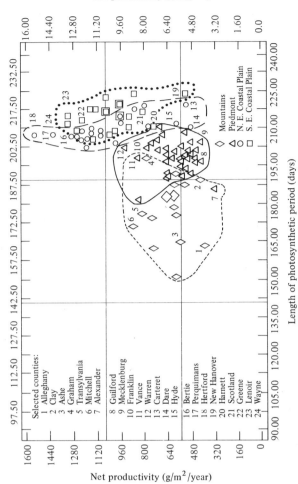

FIGURE 6–6. Estimated productivity rates of North Carolina counties compared to length of growing season. The counties are grouped into physiographic regions. Abscissa: length of vegetation period in days; ordinate: net primary productivity ($1^{t}/_{ha}$ = 100g/m²).

equivalents. This chapter summarizes the techniques and shows how they were refined in the North Carolina study. Other studies (Cottam *et al.*, 1973; see also Chapter 7) have also demonstrated that this technique can be used to relate landscape productivity patterns to major factors of the environment.

References

Art, H. W., and P. L. Marks. 1971. A summary table of biomass and net annual primary production in forest ecosystems of the world. In *Forest Biomass Studies*, H. E. Young, ed., pp. 3–34. Orono, Maine: Univ. of Maine, Life Sciences and Agriculture Experiment Station.

Bray, J. R., and L. A. Dudkiewicz. 1963. The composition, biomass, and productivity of two *Populus* forests. *Bull. Torr. Bot. Club* 90:298–308.

Cottam, G., E. Howell, F. Stearns, and N. Kubriger. 1973. Productivity profile of Wisconsin (Report of work through August 31, 1972). Deciduous Forest Biome Memo Rep. 72-142.

DeSelm, H. R., D. Sharpe, P. Baxter, R. Sayres, M. Miller, D. Natella, and R. Umber. 1971. Tennessee productivity profiles. 182 pp. (mimeogr.) US-IBP Eastern Deciduous Forest Biome Memo Rep. 71-13.

Dorman, M., Jr., G. Burleson, J. Robertson, and J. Wheeler. 1970. *Profile, North Carolina Counties*, 2nd ed. Raleigh, North Carolina: Statistical Services Sect., Budget Div., Dept. of Administration.

Duvigneaud, P., and S. Denaeyer-DeSmet. 1967. Biomass, productivity and mineral cycling in deciduous forests in Belgium. In *Symp. Primary Productivity and Mineral Cycling in Natural Ecosystems*, H. E. Young, ed., pp. 167–186. Orono, Maine: Univ. of Maine Press.

Filzer, P. 1951. *Die natürlichen Grundlagen des Pflanzenertrages in Mitteleuropa*, 198 pp. Stuttgart: Schweizerbart.

Kira, T., K. Shinozaki, and K. Hozumi. 1969. Structure of forest canopies as related to their primary productivity. *Plant Cell Physiol.* 10:129–142.

Knight, H. A., and J. P. McClure. 1966. North Carolina's timber. U. S. Forest Service Resource Bull. SE-5. Asheville, North Carolina: USDA Forest Service, Southeastern Forest Experiment Station.

Lieth, H. 1964. Versuch einer kartographischen Darstellung der Produktivität der Pflanzendecke auf der Erde. *Geographisches Taschenbuch* 1954/1965, pp. 72–80. Wiesbaden: Steiner Verlag.

————. 1972. Computer mapping of forest data. In *Proc. 51st Annu. Mtg*, pp. 53–79. Society of American Foresters, Appalachian Sect.

Madgwick, H. I. A. 1968. Seasonal changes in biomass and annual production of an old-field *Pinus virginiana* stand. *Ecology* 49:149–152.

Monsi, M. 1968. Mathematical models of plant communities. In *Functioning of Terrestrial Ecosystems at the Primary Production Level: Proc. Copenhagen Symp. 1965*, F. E. Eckardt, ed., *Natural Resources Res.* 5:131–149. Paris: UNESCO.

Nemeth, J. 1971. Dry-matter production in young loblolly (*Pinus taeda* L.) and slash pine (*Pinus elliotti*) plantations. Doctoral dissertation. Raleigh, North Carolina: North Carolina State Univ.

Odum, E. P. 1971. *Fundamentals of Ecology*, 3rd ed. 574 pp. Philadelphia, Pennsylvania: Saunders.

146

Paterson, S. S. 1956. *The Forest Area of the World and its Potential Productivity*, 216 pp. Goteborg: Royal Univ. of Goteborg.

Post, L. J. 1970. Dry-matter production of mountain maple and balsam fir in northwestern New Brunswick. *Ecology* 51:548–550.

Radford, J. R. 1971. Biological determination of the length of growing season in North Carolina: Computer mapping and environmental correlation for spring and fall plant phenophases. Master's thesis. Chapel Hill, North Carolina: Univ. of North Carolina.

Reader, J. R. 1972. *SYMAP (version 5.16A) Instruction Manual*. Triangle Park, North Carolina: Triangle Univ. Computation Center, Research Div.

———. 1973. A phenological approach to the estimation of photosynthetic period and primary productivity of the macroregion and biome levels in the Eastern Deciduous Forest biome, 11 pp. (mimeogr., with illust.) US-IBP Eastern Deciduous Forest Biome Memo Rep. 73-2.

Satoo, T. 1967. Primary production relations in woodlands of *Pinus densiflora*. In *Symp. Primary Productivity and Mineral Cycling in Natural Ecosystems*, H. E. Young, ed., pp. 62–80. Orono, Maine: Univ. of Maine.

———. 1970. A synthesis of studies by the harvest method: Primary production relations in the temperate deciduous forests of Japan. In *Analysis of Temperate Forest Ecosystems*, D. E. Reichle, ed., *Ecological Studies* 1:55–72. New York: Springer-Verlag.

Sharp, D. 1973. North Carolina productivity profile, rev. ed., 1972, 24 pp. (mimeogr.) US-IBP EDF Memo Rep. 73–4.

Snedecor, G. W., and W. G. Cochran. 1967. *Statistical Methods*, 6th ed. Ames, Iowa: Iowa State Univ. Press.

Stearns, F., *et al.* 1971. Productivity profile of Wisconsin, 35 pp. (mimeogr. with illust.) Deciduous Forest Biome Memo Rep. 71-14.

Weck, J. 1955. *Forstliche Zuwachs- und Ertragskunde*, 2nd ed. Radebeul and Berlin: De Gruyter.

Whigham, D., *et al.* 1971. The North Carolina productivity profile, 1971. 42 pp. (mimeogr) US-IBP Eastern Deciduous Forest Biome Memo Rep. 23-71.

Whittaker, R. H. 1961. Estimation of net primary production of forests and shrub communities. *Ecology* 42:177–180.

———. 1966. Forest dimensions and production in the Great Smoky Mountains. *Ecology* 47:103–121.

———. 1970. *Communities and Ecosystems*, 162 pp. New York: Macmillan.

———, and G. M. Woodwell, 1968. Dimensions and production relations of trees and shrubs in the Brookhaven Forest, New York. *J. Ecol.* 56:1–25.

———, and G. M. Woodwell. 1969. Structure, production, and diversity of the oak–pine forest at Brookhaven, New York. *J. Ecol.* 57:155–174.

Woodwell, G. M., and R. H. Whittaker. 1970. Primary production and the cation budget of the Brookhaven Forest. In *Symp. Primary Productivity and Mineral Cycling in Natural Ecosystems*, H. E. Young, ed., pp. 151–166. Orono, Maine: Univ. of Maine.

7

Methods of Assessing the Primary Production of Regions

David M. Sharpe

Most published work on primary production has been done at the local level. Estimates of production for regions have been based upon extrapolations from small samples of stand productivities. Recently, regional production rates have been studied in a more integrated manner, for example, in the Biome and Regional Analysis Program of the Eastern Deciduous Forest Biome Program, which is one of the contributions of the United States to the International Biological Program (IBP). The objective of this chapter is to review the methods used in estimating primary production rates for specific regions, and to suggest directions for improvement.

Conceptual Framework for Regional Production

Net primary productivity is defined as the difference between cumulative photosynthesis and cumulative respiration by green plants per unit time and space. Woodwell (1970) expressed this relationship by the formula

$$GP - Rs_A = NP$$

where GP is gross primary productivity or photosynthesis; Rs_A is respiration of all parts of autotrophic plants; and NP is the resultant net primary productivity.

More generally, net primary production can be conceived as an instantaneous rate, or alternatively as a cumulative amount or average, for any period of time and unit of space. Net primary production of a plant is the sum total of the production of its components; that of a hectare is the sum of the production

KEYWORDS: Regional production assessment; methods; evapotranspiration model; United States of America.

Primary Productivity of the Biosphere, edited by Helmut Lieth and Robert H. Whittaker.
© 1975 by Springer-Verlag New York Inc.

of plants occupying that area, and so on in a hierarchy of space from the leaf to the ecosphere. Each method for estimating net primary production is appropriate to some limited time and space. Exchange of carbon dioxide between plant and atmosphere can be measured with such instruments as the infrared gas analyzer only for small areas and for short periods. Harvesting is usually done on fractions of hectares. The scales chosen suggest appropriate methods; conversely, the available methods impose restrictions on the scales of time and space that can be studied. Which scales are most appropriate for interpreting primary production of regions? How adequate are current methods? How might these be further improved?

Many schemes have been devised to present the hierarchic nature of ecologic units (Novikoff, 1945; Evans, 1956; Dansereau, 1962). Rowe (1961), for example, established a hierarchy that extends from the monocene (Friederichs, 1958) through local and regional ecosystems to the ecosphere. Although conceptually appealing, these hierarchies are not based on the organization necessary for the use of the concept as a framework for measurements. Goff *et al.* (1971) formulated a hierarchy of regions based on the need for data sets and computer facilities to study regions of sizes differing by order-of-magnitude increments of linear dimensions, as shown in Figure 7–1. The ecosphere (designated R1) has a characteristic length (the circumference of the earth) exceeding 10^4 km. Biomes (R2) have characteristic lengths of 10^3–10^4 km, and so on through a succession of smaller regions (R3–R5) to research sites (R6–R8), which have characteristic lengths of 10^{-1} km or less. R9 and R10 are units of space occupied by organisms and organs, respectively.

FIGURE 7–1. Some space and time scales appropriate for regional productivity (shaded) and major periods of integration for production in context of space–time hierarchy.

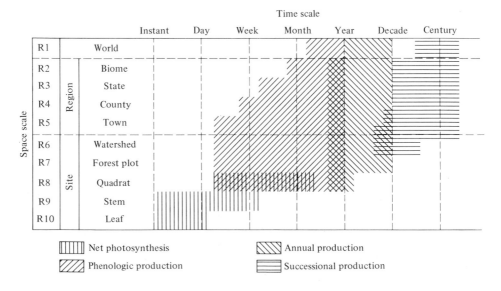

Likewise, a time hierarchy can be distinguished from the instant for rates of quantum or molecular exchange to the decades and centuries for succession and land use change. Units of time are shown at the top of Figure 7–1. The levels of the time hierarchy defined here are the instant (approximated by a second or minute), day, week, month, year, decade, and century.

Certain time scales have been argued for, implicitly or explicitly. Lieth (1970, 1971a) has promoted "the phenologic viewpoint in productivity studies" to deal with primary production over periods of days and weeks. Change in productivity through years or decades of succession has been studied by many ecologists. E. P. Odum (1960) and Golley and Gentry (1965) compared productivity on fields abandoned for various periods. Loucks (1970) hypothesized a peak in stand productivity in the late pioneer stage, which is followed by a decline as succession proceeds; Kira and Shidei (1967) found a similar trend in age series of forests. The biomass accumulation ratio was proposed as an index of stage of succession (Whittaker, 1966; Woodwell, 1967). The integration of net primary production rate for days and weeks or the year, and the trend of annual production associated with stand maturation and succession, are readily identifiable themes in production research. The net photosynthesis for periods of time shorter than encompassed by phenologic production has received attention mostly by physiologists using gas-exchange methods.

In general, space and time are linked through rates of processes, just as time and distance are related through velocity. Phenologic and annual production and the changes in annual production during succession usually involve attention over progressively larger spans of area as well as time. The horizontal dashed lines that distinguish the regions at scales of R2 to R5 from the world (R1) and site studies (R6–R10) identify categories of production relevant to regional studies. Estimates of phenologic, annual, and successional production are all important for interpreting the production of regions. The elements of the space–time hierarchy that fall within the domain of regional production are represented by the shaded area on Figure 7–1. The key point of Figure 7–1 is that studies of regional production made for markedly different space and time dimensions need differing data resolution, even though each ideally expresses an integral of photosynthesis minus respiration over an appropriate time and area.

Methods of Estimating Regional Production

Three sources of data have been used in IBP studies to provide information with different degrees of spatial resolution. The first method draws upon continuous forest-inventory plot data and allometric relations to develop the equivalent of a network of intensive plot studies with several plots per county. A second method involves the use of published data on areas in counties devoted to various land uses and data on agricultural and forest yields as provided by the *Census of Agriculture* and published forest inventories. Both methods assume that primary production can be extrapolated from these data through the use of appropriate conversion factors, but obtaining these factors is an unfinished task that is discussed subsequently. A third method relates regressions of the most appropriate index of production on environmental variables, for example, from networks of weather stations, to map production over large regions (R2 and R3) and the world.

Each of the methods considered in the next section utilizes a large existing data base, because it is virtually impossible to collect new data for so many points in the field. Unfortunately, data sets that have been developed for other purposes require major adjustments to derive regional production estimates.

Continuous forest inventories

The forest resources of the United States are censused periodically by the Forest Service in a program of continuous forest inventory (CFI). Other agencies have CFI programs in their regions, such as the Tennessee Valley Authority (TVA).

The CFI program involves sampling of forests by establishing forest inventory plots that are resurveyed at intervals (e.g., 5 or 10 years). Plots usually are located on a grid system; the Forest Service locates plots on a 3-mile- (4.8-km-) square grid, and the TVA uses an 8.5-mile- (13.7-km-) square grid. A variable-radius plot (Forest Service), or a fixed-radius plot, usually of 0.08 ha (TVA) may be established. Tallied trees on each plot are identified so that the basic record of the CFI program relates measurements to specific trees. Each tree is classified by species, and by diameter-size class as sapling, poletimber, and saw-timber. Measurements to assess both the quantity and quality of forest products in trees of commercial species and merchantable size (12.7 cm DBH and larger) are made; saplings of commercial species are censused to assess the potential for forest products; and noncommercial species are measured. The DBH (diameter breast height) of each measured tree is recorded, along with other attributes of commercial trees of merchantable size.

Records of CFI surveys are available in two forms: (1) as published summaries that provide information on forest types, merchantable standing crop, and in some cases growth to the merchantable growing stock, as discussed in the next section, and (2) as unpublished data for each tree on each plot on the CFI program, available on punched cards or magnetic tape. The plot records provided by the TVA are the data base for the study discussed here. The method is generally applicable to other CFI data as well.

Records of 224 plots in Tennessee are being used to test and revise this method. Some of these plots were installed in 1960 and were resurveyed in 1965 and 1970. Others were installed in 1966 and were resurveyed in 1970. The 1965 (or 1966) to 1970 period was used to compute average annual net primary production for each plot, and 1970 was chosen to compute biomass of the plot (DeSelm *et al.*, 1971).

The average annual net primary production and biomass of each plot are the summations of the production and biomass of each stand component. The biomass and primary production of poletimber, sawtimber, and saplings were computed; then adjustments were made to account for biomass and production of undergrowth and roots and for insect consumption. The components of the stand

Table 7–1   Outline for computing net primary production and standing
crop of Tennessee Valley Authority forest inventory plots

Components of the stand	1965–1970 net primary productivity (g/m²/year)	1970 standing crop (kg/m²)
Undergrowth (shoots and roots)	30—Average of local research	0.135—Average of local research
Saplings	Foliage only: foliage biomass × turnover rate	Bole and branch, foliage (1970 survey)
Poles and sawtimber	Bole and branch: 1970 biomass minus 1965 biomass Foliage: average biomass × turnover rate	Bole and branch, foliage (1970 survey)
Roots (saplings, poles, sawtimber)	25% of shoot production	25% of shoot biomass
Insect consumption (saplings, poles, sawtimber)	3% of shoot production	None

under consideration and the general procedure for computing net primary production and biomass standing crop of each component are shown in Table 7–1.

The biomass of each measured tree of any species and size was computed
from the recorded DBH of the tree by using allometric relations between DBH
and bole and branch biomass, and DBH and foliage biomass. The equations were
developed by Sollins and Harris (personal communication) from stem analyses
of conifers and deciduous species in Georgia, North Carolina, and Tennessee
computed by Sollins and Anderson (1971). The equations were adjusted to
remove the bias inherent in logarithmic transformation of data in regression
analysis discussed by Beauchamp and Olson (1972). The equations (with the
original in brackets, preceded by the adjustment factor) are

$$BB = (1.15)[0.119D^{2.393}]$$
$$FB = (1.37)[0.03D^{1.695}]$$

where *BB* is the biomass of bole and branches in kilograms for dry weight;
FB is the biomass of foliage in kilograms for dry weight; and *D* is the diameter
breast height (expressed as *DBH* throughout this volume) in centimeters.

The biomass of poletimber and sawtimber for 1965 (or 1966) and 1970
and for saplings in 1970 was computed as the sum of bole and branch plus
foliage biomass. Net primary production of each poletimber and sawtimber tree
was computed as (1) the difference between bole and branch biomass in 1970
and 1965 (or 1966) divided by the 5- or 4-year interval; and (2) the average
foliage biomass for 1965 (or 1966) and 1970 multiplied by the foliage turnover rate (once every year for deciduous species and an assumed 3 years for
evergreens). Because the TVA did not measure saplings until the 1970 survey,
only their foliage production could be computed as the 1970 foliage biomass

multiplied by the appropriate foliage turnover rate. Neglect of bole and branch production of saplings was compensated for by the production computed for trees that grew to poletimber size between 1965 (or 1966) and 1970. These trees had a DBH of zero in 1965 (or 1966) that was arbitrarily assigned to them by TVA, and a correspondingly high bole and branch production in this study.

Unfortunately, no forest inventory program takes into account all of the ecologically significant components of the forest stand. Saplings were not measured by the TVA in the initial survey, and the biomass and growth of seedlings, herbs, shrubs, and root systems are not measured. The adjustments shown in Table 7–1 were made to account for this additional production. Above- and belowground production of herbs and shrubs was considered as a constant 30 $g/m^2/year$. Root production of trees was assumed to be 25% of aboveground production, and animal consumption to be 3% of aboveground production.

One uncertainty of this method results from the extensive use of one set of allometric relations. Stem analyses of trees from West Tennessee are not included in the data set of Sollins and Anderson, and the difficulty of stem analysis of large trees biases the sample toward saplings and small poletimber. As more stem analyses are made we shall gain confidence in the allometric relations.

Moreover, the adjustments for undergrowth, roots, and insect consumption are recognized as arbitrary. Table 7–2 shows the estimated net primary production and biomass for the TVA CFI plots in Knox County, Tennessee. Net primary production increases generally as basal area increases and stocking improves from 36 $g/m^2/year$ for plot 360 to 1230 $g/m^2/year$ for plot 365. Basal area in this case takes account of pole- and sawtimber, but not saplings. Stocking is defined as follows (TVA, 1967):

> *Overstocked.* 100% crown closure or more than 700 seedlings and saplings per acre (1750 per hectare)
> *Good stocking.* 70–99% crown closure or 550 seedlings and saplings per acre (1360 per hectare)
> *Fair.* 40–69% crown closure or 300–549 seedlings and saplings per acre (1040–1359 per hectare)
> *Poor.* 10–39% crown closure or 100–300 well-distributed seedlings and saplings per acre (247–1039 per hectare)
> *Other.* Less than 10% crown closure or less than 100 well-distributed seedlings and saplings per acre (247 per hectare).

Plot 360 was a dense pole stand of *Pinus echinata* and *P. virginiana*, which was cut between 1960 and 1965; only an estimated 100 saplings per hectare remained in 1970. Plot 365, by contrast, is in a yellow pine–hardwood stand with a large number of rapidly growing pole- and sawtimber trees and no evidence of recent cutting.

The extremely low production for plot 360 results from assigning to each plot a constant undergrowth production of 30 $g/m^2/year$, which is more representative of closed stands in Tennessee than of abandoned and recently distributed land. Similarly, recent studies suggest that root production is significantly higher

Table 7–2   Net primary productivity (g/m²/year) and biomass (kg/m²), dry matter, for selected plots in Knox County, Tennessee

Plot	Forest type	Basal area (m²/ha)[a]	Stocking class	Net primary production (trees)			Biomass		
				DBH 2.5–12.4 cm	DBH $\geq$ 12.5 cm	Total	DBH 2.5–12.4 cm	DBH $\geq$ 12.5 cm	Total
360	391[b]	0.0	Poor	6	0	36	0.08	0	0.21
361	380[c]	1.44	Fair	213	260	503	5.02	0.92	6.08
363	380	1.73	Fair	57	278	365	2.01	1.29	3.44
365	380	27.74	Good	43	1153	1226	1.02	22.84	23.99

[a] Trees DBH $\geq$ 12.5 cm.
[b] Cedar–pine–hardwood.
[c] Yellow pine–hardwood.

than was hitherto suspected (Harris and Todd, 1972). Better measures of these frequently ignored or hard-to-measure stand components will certainly raise the estimates of forest production.

Table 7–3 shows the primary production and biomass as computed from the TVA plots located in each physiographic province (after Fenneman, 1938). If only the plots that have overstocking and good stocking are considered, estimated productivity varies from 899 g/m²/year in the Cumberland Plateau to 1419 g/m²/year in West Tennessee. If plots comprised of all stocking classes are included, computed productivity and biomass are decreased, as shown in Table 7–3. This reduction is less extreme than is shown in Table 7–2 because 134 of the 224 TVA plots in Tennessee were classified as overstocked or well stocked and 79 as having fair stocking; most had high production reflected in computed production of pole- and sawtimber.

Censuses and conversion factors

An alternative approach to the study of productivity is shown by the productivity profiles of North Carolina, Tennessee, New York, Massachusetts, and Wisconsin. These studies were conducted by four teams that coordinated their work, but that, in some respects, used different techniques and sources of data (Art et al., 1971; DeSelm et al., 1971; Stearns et al., 1971; Whigham and Lieth, 1971). The productivity profiles also tapped reservoirs of data collected by federal and state agencies, again with purposes other than regional productivity. The general strategy of each profile was to determine the area in each county that was devoted to each of a number of land-use categories, to establish an average primary production value for each land-use category, and by multiplying area by average production and summing across all land uses, to estimate county-level productivity.

Table 7–3   Tennessee productivity profile: Forest biomass and net primary productivity by physiographic region for two stocking categories in Tennessee

Region	Well stocked		All stocking classes	
	1965–1970 net primary productivity (g/m²/year)	1970 biomass (kg/m²)	1965–1970 net primary productivity (g/m²/year)	1970 biomass (kg/m²)
Appalachian Mountains	1203	19.5	1081	17.6
Great Valley	1108	15.6	940	12.5
Cumberland Plateau	899	14.0	831	12.5
Highland Rim	1001	15.4	894	13.4
Nashville Basin	1086	21.4	970	17.0
West Tennessee	1419	20.6	1074	15.3
Average:	1091	16.7	936	13.8

Land-use categories, and the area per county in each category, were determined from state or federal sources. The 1964 *Census of Agriculture* (U.S. Bureau of the Census, 1967) was a major source of information for the New York–Massachusetts profiles as was the Crop Reporting Service in North Carolina and Tennessee. Forest yields for North Carolina and Wisconsin are based on published records of net annual growth to growing stock (the increment to pole- and sawtimber, plus ingrowth to these size classes, minus losses incurred by mortality). These data have been collected by the Forest Service in the CFI program of each state. Yields are published by species in some states (e.g., Wisconsin), or by forest type (e.g., North Carolina).

In the Wisconsin and Tennessee productivity profiles no published statistical data were available on production of wetlands and water bodies, urban areas and rights-of-way, and the catchall category of open land (for abandoned farmland, nonstocked forest land, farmstead, and county roads). Consequently, indirect evidence of a limited number of ecologic studies had to be relied upon (Stearns *et al.*, 1971; DeSelm *et al.*, 1971).

The major issue of the productivity profiles has been how to convert yields of agricultural crops and net annual growth to growing stock of forests to a more complete net primary production budget for these land-use categories. This involves more than converting the units in which yield is reported, for example, bushels, to dry weight of the yield; accounting for the unharvested or uneconomic components of production that are not included in yield figures has been treated to date only as an approximation. For agricultural crops, the unharvested biomass of the plant, such as roots, stalk, husk, and leaves for corn, must be accounted for, along with any of these components that are lost during the growing season. For forests, growth in roots, unharvested portions of merchantable boles, branches, and foliage, and unmerchantable trees is needed, along with mortality of individuals and parts (e.g., branch pruning and root sloughing); and herbaceous and woody undergrowth all must be added to net annual growth to growing stock.

The following quotation details how the conversion factor for wheat was determined for the North Carolina study (Whigham and Lieth, 1971). The same logic was used for other crops and other states, but the values probably can be improved in all cases:

Extant data were given as yield in bushels per acre (bu/ac). Each bushel of wheat weighs approximately 60 pounds (lb); the data were initially multiplied by 60 to convert yield in bu/ac to yield in lb/ac. Finally, yield in lb/ac was converted to yield in t/ha. The formula:

$$\frac{\text{yield in lb/ac}}{892.2} = \text{yield in t/ha (100 g/m}^2\text{)}$$

Yield rates (t/ha) were then converted to total plant productivity rates by using a conversion factor (total plant production/plant yield). Adjusting that ratio (2.42) for water content (12%), it became 2.13 and the rate of yield × 2.13 = corrected production rate. Total county wheat production was then calculated by multiplying the corrected production rate by the hectares of wheat in the county (i.e., tons of wheat per hectare multiplied by hectares = tons of wheat).

A similar procedure was used to convert annual growth to growing stock of forests, expressed in cords or cubic feet per acre, to dry weight production per hectare. Detailed information about the conversion from yield to total productivity and the assessment of production for individual counties is given in Chapter 6 of this volume.

Conversion factors used to date to extrapolate yield data for selected crops, forest, urban, open land, and water land-use categories are shown in Table 7–4. In all cases except for hay, commercial yields are no more than 50% of total production, that is, conversion factors are 2.0 or more. The production of urban areas, open land, and water was computed as a proportion of the hay production of a county, or a constant production was assigned, except as noted in footnote c of Table 7–4. The choice of conversion factor is therefore critical to the accuracy of estimated primary production. The teams working on the productivity profiles collaborated on this issue, so the similarity of these trial conversion factors is not surprising. Conversion factors remain a major issue for estimating total production, as well as the fraction of that total that can be used by man for particular purposes.

Table 7–4 Factors used in productivity profiles to convert yields for selected agricultural and forest crops to dry matter net primary productivity[a]

Cover type	New York– Massachusetts	Wisconsin	Tennessee	North Carolina
Corn (grain)	2.14	2.12	3.68	2.03
Small grains	2.32	2.45	3.46	2.64
Hay	1.12	1.48	1.48	1.12
Forest				
Pine	—	3.5	—	2.0
Aspen, ash	—	3.0	—	2.0
Urban	—	160 g/m²/year	0.33 × hay production	25 g/m²/year
Open	—	Avg. other categories	0.5 × hay production	0.6 × hay production
Water	—	12 g/m²/year[b] 62 g/m²/year[d]	213–841 g/m²/year[c]	500 g/m²/year

[a] (primary productivity = conversion factor × yield dry weight); conversion factors and constants for urban, open, and water–land-use categories.

[b] North of tension zone (Curtis, 1959).

[c] Based on production measured for selected TVA reservoirs by M. P. Taylor (*personal communication*), Environmental Biology Branch, TVA, Norris, Tennessee.

[d] South of tension zone.

Table 7–5   Range in primary productivity for selected vegetation
(dry matter g/m²/year)

State	Maize Low[a]	Maize High[b]	Small grain Low	Small grain High	Hay Low	Hay High	Forest Low	Forest High	Average Low	Average High
New York–										
Massachusetts	260	590	120	580	280	580	—	—	230	800
Wisconsin	280	1260	360	670	590	1340	220	490	230	800
Tennessee	660	910	220	560	300	510	720[c]	1050[c]	420	900
North Carolina	490	1080	260	810	250	450	100[d]	430[d]	150	500
							765[e]	1500[e]		

[a] Production of county with lowest production.
[b] Production of county with highest production.
[c] Based on TVA plot data.
[d] Based on Forest Service statistics, Whigham and Lieth (1971).
[e] Based on plantation survey data; see Table 6–4.

In spite of these uncertainties, two practical types of information came from the productivity profiles. The first type is variation in primary production of selected crops from one state to another. Table 7–5 shows, for each state, the productivity of the least productive and most productive county for selected land-use categories and for the average of all land uses. For example, the lowest average corn production for the least productive county for corn in the New York–Massachusetts profile (Franklin County, Massachusetts) was 260 g/m²/year, and the highest (for Chautauqua County, New York) was 590 g/m²/year. The lack of any sharp distinction between agricultural and forest production, except in Tennessee, should be noted. The higher forest production in Tennessee, computed from CFI plot data as discussed previously, is probably more a result of conservative initial conversion factors for forests in North Carolina and Wisconsin, than of higher actual productivity in the forests of Tennessee.

Another comparison of interest is the ranking of production by land-use category, as shown for Tennessee in Table 7–6. Forests rank highest, perhaps because of the close (but still incomplete) accounting made of forest production. Row crops can produce nearly as much as forests, but only by investing more management input. The computed production of 110 g/m²/year for urban land may be very conservative. The aboveground production of one residential landscape studied in Madison, Wisconsin, exceeds the production of adjacent woodland per unit area of vegetated surface (perhaps the result of the inputs of fertilizer and supplemental irrigation and of decreased competition), and is equal when paved and roofed surfaces are included (Lawson et al., 1972). This may be found to be true generally (except, of course, for urban cores) when urban vegetation is subjected to the same close accounting as natural vegetation.

Table 7–6   Median and range in primary productivity by
land-use categories for counties in Tennessee
(dry matter g/m²/year)

Land-use category	Productivity		
	Low	Median	High
Forest	720	900	1050
Agricultural row crops	660	750	920
Hay crops	300	400	510
Small grains	220	390	560
Pasture	200	250	300
Urban	080	110	190
Open land	120	160	280
Lakes and rivers	220	560	900

Relating Productivity to Environment

Weather records constitute another widespread source of data that can be brought to bear on the problems of regional productivity. A number of models for agricultural and natural vegetation are available from the work of agronomists, silviculturists, climatologists, and ecologists (see Chang, 1968a; Lowry, 1969; Munn, 1970), so that the task becomes one of selecting models that satisfy the needs of regional ecosystem productivity analysis. Some of these are (1) the weather data needed must be available for a large number of stations, and of uniform quality; (2) the model derives production for a variety of plant communities in a region, and is not restricted to single crops or species; and (3) the model reflects major changes in production at a particular scale, for example, differences in annual production between such R3 regions as states.

Most models relate to a limited number of species (e.g., Currie and Peterson, 1966; Zahner and Stage, 1966; Albrecht, 1971); rely on data with high resolution (e.g., deWit, 1958; Monteith, 1965); ignore seasonality of energy or moisture resources, which becomes important in interregional comparisons (e.g., Drozdov, 1971; Lieth, 1971b); or assume either energy or moisture to be in adequate supply in all climatic conditions (e.g., Chang, 1968b, 1970).

Models that relate growth to a component of a water balance, usually actual evapotranspiration but sometimes deficit as well, avoid these shortcomings. A water balance is a budget of water in response to an estimated demand for moisture imposed by an energy load versus precipitation and available supply of soil moisture. The accounting period may vary from a day to a month. Many water-balance schemes have been devised, but the simplest of these requires only air temperature and precipitation data, the geodetic coordinates of the weather station, and a measured or assumed moisture-storage capacity for the root zone (Thornthwaite and Mather, 1957).

Actual evapotranspiration has been related to yields of agricultural crops (Arkley and Ulrich, 1962; Arkley, 1963), diameter growth of trees (Zahner

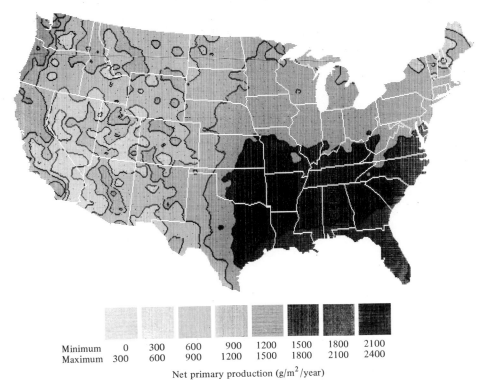

Minimum	0	300	600	900	1200	1500	1800	2100
Maximum	300	600	900	1200	1500	1800	2100	2400

Net primary production (g/m²/year)

FIGURE 7–2. Average annual net primary production in conterminous United States after C. W. Thornthwaite Memorial model developed by Lieth and Box, and average annual water balances computed by C. W. Thornthwaite Associates.

and Stage, 1966; Zahner and Donnelly, 1967; Manogaran, 1972), and to the net primary production of ecosystems (Rosenzweig, 1968; Lieth and Box, 1972). Maps of the primary production of the United States were developed using the Lieth–Box and Rosenzweig models, as shown in Figures 7–2 and 7–3. The Rosenzweig and Lieth–Box models have the same general logic; each considers primary production as a function of actual evapotranspiration. However, there are differences between the data sets of measured primary production and of actual evapotranspiration upon which each is based, and the mathematical function chosen to relate production to evapotranspiration. This has been discussed in detail (Lieth and Box, 1972; see also Chapter 6, this volume); some general comments on the methods used will clarify disparities between the two maps.

Each model uses measured values of primary production as a data set; Lieth and Box use a data set of about 50 values of aboveground and belowground production from North America, South America, Eurasia, and Africa. Rosenzweig's data set of 25 values of aboveground production only derives largely from the Great Smoky Mountains in Tennessee (15 points from Whittaker, 1966) with a selection of other data from desert, grassland, tundra, and tropical

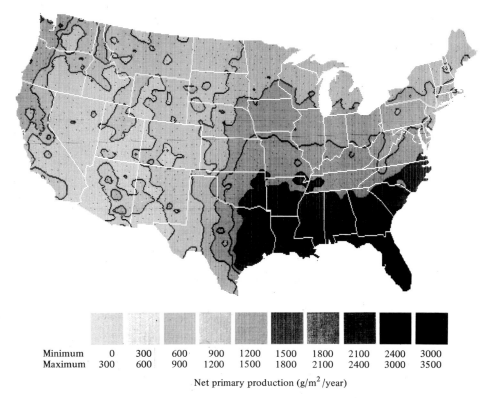

Minimum	0	300	600	900	1200	1500	1800	2100	2400	3000
Maximum	300	600	900	1200	1500	1800	2100	2400	3000	3500

Net primary production (g/m^2/year)

FIGURE 7–3. Average annual net primary production in conterminous United States after M. L. Rosenzweig and average annual water balances computed by C. W. Thornthwaite Associates.

forest. The maximum net primary production value in each data set is about 2900 g/m^2/year. Evapotranspiration values in Rosenzweig's model are from published water budgets computed by the Thornthwaite method (C. W. Thornthwaite Associates, 1964). Evapotranspiration values in the Lieth–Box data set are derived from the map *Annual Effective Evapotranspiration* (scale 1 : 30,000,000) (Geiger, 1965) by estimating evapotranspiration for each site in the production data set.

Each model results from a least-squares fit of a curve to the respective data sets. Rosenzweig's model is a linear regression of the logarithms of the variables. The Lieth–Box model is a saturation curve with 3000 g/m^2/year as asymptote. The equations are

$$NPP(\text{aboveground}) = 0.0219E^{1.66} \text{ (Rosenzweig)}$$
$$NPP(\text{total}) = 3000[1 - e^{-0.0009695(E-20)}] \text{ (Lieth–Box)}$$

where *NPP* is net primary production (g/m^2/year) and *E* is actual evapotranspiration (millimeters per year).

The maps in Figures 7–2 and 7–3 were produced using these equations and a common data base of evapotranspiration derived from average annual water budgets for about 1100 weather stations computed by the Thornthwaite method. The maps show similar trends in production across the United States with maximum values in the Southeast, and a minimum in the Intermountain West. The estimates are most similar in the Southern Appalachian Mountains, and diverge toward places having higher and lower production.

The evapotranspiration data base used for these maps and the logic of each model account for their differences. Brief review of the Geiger map indicates that the Thornthwaite estimates of evapotranspiration in the eastern United States are higher than the Geiger map shows. An adjustment to account for the overestimate of evapotranspiration (as viewed from the perspective adopted for the Lieth–Box model) would reduce the Lieth–Box production estimate. Rosenzweig's data set accounts for aboveground production only; adjustment of the Rosenzweig or Lieth–Box model, so that each accounts for the same stand components in the production estimate, would increase Rosenzweig's estimates or decrease the Lieth–Box estimates.

The sharp rise in net primary production between North Georgia and the Florida panhandle in Figure 7–3 (Rosenzweig's model), which is not shown in Figure 7–2 (Lieth–Box model) identifies the major difference between the two models. The data set of production values used by Rosenzweig is generally more conservative than the Lieth–Box model for given values of evapotranspiration, and the maximum values of production and evapotranspiration are nearly coincident. Yet Rosenzweig considers primary production to be a power function of evapotranspiration, which imposes no limit on production as evapotranspiration increases, whereas the Lieth–Box model is a saturation curve that imposes an upper limit to production of 3000 g/m²/year. Tests of alternative curves on Rosenzweig's data show that a simple linear regression of production on evapotranspiration has the same correlation coefficient ($r = 0.95$) as the linear regression of the logarithms of the variables. A logistic curve with an asymptote of 3000 g/m²/year has a slightly poorer least-square fit. Alternative curve forms could be fitted to the Lieth–Box data sets, as well, perhaps with some decrease in the goodness-of-fit.

Lieth and Box (1972) justify their use of the saturation curve on the ground that it conforms to Mitscherlich's yield law. Other lines of evidence indicate a need for a ceiling on primary production. Stanhill (1960), Black (1966), and Chang (1968b) point to the increasing toll taken by respiration on the gross photosynthesis of agricultural crops, pasture, and forests as temperatures increase when moisture is in adequate supply. Drozdov (1971) models primary production as a saturation-curve function of net radiation in subhumid and humid environments. The ecologic reasoning expressed in the Lieth–Box model, appears to be superior to that of the Rosenzweig model. The values of production associated with evapotranspiration and the specific form of the curve are likely to change as more measurements of production and studies of the physiology of net photosynthesis provide further insight into plant–environment relationships in production.

Discussion

Two goals for studies of regional production are confidence in the methods and results of each study, and as a corollary, convergence of the results of different studies. Considerable progress toward this goal is being made in the local ecosystem analysis programs of IBP; the methods and results of the study of regional production discussed here deserve less confidence. Comparison of several results illustrates this point.

The Lieth–Box production map shows production for Tennessee in the range of 1500–1800 g/m²/year; the summary for Tennessee in Table 7–6 shows production for forests as computed from CFI data as 720–1050 g/m²/year, and an agricultural production of 660–920 g/m²/year for row crops and 300–510 g/m²/year for hay. By contrast, the production for Wisconsin in the Lieth–Box map is 1200–1500 g/m²/year, whereas Table 7–5 shows a forest production of 220–490 g/m²/year, a corn production of 280–1260 g/m²/year, and a hay production of 590–1340 g/m²/year. The trend for Tennessee is: Lieth–Box > forest ≧ row crops > pasture; for Wisconsin it is: Lieth–Box ≧ hay ≧ corn ≧ forest. Does this ordering indicate real differences in the absolute and relative production of these land–use categories within and between Tennessee and Wisconsin, or does it indicate inaccuracies in the methods? The Wisconsin forest production values are quite low in relation to production estimates for a range of eastern forests, some of them in cool mountain climates, in the Great Smoky Mountains (Whittaker, 1966). No firm answer can be given, however, and the uncertainties that underlie these orderings indicate some themes for further study. These include

1. The reason for the consistently high values in the Lieth–Box model: Each region represented in the Lieth–Box data set is likely to have a spectrum of productivities created by topographic and edaphic diversity and successional and land–use patterns. The situation of each site study in this spectrum is unknown; my conjecture is that poorly stocked, inaccessible, and disturbed stands are avoided, and that the Lieth–Box model represents the productivity of well-stocked stands on the accessible, better sites of a region. Clarification of this for a region may come from using the sampling data from CFI and agricultural census programs to ascertain whether the production defined by the Lieth–Box model or a successor to it defines a maximum production that will be approached by other methods as they become more accurate, or whether it has some alternative significance as an index of regional production.
2. The reason for the shifts in the ordering of production by land-use categories from one state to another: These shifts may result from real differences in production or they may be artifacts of method, especially of the conversion factors, constants, and assumptions used to extrapolate from CFI and agricultural census data. The confidence that can be placed in these categories has been explored for each method. Greater stress in

production studies on landscapes of large extent—agricultural, urban, and exploited forest—will enable us to develop methods to unravel the complexities of the interaction of human and environmental inputs to production.

3. The relationships among annual production, phenologic production, and trends in production through succession and land-use change: Annual production has been stressed in the regional production studies discussed here, although Figure 7–1 identifies phenologic and successional production as well. Annual production is the summation of phenologic production for stand components and phenophases; accurate estimates of both phenologic and annual production will support each other. Change of annual (and phenologic) production over years and decades is characteristic of succession. Further emphasis on these shorter- and longer-term processes will clarify the dynamics of production in ways appropriate to the regional studies.

4. The relationship between primary production and environment: An implication of the space–time hierarchy for regional production is that it would be useful to have a variety of models, each expressing the environmental effects on production most significant for a given scale. The independent variable in the Lieth–Box model—actual evapotranspiration—expresses broad regional patterns of the interaction of precipitation and solar energy. It does not consider how evapotranspiration might change locally with soils and topography. The emphasis in IBP toward relating processes for particular ecosystems suggests the need for models relating production to environment on a local scale. Intensive studies in the local ecosystem-analysis programs of IBP may well provide new and larger data sets to both clarify local production relationships and enhance the reliability of regional production studies.

Acknowledgments

This research was supported by the Deciduous Forest Biome Project, International Biological Program, funded by the National Science Foundation under Interagency Agreement AG–199, 40–193–69 with the Atomic Energy Commission, Oak Ridge National Laboratory, Oak Ridge, Tennessee.

References

Albrecht, J. C. 1971. A climatic model of agricultural productivity in the Missouri River Basin. *Publ. Climatol.* 24(2):1–107. Centerton (Elmer), N.J.: Thornthwaite Lab. Climatol.

Arkley, R. J. 1963. Relationships between plant growth and transpiration. *Hilgardia* 34:559–584.

————, and R. Ulrich. 1962. The use of calculated actual and potential evapotranspiration for estimating potential plant growth. *Hilgardia* 32:443–468.

Art, H. W., P. L. Marks, and J. T. Scott. 1971. Progress report—Productivity profile of New York. 30 pp. (mimeogr.) US-IBP EDFB Memo Rep. 71-12.

Beauchamp, J. J., and J. S. Olson. 1972. Estimates for the mean and variance of a lognormal distribution where the mean is a function of an independent variable. 22 pp. (mimeogr.) US-IBP EDFB Memo Rep. 71-101.

Black, J. N. 1966. The utilization of solar energy by forests. *Forestry Suppl.* 98–109.

Chang, J.-H. 1968a. *Climate and Agriculture: An Ecological Survey,* 304 pp. Chicago, Illinois: Aldine.

———. 1968b. The agricultural potential of the humid tropics. *Geograph. Rev.* 58: 333–361.

———. 1970. Potential photosynthesis and crop productivity. *Ann. Assoc. Amer. Geogr.* 60:92–101.

Currie, P. O., and G. Peterson. 1966. Using growing-season precipitation to predict crested wheatgrass yields. *J. Range Mgt.* 19:284–288.

Curtis, J. T. 1959. *The Vegetation of Wisconsin,* 657 pp. Madison, Wisconsin: Univ. of Wisconsin Press.

Dansereau, P. 1962. The barefoot scientist. *Colo. Quart.* 12:101–115.

DeSelm, H. R., D. Sharpe, P. Baxter, R. Sayres, M. Miller, D. Natella, and R. Umber. 1971. Final report; Tennessee productivity profile; 182 pp. (mimeogr.) US-IBP EDFB Memo Rep. 71-13.

deWit, C. T. 1959. Potential photosynthesis of crop surfaces. *Neth. J. Agr. Sci.* 7:141–149.

Drozdov, A. V. 1971. The productivity of zonal terrestrial plant communities and the moisture and heat parameters of an area. *Sov. Geogr. Rev. Transl.* 12:54–59.

Evans, F. C. 1956. Ecosystem as the basic unit in ecology. *Science* 123:1127–1128.

Fenneman, N. A. 1938. *Physiography of the Eastern United States*, 714 pp. New York: McGraw-Hill Book Co.

Friederichs, K. 1958. A definition of ecology and some thoughts about basic concepts. *Ecology* 39:154–159.

Geiger, R. 1965. *The Atmosphere of the Earth* (12 wall maps and text). Darmstadt, Germany: Justus Perthes.

Goff, F. G., F. P. Baxter, and H. H. Shugart, Jr. 1971. Spatial hierarchy for ecological modeling, 12 pp. (mimeogr.) US-IBP EDFB Rep. 71-41.

Golley, F. B., and J. B. Gentry. 1965. A comparison of variety and standing crop of vegetation on a one-year and a twelve-year abandoned field. *Oikos* 15:185–199.

Harris, W. F., and D. E. Todd. 1972. Forest root biomass production and turnover, 17 pp. (mimeogr.) US-IBP Memo Rep. 72-156.

Kira, T., and T. Shidei. 1967. Primary production and turnover of organic matter in different forest ecosystems of the western Pacific. *Jap. J. Ecol.* 17:70–87.

Lawson, G. J., G. Cottam, and O. Loucks. 1972. Structure and primary productivity of two watersheds in the Lake Wingra Basin, 51 pp. (mimeogr.) US-IBP EDFB Memo Rep. 72-98.

Lieth, H. 1970. Phenology in productivity studies. In *Analysis of Temperate Forest Ecosystems,* D. E. Reichle, ed., *Ecological Studies* 1:29–46. New York: Springer-Verlag.

———. 1971. The phenological viewpoint in productivity studies. In *Productivity of Forest Ecosystems: Proc. Brussels Symp. 1969*, P. Duvigneaud, ed., *Ecology and Conservation*, Vol. 5, 71–84. Paris: UNESCO.

———. 1973. Primary production: Terrestrial ecosystems. *Human Ecol.* 1:303–332.

———, and E. Box. 1972. Evapotranspiration and primary productivity: C. W. Thornthwaite Memorial Model. In *Papers on Selected Topics in Climatology*, J. R. Mather, ed. (*Thornthwaite Mem.* 2:37–46.) Elmer, New Jersey: C. W. Thornthwaite Associates.

Loucks, O. L. 1970. Evolution of diversity, efficiency and community stability. *Amer. Zool.* 10:17–25.

Lowry, W. P. 1969. *Weather and Life: An Introduction to Biometeorology*, 305 pp. New York: Academic Press.

Manogaran, C. 1972. Climatic limitations on the potential for tree growth in southern forests. Unpublished dissertation. Carbondale, Illinois: Southern Illinois Univ.

Monteith, J. L. 1965. Light distribution and photosynthesis in field crops. *Ann. Bot. N.S.* 29:17–37.

Munn, R. E. 1970. *Biometeorological Methods*, 336 pp. New York: Academic Press.

Novikoff, A. B. 1945. The concept of integrative levels and biology. *Science* 101: 209–215.

Odum, E. P. 1960. Organic production and turnover in old field ecosystems. *Ecology* 41:34–49.

Rosenzweig, M. L. 1968. Net primary productivity of terrestrial communities: Prediction from climatological data. *Amer. Natur.* 102:67–74.

Rowe, J. S. 1961. The level-of-integration concept and ecology. *Ecology* 42:420–427.

Sharp, D. D., H. Lieth, and D. Whigham. Chapter 6, this volume.

Sollins, P., and R. M. Anderson. 1971. Dry-weight and other data for trees and woody shrubs of the Southeastern United States. ORNL-IBP-71-6, 80 pp. Oak Ridge, Tennessee: Oak Ridge National Laboratory.

Stanhill, G. 1960. The relationship between climate and the transpiration and growth of pastures. In *Proc. 8th Int. Grassland Congress*, Tel Aviv: 293–296.

Stearns, F., N. Kobriger, G. Cottam, and E. Howell. 1971. Productivity profile of Wisconsin, 82 pp. (mimeogr.) US-IBP EDFB Memo Rep. 71-14.

Tennessee Valley Authority. 1967. TVA forest inventory field manual for county-wide units and watersheds in the Tennessee Valley, 20 pp. (mimeogr.) Norris, Tennessee: Forest Survey Sect., Forest Products Branch, Div. of Forestry, Fisheries and Wildlife Development.

Thornthwaite, C. W., and J. R. Mather. 1957. Instructions and tables for computing potential evapotranspiration and the water balance. *Publ. Climatol.* 10(3):181–311. Centerton, N.J.: Thornthwaite Lab. Climatol.

Thornthwaite, C. W. Associates. 1964. Average climatic water balance data of the continents, Part 7: United States. *Publ. Climatol.* 17(3):415–615. Centerton, N.J.: Thornthwaite Lab. Climatol.

U. S. Bureau of the Census, Census of Agriculture. 1967. Statistics for the State and Counties, Tennessee. Washington, D. C.: U. S. Government Printing Office.

Whigham, D., and H. Lieth. 1971. North Carolina Productivity Profile 1971, 143 pp. (mimeogr.) US-IBP EDFB Memo Rep. 71-9.

Whittaker, R. H. 1966. Forest dimensions and production in the Great Smoky Mountains. *Ecology* 47:103–121.

Woodwell, G. M. 1967. Radiation and the patterns of nature. *Science* 156:461–470.

———. 1970. The energy cycle of the biosphere. *Sci. Amer.* 223:64–74.

Zahner, R., and J. R. Donnelly. 1967. Refining correlations of water deficits and radial growth in young red pine. *Ecology* 48:525–530.

———, and A. R. Stage. 1966. A procedure for calculating daily moisture stress and its utility in regressions of tree growth on weather. *Ecology* 47:64–74.

Part 3

Global

Productivity

Patterns

The primary production of the earth is the prime concern of this book. Its assessment is the result of intensive work of ecologic research groups all over the world. In general, research on productivity has different emphases in terrestrial and aquatic ecosystems. Among aquatic systems, production problems differ from oceans to lakes and streams. The aquatic section in this book is, therefore, separated into treatments of marine and freshwater ecosystems. Although the oceans cover about 70% of the earth surface, their contribution to world production is much less than that of land communities. Marine communities differ greatly from land communities in structure, nutrient relationships, and appropriate research approaches. The plankton communities of the open oceans are less variable in their range of productivities and other characteristics than are land communities or freshwater communities (Chapter 8).

The freshwater bodies of the world cover only a small portion of the total land area. Consequently, their contribution to the primary production of the earth is very limited. Nevertheless, man is concerned with the productivity of freshwater bodies as a source of food, and he is also concerned about cultural eutrophication of these bodies by overfertilization. The brackish and inland saltwater bodies are not covered separately in this volume. The coastline areas— the "interface" of land and sea—include communities that are highly productive and important to man as a source for the major food species of fish. Inland salt water bodies (some of which are highly productive, but not productive of fish) as well as freshwater ecosystems are treated in this section (Chapter 9).

The terrestrial ecosystems are dealt with in Chapter 10, which compares and sums their production for the entire world. Knowledge of terrestrial productivity is rapidly increasing, and while this book is in press new summaries are being prepared. Although later work is not expected to alter significantly the calculations for the temperate ecosystems of the world, corrections may be needed in tropical areas. Therefore, included in this book is a current evaluation of productivity in tropical ecosystems (Chapter 11).
It is hoped that the four contributions in this section come as close to the real production pattern on earth as possible at the moment. We hope that, in any case, our summary of the extensive research on productivity up to this time may remain a benchmark in our knowledge of the primary production of the earth.

8

Primary Productivity of Marine Ecosystems

John S. Bunt

With current effort, the sea is yielding roughly 60 million tons of fish annually and until as recently as 1969 the catches were increasing steadily. Can harvests of this intensity be sustained? Can they be raised? To answer these questions, reliable knowledge of marine primary production is needed. This chapter deals with estimations of marine production and with the difficulties and uncertainties to which they are subject.

The Ocean Environment and Its Plant Populations

The sea has a total area of roughly 367×10^6 km² and occupies a little more than 70% of the earth's surface. With an average depth of ~ 4000 m, only the superficial, illuminated layers are capable of supporting plant growth. This productive zone, however, varies remarkably in character. It includes habitats as diverse as the polar pack ice, the shallow warm waters and sediments of mangrove-fringed tropical estuaries, surf-beaten intertidals, coral reefs and seaweed beds, as well as coastal waters, the vast stretches of the open ocean, and zones characterized by an upwelling of nutrient-rich waters from layers far below the surface.

As a milieu for the support of life, the sea differs basically from the land in its fluid mobility and instability, its transparency, limited capacity to supply plant nutrients, and comparative thermal stability. Only marine sediments and solid substrata that receive solar radiation are comparable, in some respects, to sub-aerial formations. The incidence of solar radiation varies with latitude and with

KEYWORDS: Net primary production; marine ecosystems; ecology; global overview; global pattern.

Primary Productivity of the Biosphere, edited by Helmut Lieth and Robert H. Whittaker.
© 1975 by Springer-Verlag New York Inc.

season, and a complex of factors influence light penetration into the sea. For phytoplankton populations, water stability and nutrient status have more critical effects on productivity, and these have an interrelationship that varies with latitude. In some areas thermal stratification, with warmer less dense surface water above colder denser deep water, largely prevents local movement of water between the surface and the depths. Sinking organisms and their dead remains carry nutrients needed for plant growth downward; and when the water is stratified, these nutrients are gradually depleted in the lighted surface waters. Thermal stratification, with consequent nutrient impoverishment of the surface layers, is most pronounced in the tropics. In temperate-zone waters, stratification is typically seasonal—most pronounced and associated with nutrient depletion during the summer months. Stratification in polar waters normally is weak and transient. In the zone of fast ice and pack ice, the ice layer itself effectively protects algal cells in it from sinking.

Within the vast fluid space of the seas, the bulk of the photosynthesis is by microscopic algae—the simple, but often remarkably beautiful, taxonomically and metabolically diverse cells of the phytoplankton. Related species of algae live on sediments and a variety of other surfaces including, as a truly exotic habitat, the hides of whales. The familiar and often rapidly growing seaweeds are conspicuous in the intertidal zone and may extend to some depth on the lighted substrate below the intertidal, especially in highly transparent tropical waters. A limited group of angiosperm species are important in the productivity of special habitats.

Measurement of Primary Productivity

In principle, it should be possible to measure primary productivity from observed changes in the environmental concentrations of any of the raw materials involved in photosynthesis. Once popular, this approach, which is centered on major nutrients or dissolved oxygen, is little used nowadays because it is impractical to determine the influences of water movement with sufficient precision. Planktonic productivity is normally measured by exchange of respiratory gases—either O_2 or CO_2—between plankton cells and the water in small enclosed samples. Comparable procedures sometimes are used for benthic organisms, although for these it may be more convenient to measure biomass changes with time. No technique is free of difficulty or uncertainty.

According to Strickland (1965), the Winkler method enables the reliable determination of changes in dissolved oxygen as small as 0.02 ml/liter. For samples of reasonable volume, and for incubation periods of acceptably short duration, this level of sensitivity can provide worthwhile data only in exceptionally productive waters. Winkler analysis does not allow continuous observation of changing O_2 concentrations. Oxygen electrodes can provide this sort of information, but unnatural concentrations of organisms would be necessary in most circumstances if instantaneous rates were to be read. Release of oxygen into the water in the light expresses net photosynthetic activity. Traditionally, gross productivity is estimated by adding rates of oxygen removal from the water in the dark to rates of oxygen increase in the water in the light. As explained in a review by Jackson and Volk (1970), this practice is not acceptable. Moreover, data on oxygen exchange are difficult to interpret unless reliable information is available on photosynthetic quotients.

The sensitivity needed to measure photosynthetic activity in less productive waters can be achieved through observing incorporation of radiocarbon supplied as bicarbonate. This method was introduced to oceanography by Steemann Nielsen (1952) and is now the method of choice in routine productivity studies throughout the world. Very briefly, the technique calls for the incubation of a sample of seawater with plankton in a bottle or transparent container to which a measured amount of $NaH^{14}CO_3$ has been added. The phytoplankton cells take up the tagged $^{14}CO_2$, and during photosynthesis they incorporate the radioactive ^{14}C in organic compounds. Subsequently, the plankton cells and other particulate materials are recovered by filtration (or any other appropriate procedure). The radioactivity of the samples, established by planchet or liquid scintillation counting, indicates uptake of radiocarbon in photosynthesis. There is no universally accepted procedure, but the technique for analysis of plankton is described in detail by Strickland and Parsons (1965). Measurement of photosynthesis in sediments and macrophytes cannot be undertaken by means of any single procedure. The types of problems that arise and some of the solutions are described in the third UNESCO *Monograph on Oceanographic Methodology* (Anon., 1973); see also Chapter 3.

Various difficulties burden the interpretation and extrapolation of photosynthetic rate data based on such processes as O_2 exchange and ^{14}C fixation. The question, at one period actively debated, of whether ^{14}C uptake gives a measure of net or gross photosynthesis or some intermediate value, has never been resolved. All earlier arguments advanced by workers such as Ryther (1956) and Steemann Nielsen and Hansen (1959) were based in part on the assumption that dark respiration continues unaltered in the light. Bunt (1965), however, presented evidence that dark respiration is partially or completely inhibited in the light, and some of the information reviewed by Jackson and Volk (1970) supports that finding. Processes associated with photorespiration are attracting widespread interest, especially among crop physiologists. The subject has been rather neglected with regard to the sea, although a 1973 expedition to the Great Barrier Reef by R/V Alpha Helix was devoted entirely to this topic. The results of these investigations have not yet appeared.

A further complication in measuring marine primary productivity centers on the fact that algae exposed to $^{14}CO_2$ commonly excrete some of their labeled photosynthetic products into the surrounding medium. Sieburth and Jensen (1969) report that exudation in *Fucus vesiculosus* can amount to 40% of the carbon fixed. Thomas (1971) has found that excretion, as a percentage of total photosynthetic fixation, increased seaward from 7% in Georgia estuaries to $\sim 13\%$ in coastal waters, and approached 44% in the western Sargasso Sea. Although percentages of release were low in the estuaries, in absolute amounts excretion by the estuarine phytoplankton was estimated as high as 40 mg $C/m^3/$ day. These findings will be important in the subsequent discussion.

The extrapolation in time and space of data obtained with small samples subjected to short exposure to radiocarbon is uncertain at best. The reasons are technical as well as biologic. Photosynthetic activity is not constant during the day and may vary with degree of shade adaptation at different levels in the water column. Moreover, it is often impractical, and may be biologically inadvisable, to continue incubations for more than a few hours. Survey activities over wide areas commonly prevent adequate replication. As a further compromise, it is frequently necessary to measure carbon fixation with the artificial light on deck incubators. When possible these are operated at a range of light intensities, with measurement or estimation of extinction coefficients in the water column as a basis for estimating production at different depths. Deck incubators normally are operated at the temperature of the seawater intake; this is unfortunate in areas with significant temperature gradients in the water column.

In principle, it would be desirable to make *in situ* incubations in the water column, with the bottled seawater and plankton at the different depths, temperatures, and light intensities for which photosynthetic rates are to be measured. For research vessels with large regions to investigate, this usually is not feasible. Further problems result from raising and lowering water samples and from the fact that the enclosed phytoplankton is held at fixed depths, whereas the natural community is likely to be in constant motion. Some of these problems may not arise in benthic investigations. However, satisfaction of the nutrient requirements of attached algae may be partly dependent on constant water movement: to enclose samples for any length of time interferes with their metabolic activity.

Net primary productivity on land can be determined most directly by terminal or periodic harvest of plant growth. The harvest method can be used in the marine environment for attached algae and vascular plants such as eelgrass, *Thalassia testudinum*. This tactic is most reliable if the harvest can coincide with natural cycles of growth or if the production of new material and the loss of old can be estimated by marking without interfering with the standing stock (Mann, 1972, 1973). Such measurements may not be feasible, and if they are, may not take account of grazing effects and loss of dissolved organic matter into the water (or possible uptake of dissolved organic matter).

Faced with so many difficulties, it is not surprising that some researchers have attempted to formulate equations for estimating primary productivity from a few easily measured variables. To meet requirements, empirical equations of the types discussed by Strickland (1965) must have substantial and frequently rough built-in assumptions. For example, it is common to assume uniform distribution of phytoplankton in the water column and to accept a single standard response curve of photosynthesis versus light intensity. Steele (1969) has discussed some of the difficulties associated with modeling aquatic productivity. Figure 8–1 gives a current global model provided by H. Lieth, which shows the generalized distribution of productivity in the world oceans.

Determinants of Productivity

The features of the environment that control plant growth on land are well understood and can be controlled and manipulated with remarkable success. The primary producers of the sea have the same basic needs, even though the marine environment creates special difficulties that normally do not arise on land. Although some of the major needs of phytoplankton communities are evident, a long and continuing effort (see Provasoli *et al.*, 1957; Hutner and Provasoli, 1964) to satisfy the needs of any given individual species of wild phytoplankton in the laboratory has not had outstanding success.

With the obvious exception of nighttime hours (which become seasonal toward the Poles and further prolonged by sea ice), photosynthesis in the upper layers of the sea need not be limited by light. However, plants near the water surface may be exposed to supraoptimal light intensities, whereas little or no growth is possible for those well below the surface at suboptimal intensities. Changes in light quality with depth may also be important, especially if effective utilization requires a particular combination of wavelengths. The way light intensity and spectral quality change with depth is complex; it depends in part on the concentration and nature of dissolved living and dead organic as well as inorganic materials in the water column. It is generally assumed that the photic zone extends to the depth at which the light intensity is reduced to 1% of the value at the surface. The 1% level should not be taken too literally, for growth by photosynthesis has been shown at much lower intensities. For example, Bunt (1968) has measured the growth of Antarctic diatoms at light intensities as low as 0.0002 ly/min.

A brief consideration of other factors influencing productivity includes the following. Temperature, in general, does not seem to be a major factor in controlling productivity. Rates comparable in magnitude have been obtained in widely separated latitudes. Data collected by Bunt and Lee (1970) in Antarctic Sea ice constitute, however, an example of limitation at the extremes. A wide range of organic as well as inorganic nutrients and metabolites can affect the growth of marine algae. Some organic substances, including vitamins, are essential for some species of algae; others are not essential but are stimulatory; still others are toxic or inhibitory. Excellent accounts of this topic have been prepared by Provasoli (1963). As another determinant of productivity, grazing (e.g., McAllister, 1970), should at least be mentioned. We also recognize the diversity and complexity of controlling influences in estuaries (Woodwell *et al.*, 1973).

Whatever the effects of organic materials on algae, it is clear that in contrast to the land, reserves of the major inorganic nutrients (nitrogen and phosphorus) are commonly limited or almost nonexistent in the sea. It is often considered that over much of the ocean's area the supply of available phosphorus is the more critical; but both nitrogen and phosphorus may be limiting in some areas, and in coastal waters nitrogen may exert primary control. Various trace elements are known to be necessary for algal growth, but it may be the states of these substances in seawater rather than their concentrations that exert control. It should be stressed that specific information on the inorganic nutritional requirements and nutrient uptake kinetics of the marine algae is quite limited. Furthermore, it is evident from the results of various investigators (e.g., Eppley *et al.*, 1969) that the physiology of algal nutrition is complex. The subject of nutrient limitation in the sea has been considered recently by several authors including Dugdale (1967), Barber *et al.* (1972), MacIsaac and Dugdale (1969), Eppley

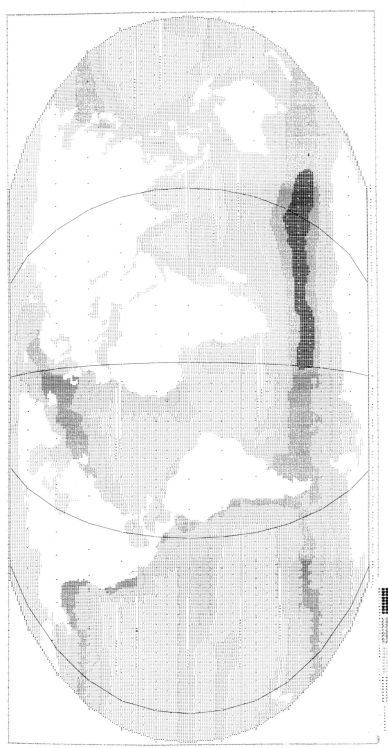

FIGURE 8–1. Net primary productivity pattern of the oceans. The ranges mapped are in dry matter g/m²/year, and a summation for the map (using mean levels in Table 13–5) gives a total production of 41.6×10^9 t/year for the world's oceans. This is about 20% short of the value given in Table 8–2. The discrepancy stems, in part, from the 2% larger ocean area in Table 8–2 versus that in Table 13–5, and in part from higher mean values set by Koblentz-Mishke *et al*., as used in Table 8–2, versus the accounting from this map by Box (Table 13–5). The map was produced by UNC graduate students as term paper.

and Strickland (1968), and, with regard to pollution, by Ryther and Dunstan (1971).

Exposures of phytoplankton cells to light, nutrients, and acceptable temperatures—which together make for optimal productivity—are dependent on the hydrodynamic characteristics of the water column and the characteristics of algal cells that enable them to remain in suspension, or at least to sink only slowly. Some of the most productive situations are those in which nutrient-rich deep waters upwell and spread out at the surface; some of the least productive are those in which stratification is so stable at the surface that nutrients become depleted and are not replaced. Smayda (1970) has examined the biologic problems of suspension in some detail, and there is an old and detailed literature on the question of water stability and upwelling (see, e.g., Ryther, 1963; Redfield *et al.*, 1963). Detailed attention is being given to upwelling phenomena in currently active programs under sponsorship of The International Decade of Ocean Exploration (IDOE).

Observed Productivity

Taking data then available from ^{14}C uptake or comparable techniques and with their shortcomings in mind, Ryther (1963), attempted a comparative analysis of global marine productivity. Between and within the major regions considered, there was both wide variation and substantial uncertainty. No global total was suggested. Later, Ryther (1969) accepted 15–18×10^9 t C/year as the most likely level of open ocean primary production.[1] Dividing the oceans into three provinces, he suggested mean productivity values of 50, 100, and 300 g C/m^2/year for the open ocean, coastal zones, and upwelling areas, respectively (see Table 8–1; multiply by ~ 2.2 for equivalent dry-matter productivity). The lower ranges prevail over the greater part of the oceans (Fig. 8–1).

Ryther's (1969) conclusions on productivity for the sea, a slight increase over figures of Steemann Nielsen and Jensen (1957), were influenced by a large body of data then being prepared for publication by Russian workers. Now

---

[1] t = metric ton = 10^6g.

Table 8–1   Global planktonic primary production as estimated by Ryther (1969)

Province	Percentage of ocean	Area (km^2)	Mean productivity (g C/m^2/year)	Total production (10^9t C/year)
Open ocean	90.0	326.0×10^6	50	16.3
Coastal zone^a	9.9	36.0×10^6	100	3.6
Upwelling areas	0.1	3.6×10^5	300	0.1
Total:		362.4×10^6		20.0

^{a} Includes offshore areas of high productivity.

Table 8–2   Global planktonic primary production as estimated by Koblentz-Mishke *et al.* (1970)

Type of water	Primary productivity level (mg C/m²/day)		Areas of each type of water in different oceans			Summary of annual production of the world ocean for each type of water (10⁹ t C/year)
	Mean value	Limits of fluctuation	Area[a]	× 10³ km²	World ocean (%)	
Oligotrophic waters of the central parts of subtropical halistatic areas	70	< 100	PO	90,105	24.6	3.79
			IO	19,599	5.3	
			AO	30,624	8.3	
			OW	8,000	2.2	
			WO	148,329	40.4	
Transitional waters between subtropical and subpolar zones; extremity of the area of equatorial divergences	140	100–150	PO	33,357	9.1	4.22
			IO	23,750	6.5	
			AO	22,688	6.2	
			OW	3,051	0.8	
			WO	82,847	22.6	
Waters of equatorial divergence and oceanic regions of subpolar zones	200	150–250	PO	31,319	8.5	6.31
			IO	18,886	5.2	
			AO	32,650	8.9	
			OW	3,642	1.0	
			WO	86,498	23.6	
Inshore waters	340	250–500	PO	10,422	2.8	4.80
			IO	7,944	2.2	
			AO	14,183	3.9	
			OW	6,184	1.7	
			WO	38,735	10.6	
Neritic waters	1,000	> 500	PO	243	0.07	3.90
			IO	5,289	1.4	
			AO	2,717	0.74	
			OW	2,433	0.66	
			WO	10,683	2.9	
Total, all waters:			WO	367,092	100	23 × 10⁹ t C/year

[a] PO, Pacific Ocean; IO, Indian Ocean; AO, Atlantic Ocean; OW, other waters (North Pole ocean, Indonesian seas, Mediterranean, Black, Asov, White, Okhotsk, Bering, Japan, China, Yellow seas); WO, summary value for the whole World Ocean.

generally available, the account by Koblentz-Mishke *et al.* (1970) bears examination (Table 8–2). Based on data from over 7000 stations, they divided the waters of the global ocean into five "types" with daily rates of productivity ranging from a mean of 70 mg C/m^2 in oligotrophic (nutrient-poor) waters in the central subtropics to a mean of 1000 mg C/m^2 from open coastal waters. The sum of their data provides a global estimate of 23×10^9 t C/year, somewhat above Ryther's (1969) range of values. None of these estimates takes into account benthic production, the magnitude of which has obvious interest.

Data from a variety of characteristic sites have been drawn together by West-lake (1963) and range, respectively, between 5.5 and 13 g dry organic matter synthesized per square meter per day for *Ascophyllum nodosum* in Nova Scotia and green algae on a tropical Pacific coral reef. In a similar compilation, Ryther (1959) listed net data from a coral reef and from a turtle grass flat, respectively, of 9.6 and 11.3 g dry organic material per square meter per day. Table 8–3 offers more recent information from a range of habitats. Although some high values are shown, considerable variation is also evident. Note, in particular, that available data from sediments bare of macrophyte vegetation are consistently low.

It should also be remembered that the figures given in Table 8–3 may be

Table 8–3 Experimentally determined rates of primary productivity in benthic habitats (g $C/m^2/day$)[a]

Site	Reference	Rate
Tidal fish pond, Hawaii	Hickling (1970)	1.22
Laminaria and *Agarum*, Nova Scotia	Mann (1972)	1.65
Laminaria hyperborea	Bellamy et al. (1968)	3.37
Laminaria sp.	Bellamy et al. (1973)	7.90 (2 m) 3.00 (10 m)
Intertidal seaweeds	Kanwisher (1966)	20.00
Cytoseira, Canary Islands	Johnston (1969)	10.50
Sea grasses, Laccadives	Qasim and Bhattathiri (1971)	5.80
Calcareous red algae, Eniwetok	Marsh (1970)	0.66
Reef corals, Florida	Kanwisher and Wainwright (1967)	2.70–10.20 (gross)
Intertidal blue-greens, Eniwetok	Bakus (1967)	0.65–2.15
Mangroves, Florida	Heald (1971)	1.20
Codium fragile, Long Island Sound	Wassman and Ramus (1973)	12.90
Benthic microflora, northern U.S. estuaries	Marshall et al. (1973)	0.08–0.53
Benthic microflora, tropical sediments	Bunt et al. (1972)	0.02–0.22
Benthic microflora, Scottish sediments	Steele and Baird (1968)	0.01–0.03

[a] See text for further details.

compared only with caution. They stem from a variety of procedures, all with possible shortcomings, applied at different times of year over nonuniform periods of incubation and depths, and with varying degrees of replication. For these reasons, and because sufficiently detailed geographic information is lacking, it is no surprise that global or even regional estimates of benthic production are not available. It is also clear that several types of benthic community are highly fertile and must make substantial local and regional contributions to marine production.

Theoretical Treatment

Some indication of the acceptability of estimates of global primary production may be derived from theoretical considerations. This approach has been taken by various reviewers including Rabinowitch (1945), Russell-Hunter (1970), and Vishniac (1971). Ryther (1959) made a series of deliberate assumptions to arrive at a value for maximum probable primary productivity beneath a unit area of sea surface, using only measures of incident radiation. Vishniac (1971) suggested that with an estimated 2.5×10^{21} cal/year available for marine photosynthesis, and an efficiency of 1 g carbon per 1.3×10^4 cal, an annual production of 190×10^9 t would be theoretically possible with freedom from nitrogen, phosphorus, and other resource limitations.

Russell-Hunter (1970, pp. 230–231) allowed "some 20 percent absorption (of light) by nonphotosynthetic areas and reflection by ice and snow." Of the remaining (calculated) incoming radiation, another 25% was subtracted for "additional absorption and reflection losses." An average 2% efficiency was allowed for photosynthesis under field conditions, and 9.5×10^3 cal were regarded as necessary to fix 1 g of carbon. On this basis, Russell-Hunter (1970) estimated 260×10^9 t C/year, equivalent to a mean very close to 2 g C/m^2/day. This quantity is more than 10 times higher than the Koblentz–Mishke *et al.* (1970) analysis.

Russell-Hunter's caloric requirement of 9.5×10^3 is lower than Vishniac's (1971) 13×10^3 cal per g carbon, although much higher than the 5.5×10^3 cal used by many other writers on the basis of energy yields when organic materials are combusted. The figure selected for field photosynthetic efficiency seems excessive for actual algal communities, as distinguished from cultures in optimal laboratory conditions. Wassink (1959) adopted a value of 0.11% for the world ocean. Losses caused by absorption, reflection, etc., appear to be uncertain; however, Ryther (1959) quoted other workers who found marine surface losses caused by reflection to be no more than 3–6%. It is not certain how much of the light that penetrates the surface of the sea is absorbed by photosynthetically active tissue, although Ryther (1959) assumed that in clear water the percentage could be quite high. Taking these possibilities into account, it is a simple matter to arrive at theoretic lower and upper levels of production lying between the wide limits of 12×10^9 and 488×10^9 t C/year. These estimates are the equivalent of 0.09–3.74 g C/m^2/day and, in fact, overlap slightly with the estimates from worldwide field measurements. Individual daily rates quoted by Ryther (1959) in some cases exceed the upper theoretical mean given here.

Obviously, there is nothing conclusive about theoretically calculated productivities, especially when their derivation involves assumptions that are scarcely better than guesses. Nonetheless, they can provide a useful perspective provided one is not seduced by an agreement with field data that may be circumstantial. Factors not allowed for in a theoretic treatment must affect actual productivity. Because these factors are primarily limiting factors, theoretical treatments are biased toward overestimate. The assumptions in Russell-Hunter's 260×10^9 t C/year and our high limit of 488×10^9 t C/year are probably so unrealistic that these estimates are without real value, but this does not mean that the low limit of 12×10^9 t C/year and estimates from field data are without some bias in the reverse direction. In this regard, it is instructive to consider some further alternatives.

It is an almost universal assumption that effective yields from photosynthesis are possible only for depths at which the incident illumination equals or exceeds 1% of the surface value. However, the rich populations of microalgae found in sea ice are known to be capable of developing autotrophically at much lower intensities (Bunt, 1963; Bunt and Lee, 1970), and it appears that this capability may be expanded in a more general sense (e.g., see Anderson, 1969). If we assume that the 1% of light penetrating below the accepted limit of the global photic zone were used with 9% efficiency as calculated for sea ice by Bunt and Lee (1970), and allow 9.5 kcal to fix 1 g carbon, this could provide up to 13.5×10^9 t C/year. Now if the 99% of incident light available in the "photic zone" were utilized with 0.11% efficiency as suggested by Wassink (1959), this would produce close to 19×10^9 t C/year. The total estimated production of 32.5×10^9 t C/year would be in reasonable agreement with the Koblentz–Mishke *et al.* (1970) analysis, unlike the much higher Russell-Hunter (1970) figure; this might raise serious questions over the spatial distribution of the production.

It remains to comment on the estimate of carbon production 23×10^9 t/year, based on field measurements. This is scarcely the place for detailed examination of the logistic, technical, and other potential shortcomings of currently accepted practices of productivity measurement. Some of the problems are discussed by Koblentz–Mishke *et al.* (1970) and in Chapter 3, this volume. Essentially, Koblentz–Mishke *et al.* recommend their estimate on the basis of the large number of separate measurements and their broad seasonal and global coverage. Of course, large numbers of measurements do not escape the essential limitations of the measurement procedures.

It is especially important to realize that the figure 23×10^9 t C/year is an estimate of fixation for the particulate components of the phytoplankton. It does not include photosynthetic products excreted into the seawater, the amount of which may be substantial. It also omits benthic production. To the best of my knowledge no one has yet attempted an estimate for worldwide benthic primary production. This would have to include not only the seaweeds, many of them known to be highly productive, but all corals containing symbiotic algae, and marine vascular plants as well as epiphytic and benthic microalgae. Such an evaluation would be both difficult because of the diversity of benthic habitats and premature because so few measurements have been attempted. However, tentative theoretical limits of benthic production may be suggested.

The extent of the benthic environment lying within the photic zone is difficult to assess but may be associated reasonably with the coastal fringe. According to Karo (1956), the total length of the global coastline may be taken as 280,000 statute miles (450,800 km). Setting the mean width of the benthic photic zone within the arbitrary but probably reasonable limits of 1 and 10 km, one obtains plane areas between 0.45 and 4.5×10^6 km². The yield of plant carbon derivable annually from these areas could amount to $0.65–6.5 \times 10^9$ t based on theoretical production potentials suggested by Ryther (1959) for an average radiation incidence of 200 g cal/cm²/day. The lower estimate represents $\sim 2\%$ of the current assessment of 23×10^9 t carbon for the world ocean and is probably conservative.

Conclusion

For total oceanic primary production the estimate 23×10^9 t C/year would need revision upward. The great photosynthetic industry of the sea that this, or even a somewhat higher figure, expresses does not, however, imply great reserves of unused but useful food for man (Ryther, 1969). Only very small fractions of primary production can be harvested through the secondary production of fish which are high in trophic pyramids; man's catch is further limited to certain fish populations that are large enough and concentrated enough to make harvest economically feasible. As an example of the disparity between primary production and yield of fish, a few simple calculations show that if the 10 g plant carbon/m² recovered by Bunt and Lee (1970) from annual Antarctic Sea ice is representative, the yearly production in this extreme environment could amount to 0.26×10^9 t. As trivial as this quantity is, compared with estimated global production, it exceeds by a factor of about 30 current world catches of fresh fish reported by FAO (1971). In this regard, it must be stressed that, as far as present human food needs are concerned, global primary production is of less interest than is production that directly influences fish stocks economically accessible to man. The distribution of harvestable fish stocks over the area of the ocean is very uneven; these stocks are strongly concentrated in the limited areas of inshore waters and upwelling where primary productivity is relatively high. Because of their concentration, these fish populations may be vulnerable to overharvest, from the effort to catch ever more, and to the effects of pollution (of inshore waters especially) on the fish and the food bases supporting them.

In conclusion (1) the estimate of marine primary production by Koblentz–Mishke *et al.* (1970) as 23×10^9 t C (about 50×10^9 t dry matter) per year is the best now available from field evidence; (2) this estimate is somewhat too low if benthic production, loss of dissolved organic matter from plankton cells, and photosynthesis at the low light intensities are considered; and (3) neither the lower nor a higher estimate offers much encouragement to the hope for a more abundant harvest of food from the sea. Only a more detailed comprehension of marine trophic dynamics than is now available will determine how much longer the sea can maintain its present contribution to human needs. However, the argument of Vishniac (1971) and others that we must be approaching the limit set by incoming radiation and feasibility of harvest seems inescapable. The possibility of even maintaining the present level of harvest from the seas may depend on how much of an effort modern society is prepared to expend in maintaining the quality of the coastal environment.

Acknowledgment

This chapter was prepared with partial support from Grant GA–25440 from the National Science Foundation.

References

Anderson, G. C. 1969. Subsurface chlorophyll maximum in the northeast Pacific Ocean. *Limnol. Oceanogr.* 14:386–391.

Anon. 1973. A guide to the measurement of marine primary production under some special conditions. *Monographs on Oceanographic Methodology*, No. 3, 73 pp. Paris: UNESCO.

Bakus, G. J. 1967. The feeding habits of fishes and primary production at Eniwetok, Marshall Islands. *Micronesica* 3:135–149.

Barber, R. T., R. C. Dugdale, J. J. MacIsaac, and R. L. Smith. 1972. Variations in phytoplankton growth associated with the source and conditioning of upwelling water. *Invest. Pesquera* 35:171–193.

Bellamy, D. J., D. M. John, and A. Whittick. 1968. The "kelp forest ecosystem" as a "phytometer" in the study of pollution of the inshore environment. *Underwater Assoc. Rep.* 79–82.

————, A. Whittick, D. M. John, and D. J. Jones. 1973. A method for the determination of seaweed production based on biomass estimates. *Monographs on Oceanographic Methodology*, No. 3, pp. 27–33. Paris: UNESCO.

Bunt, J. S. 1963. Diatoms of Antarctic sea ice as agents of primary production. *Nature (London)* 199:1255–1257.

————. 1965. Measurements of photosynthesis and respiration in a marine diatom with the mass spectrometer and with carbon-14. *Nature (London)* 207:1373–1375.

————. 1968. Some characteristics of microalgae isolated from Antarctic sea ice. *Antarctic Res. Ser.* 11:1–14.

————, and C. C. Lee. 1970. Seasonal primary production in Antarctic sea ice at McMurdo Sound in 1967. *J. Marine Res.* 28:304–320.

Bunt, J. S., C. C. Lee, and E. Lee. 1972. Primary productivity and related data from tropical and subtropical marine sediments. *Marine Biol.* 16:28–36.

Dugdale, R. C. 1967. Nutrient limitation in the sea: Dynamics, identification, and significance. *Limnol. Oceanogr.* 12:685–695.

Eppley, R. W., J. N. Rogers, and J. J. McCarthy. 1969. Half-saturation constants for uptake of nitrate and ammonium by marine phytoplankton. *Limnol. Oceanogr.* 14:912–920.

————, and J. D. H. Strickland. 1968. Kinetics of marine phytoplankton growth. *Advan. Microbiol. Sea* 1:23–62.

FAO. 1971. *Yearbook of Fishery Statistics,* Vol. 30: *Catches and Landings,* 469 pp. Rome: Food and Agriculture Organization of the United Nations.

Heald, E. J. 1971. Estuarine and coastal studies: The production of organic detritus in a south Florida estuary. *Sea Grant Tech. Bull.* No. 6, pp. 1–105. Coral Gables, Florida: Univ. of Miami.

Hickling, C. F. 1970. Estuarine fish farming. *Advan. Marine Biol.* 8:119–213.

Hutner, S. H., and L. Provasoli. 1964. Nutrition of algae. *Annu. Rev. Plant Physiol.* 15:37–56.

Jackson, W. A., and R. J. Volk. 1970. Photorespiration. *Annu. Rev. Plant Physiol.* 21:385–432.

Johnston, C. S. 1969. The ecological distribution and primary production of macrophytic marine algae in the Eastern Canaries. *Int. Rev. Ges. Hydrobiol.* 54: 473–490.

Kanwisher, J. W. 1966. Photosynthesis and respiration in some seaweeds. In *Some Contemporary Studies in Marine Science,* H. Barnes, ed., pp. 407–420. London: Allen and Unwin.

————, and S. A. Wainwright. 1967. Oxygen balance in some reef corals. *Biol. Bull.* 133:378–390.

Karo, H. A. 1956. World coastline measurements. *Int. Hydrograph. Rev.* 33:131–140.

Koblentz-Mishke, O. J., V. V. Volkovinsky, and J. G. Kabanova. 1970. Plankton primary production of the world ocean. In *Scientific Exploration of the South Pacific,* W. S. Wooster, ed., pp. 183–193. Washington, D.C.: National Academy of Science.

McAllister, D. C. 1970. Zooplankton rations, phytoplankton mortality and the estimation of marine production. In *Marine Food Chains,* J. H. Steele, ed., pp. 419–457. Berkeley: Univ. of California.

MacIsaac, J. J., and R. C. Dugdale. 1969. The kinetics of nitrate and ammonia uptake by natural populations of marine phytoplankton. *Deep-Sea Res.* 16:45–58.

Mann, K. H. 1972. Ecological energetics of the sea-weed zone in a marine bay on the Atlantic coast of Canada. II. Productivity of the seaweeds. *Marine Biol.* 14:199–209.

————. 1973. Seaweeds: Their productivity and strategy for growth. *Science* 182: 975–981.

Marsh, J. A. 1970. Primary productivity of reef-building calcareous red algae. *Ecology* 51:255–263.

Marshall, N., D. M. Skauen, H. C. Lampe, and C. A. Oviatt. 1973. Primary production of benthic microflora. *Monographs on Oceanographic Methodology,* No. 3, pp. 37–44. Paris: UNESCO.

Provasoli, L. 1963. Organic regulation of phytoplankton fertility. In *The Sea,* M. N. Hill, ed., Vol. 2, pp. 165–219. New York: Wiley (Interscience).

————, J. J. A. McLaughlin, and M. R. Droop. 1957. The development of artificial media for marine algae. *Arch. Mikrobiol.* 25:392–428.

Qasim, S. Z., and P. M. A. Bhattathiri. 1971. Primary production of a sea grass bed on Kavaratti Atoll (Laccadives). *Hydrobiologia* 38:29–38.

Rabinowitch, E. I. 1945. *Photosynthesis.* Vol. I. New York: Wiley (Interscience).

Redfield, A. C., B. H. Ketchum, and F. A. Richards. 1963. The influence of organisms on the composition of sea water. In *The Sea*, M. N. Hill, ed., Vol. 2, pp. 76–77. New York: Wiley (Interscience).

Russell-Hunter, W. D. 1970. *Aquatic Productivity*, 306 pp. London: Macmillan.

Ryther, J. H. 1956. Interrelation between photosynthesis and respiration in the marine flagellate, *Dunalielle euchlora. Nature (London)* 178:861–862.

———. 1959. Potential productivity of the sea. *Science* 130:602–608.

———. 1963. Geographic variations in productivity. In *The Sea*, M. N. Hill, ed., Vol. 2, pp. 347–380. New York: Wiley (Interscience).

———. 1969. Photosynthesis and fish production in the sea. *Science* 166:72–76.

———, and W. M. Dunstan. 1971. Nitrogen, phosphorus and eutrophication in the coastal marine environment. *Science* 171:1008–1013.

Sieburth, J. M., and A. Jensen. 1969. Studies on algal substances in the sea. II. The formation of gelbstoff by exudates of Phaeophyta. *J. Exp. Marine Biol. Ecol.* 3:275–289.

Smayda, T. J. 1970. The suspension and sinking of phytoplankton in the sea. *Oceanogr. Marine Biol. Annu. Rev.* 8:353–414.

Steele, J. H. 1969. Notes on some theoretical problems in production ecology. In *Primary Productivity in Aquatic Environments*, C. R. Goldman, ed., pp. 383–398. Los Angeles: Univ. of California Press.

———, and I. E. Baird. 1968. Production ecology of a sandy beach. *Limnol. Oceanogr.* 13:14–25.

Steemann Nielsen, E. 1952. The use of radioactive carbon (^{14}C) for measuring organic production in the sea. *J. Cons. Perm. Int. Explor. Mer.* 18:117–140.

———, and V. K. Hansen. 1959. Measurements with the carbon-14 technique of the respiration rates in natural populations of phytoplankton. *Deep-Sea Res.* 5:222–233.

———, and E. Aabye Jensen. 1957. Primary oceanic production. The autotrophic production of organic matter in the oceans. *Galathea Rep.* 1:49–135.

Strickland, J. D. H. 1965. Production of organic matter in the primary stages of the marine food chain. In *Chemical Oceanography*, J. P. Riley and G. Skirrow, eds., Vol. 1, pp. 477–610. New York: Academic Press.

———, and T. R. Parsons. 1965. A manual of seawater analysis. *Fish. Res. Bd. Can.* No. 125.

Thomas, J. P. 1971. Release of dissolved organic matter for natural populations of marine phytoplankton. *Marine Biol.* 11:311–323.

Vishniac, W. 1971. Limits of microbial productivity in the ocean. In *Microbes and Biological Productivity*, D. E. Hughes and A. H. Rose, eds., *Symp. Soc. Gen. Microbiol* 21:355–366.

Wassink, K. E. C. 1959. Efficiency of light energy conversion in plant growth. *Plant Physiol.* 34:356–361.

Wassman, E. R., and J. Ramus. 1973. Primary production measurements for the green seaweed *Codium fragile* in Long Island Sound. *Marine Biol.* 21:289–298.

Westlake, D. F. 1963. Comparisons of plant productivity. *Biol. Rev.* 38:385–425.

Woodwell, G. M., P. H. Rich, and C. A. S. Hall. 1973. Carbon in estuaries. In *Carbon and the Biosphere*, G. M. Woodwell and E. V. Pecan, eds. *Brookhaven Symp. Biol.* 24:221–240. Springfield, Va.: Tech. Inf. Center and Atomic Energy Comm. (CONF–720510).

9 Primary Production of Inland Aquatic Ecosystems

Gene E. Likens

Given the world's expanding human population, it is important to evaluate the net primary production of different ecosystems that can provide food. The inland aquatic ecosystems comprise less than 1% of the Earth's surface, but often are among the most productive areas. Many of these aquatic ecosystems have undergone dramatic changes in recent years as a result of man's activities. In some cases the change has been beneficial to man's short-term desires and requirements, but often the changes have been detrimental (e.g., polluted water supplies) because man has used water bodies widely as an inexpensive receptacle for waste products. Other responses and their implications were initially less obvious; for example, even though some aquatic ecosystems have been fertilized artificially by man's activities, thereby increasing productivity (cultural eutrophication), in many cases this productivity has been shifted to species less suitable for human consumption (e.g., Beeton, 1969; Beeton and Edmondson, 1972).

Inland Water Bodies

Among the inland bodies of water are an infinite variety of fresh and saline lakes, ponds, rivers, brooks, swamps, and marshes. This chapter groups the inland seas (e.g., Caspian) with the fresh and saline lakes in the treatment of productivity of inland aquatic ecosystems.

The dimensions of the surface inland waters of the Earth are not known precisely (cf. Hutchinson, 1957; Nace, 1960; Penman, 1970). Most of the

KEYWORDS: Primary production; freshwater ecosystems; eutrophication; global overview.

Primary Productivity of the Biosphere, edited by Helmut Lieth and Robert H. Whittaker.
© 1975 by Springer-Verlag New York Inc.

surface freshwater exists as ice and snow in glaciers and polar ice caps (some 25.5×10^6 km³). Freshwater lakes and streams cover some 0.2% of the Earth's surface and have a volume of at least 2.04×10^5 km³. A few lakes may be exceedingly deep (Lake Baikal, U.S.S.R., 1741 m), but the average depth of lakes of the world is only about 10 m. Lake Baikal contains 11% of the Earth's surface freshwaters, and as such represents the largest single reservoir of liquid freshwater. In addition, some 20% of the liquid-surface freshwaters are held by the five Laurentian Great Lakes of North America.

Saline lakes and inland seas have a somewhat smaller area and volume than freshwater lakes, but the order of magnitude is the same. A total area for inland waters of 2×10^6 km² is assumed. Freshwater marshes and swamps comprise an additional area of about 2×10^6 km².

Carbon Fixation

Man's attention usually has focused on the open water of lakes and rivers for his commercial harvest of food; however, weedy shorelines, swamps, and marshes may be the sites of greatest primary productivity. Herein lies a major difficulty when one attempts to assess the primary production of inland aquatic ecosystems. Most studies have estimated primary productivity solely from measures of phytoplanktonic photosynthesis. Even in purported studies of lake ecosystems, the primary production contributed by periphyton and rooted macrophytes often is not measured or included.

Photosynthetic fixation of carbon in inland aquatic ecosystems may occur by various communities (Fig. 9–1). These communities may be grouped and identi-fied conveniently by the type of producer organism, that is, phytoplankton, macrophytes, and periphyton. The phytoplankton represents the algal commu-nity of the open water; macrophytes are macroscopic vascular plants that are submerged or emergent, rooted or floating; and periphyton is the community of plants, other than macrophytes, that grows on submerged substrates. In many cases diatoms are dominant in the periphyton, and with other microorganisms they form a film on the surface of mud, rocks, or sand (and also on the sur-faces of the macrophytes).

In the majority of aquatic ecosystems carbon is most abundant in the in-organic form: $[\Sigma \text{ CO}_2] >$ DOC + POC detritus > POC living (Wetzel and Rich, 1973). Of the detrital fraction, the dissolved organic carbon (DOC) is usually an order of magnitude more abundant than the particulate organic carbon (POC). Only a small fraction of the total organic carbon pool is incor-ported in living organisms at any given time; but this small fraction creates the organic carbon that accumulates in other fractions and determines the functional characteristics of the ecosystem.

Very few good quantitative studies have been done on the photosynthetic and respiratory rates of the macrophyte and periphyton communities. Yet from data recently summarized by Westlake (1963, 1966), it is apparent that rooted aquatic macrophytes are particularly productive on fertile sites (Table 9–1). Apparently, as rooted emergent macrophytes colonize the sediments of a shallow

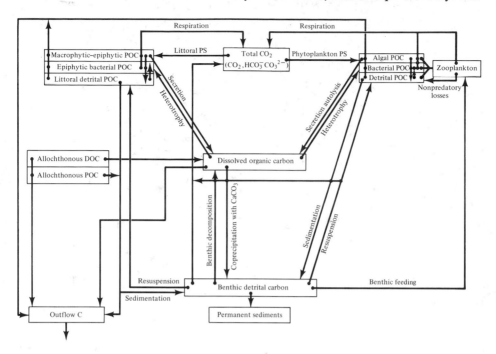

FIGURE 9–1. Diagrammatic model of the functional relationships of organic carbon in a hardwater lake. DOC, Dissolved organic carbon; POC, particulate organic carbon; PS, photosynthesis. (From Wetzel and Rich 1973.)

nutrient-rich lake during succession, biologic productivity may reach the highest level for a given site—higher than it was earlier, when the site was open water, and higher than it will be later, when the site becomes dry land. Hutchinson (1970) has speculated on why this is so. It may be that with adequate amounts of available nitrogen and phosphorus, an optimum habitat for a plant is to be rooted in a liquid medium, but to expose stems and leaves to the air, where

Table 9–1   Net primary productivity values for aquatic communities on fertile sites[a]

Producer community	mg C/m²/day[b]	g dry organic matter/m²/year
Lake phytoplankton	100–1,200	100–900
Freshwater macrophytes		
Submerged	500–2,700	400–2,000
Emergent	4,100–12,000	3,000–8,500

[a] Modified from Westlake (1963, 1966).
[b] Based on the assumption that organic matter is 50% carbon and averaged over 366 days.

188
Part 3: Global Productivity Patterns

differential replacement of carbon dioxide at the plant surface can be much faster. Plants such as *Typha* and *Phragmites* apparently make the best of both worlds.

   The relative importance of these three groups of primary producers in individual aquatic ecosystems is highly variable and largely unknown. In very large and deep lakes, phytoplankton is undoubtedly the major primary producer. However, in the shallow shoreline areas of deep lakes and especially in shallow lakes, swamps, marshes, and running-water ecosystems the contribution by periphyton and macrophytes certainly becomes more important if not dominant (see Allen, 1971). The shallows of lakes are often exceptionally productive on an areal basis and, therefore, are of special ecologic interest. One thorough study in Lawrence Lake, Michigan is very informative in this regard (Table 9–2). In this lake macrophytes were the principal primary producers and accounted for 49% of the yearly net photosynthetic total; the macrophytes, together with the periphyton on their surfaces, were responsible for 69% of the lake's total autochthonous productivity. In contrast 87% of the annual net photosynthesis was provided by phytoplankton in a small nutrient-poor lake, Mirror Lake, in northern New Hampshire (Table 9–2). Similar studies must be done for a

Table 9–2   Annual organic carbon fluxes for Lawrence Lake, Michigan and Mirror Lake, New Hampshire[a]

	Mirror Lake (g C/m²/year)	Lawrence Lake (g C/m²/year)
Inputs		
Production (net[b])		
Phytoplankton	47.0	52.6
Periphyton	1.3	41.9
Macrophytes	1.7	91.4
Bacteria (chemosynthesis)	4.0	7.1
Total autochthonous	54.0	193.0
Particular matter	6.6	4.1
Dissolved matter	11.3	21.0
Total allochthonous	17.9	25.1
Total inputs:	71.9	218.1
Outputs		
Respiration[c]	54.0	159.7
Sedimentation	7.6	16.8
Dissolved matter in outflow	9.2	35.8
Particulate matter in outflow	1.0	2.8
Total outputs:	71.8	215.1

[a] Derived from Wetzel et al. (1972) and Jordan and Likens (1975).
[b] Carbon-14 method.
[c] Does not include plant respiration.

variety of water bodies before an adequate knowledge of the sources, distribution, and trends of primary productivity within aquatic ecosystems can be gained. Such information is vital for the intelligent use and management of aquatic resources.

Carbon fixation by photosynthetic and chemosynthetic bacteria may be significant in a few rather specialized ecosystems, such as meromictic lakes or reservoirs (e.g., Sorokin, 1966; Culver and Brunskill, 1969); but the bacteria probably can be ignored for regional production estimates. However, the utilization of detrital carbon, whether autochthonous (originating within the lake) or allochthonous (originating outside the lake), may be important in the total metabolic system of lakes and streams (Fig. 9–1); and microorganisms often play a vital role in making this carbon available to consumers. For example, some 25% and 12% of the total organic carbon inputs for Mirror Lake and Lawrence Lake, respectively, came from allochthonous sources (Table 9–2). The total amount of allochthonous input is similar for both lakes, and most of this organic carbon is transported as dissolved material (Table 9–2). Surprisingly, organic carbon in direct precipitation (rain and snow) averaged about 3.1 mg/liter and represented 16% of the total allochthonous inputs for Mirror Lake (Jordan and Likens, 1975). Lund *et al.* (1963) estimated that only about one-third of the total carbon input for the North Basin of Lake Windermere could be attributed to autochthonous carbon fixation by phytoplankton. The other two-thirds, potentially available to animals and saprobes (and also possibly to algae), was attributed to external sources, although periphyton and macrophyte inputs were not evaluated carefully. Therefore, allochthonous carbon inputs can contribute significantly to the total metabolism of aquatic ecosystems through the same pathways (primarily grazing and decomposition) as carbon fixed autochthonously, and functionally should be considered as a part of the total "primary production" in evaluating the overall metabolism of aquatic ecosystems. In this regard Likens (1972a) has suggested that all reduced carbon compounds that can provide energy for consumers, from both autochthonous and allochthonous sources, be termed "ecosystem source carbon."

Measurement

Anyone familiar with the literature on primary production in aquatic ecosystems knows that there are serious difficulties in attempting to summarize it (see Chapter 3, this volume). The units of measurement are often confusing and difficult to compare, e.g. milligrams dry weight, milligrams ash-free dry weight, milligrams glucose, milligrams carbon, milligrams O_2, millimoles O_2 or CO_2, kilocalories, etc. No less confusing are the techniques. Harvest, gas exchange, change in pH, or nutrients, radioisotopic tracer, and other methods all have been used to estimate carbon fixation by photosynthesis. Unfortunately, each of these procedures provides somewhat different insights into the photosynthetic process making comparison and interpretation very difficult. Not to be ignored are the difficulties introduced when authors fail to state whether they are reporting gross or net primary production values, and what assumption they have followed in this regard. Frequently data needed for conversions from volumetric to areal units are not provided.

Also contributing to this problem are the technical and interpretative difficulties associated with the indirect measurement of carbon fixation from isolated samples of aquatic ecosystems in small glass bottles, and, particularly those associated with the widely used ^{14}C method. These precautions and difficulties have been detailed elsewhere (e.g., Goldman, 1968; Vollenweider, 1969), and include artificiality of environment, effects of incubation time and conditions, effects of light or temperature shock during sample manipulations, formation and measurement of extracellular products, incorporation or extracellular deposition of ^{14}C unrelated to photosynthesis, and calibration of sources and counting equipment. Therefore, the high sensitivity of the ^{14}C technique is often offset by the extreme precautions needed for reliable results, especially because it still is debatable as to whether the ^{14}C method measures net or gross primary productivity (e.g., Fogg, 1969).

Of more importance for ecosystem analysis is the problem of extrapolation of results from a limited number of small samples to an entire ecosystem. Photosynthesis (primary production) may be distributed unevenly in the water column of lakes. Hence, it is important to differentiate between the productivity of a lake and its water (Hutchinson, 1973), that is, productivity on an areal basis versus a volumetric basis. However, it is generally much more appropriate to compare the production of lakes on an areal basis. Findenegg (1964) has classified lake types on the basis of primary production profiles in alpine lakes (Fig. 9–2). Similar groupings were observed in the Experimental Lakes Area of Canada (Schindler and Holmgren, 1971). Following this scheme a class 1 lake has the maximum rate of productivity (mg C/m^3/time) near the surface of the water column where light intensities are high (light inhibition may occur at the surface), with a rapid decline in productivity with depth. Class 2 lakes do not have a distinct maximum, and low productivity rates prevail throughout the water column, where light intensities are adequate and apparently not limiting. Class 3 lakes usually have double maxima (near surface and mid-depth) with the highest rate of productivity at depths where light intensities are only 30–50% of the surface value (e.g., see Findenegg, 1964). In some lakes maximum rates of productivity within the water column (volumetric) may not produce maximum rates on an areal basis, whereas class 3 lakes usually have a maximum rate of production per unit area (Fig. 9–2). These results will depend on a variety of factors, including depth of trophogenic zone, algal species, light penetration, vertical distribution of nutrients, turbulence, and mixing within the water column, etc. Thus, there are often major temporal (annual, seasonal, and daily) and spatial (vertical and horizontal) differences in rates of primary productivity within an aquatic ecosystem (e.g., Rodhe, 1958; Findenegg, 1964). There are only a few careful studies in which frequent measurements made during all seasons allow for a reasonable estimate of annual primary productivity. In most cases, it is sheer guesswork to extrapolate a few daily values from short-term incubations to annual values.

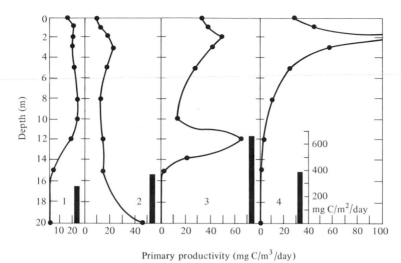

FIGURE 9–2. Primary productivity (^{14}C) patterns in alpine lakes:
(1) Lake Millstatt, (2) Lake Klopein, (3) Lake Wörth, (4) Lower
Lake of Lake Constance. Vertical bar in lower right corner of each
panel shows production per unit area of lake surface. (From
Findenegg, 1964.)

Consequently, there is a real paucity of reliable, quantitative data on annual
primary productivity. For rivers, particularly tropical rivers, such data are prac-
tically nonexistent. For example, there are no published data by which the
annual primary productivity of our greatest river systems—the Amazon, the
Congo, the Nile, and the Mississippi River—can be characterized. This lack,
although unfortunate, is understandable considering the difficulty in studying
these immense and highly complex systems of flowing water. In addition to the
problems outlined above, a parcel of water at one location in a river is constantly
changing as it moves downstream.

Amounts of Production

Because it is important to determine at least the orders of magnitude involved, I have assembled some ranges for net primary productivity values in
inland aquatic ecosystems (Table 9–3). All values are expressed in terms of net
productivity, since this is the most critical parameter in providing energy to the
heterotrophic organisms of the ecosystem. In many ways such an attempt, based
upon our current knowledge, is unrealistic if not potentially misleading. For
example, it is difficult to correct for respiration in most gas-exchange studies,
for the values may be highly variable, and measurements using opaque bottles
almost always reflect community respiration, including bacterial, zooplanktonic,
etc., rather than just algal respiration. Peterka and Reid (1968) reported
respiration values averaging 34%, but ranging between 0% and 95% of gross

Table 9–3 Net primary productivity values
for regional aquatic ecosystems

Water system	mg C/m^2/day	g C/m^2/year[a]
Tropical lakes	100–7600	30–2500
Temperate lakes	5–3600	2–950[b]
Arctic lakes	1–170	< 1–35
Antarctic lakes	1–35	1–10
Alpine lakes	1–450	< 1–100
Temperate rivers	< 1–3000	< 1–650
Tropical rivers	< 1–?	1–1000?

[a] In most cases, averaged over estimated "growing season."
[b] Naturally eutrophic lakes may reach a maximum of 450.

productivity, in Lake Ashtabula Reservoir, North Dakota. Hogetsu and Ichimura (1954) found that respiration ranged between 45% and 120% of gross productivity in Lake Suwa, and Aleem and Samaan (1969) found that respiration averaged about 30–40% of gross productivity for the water column in highly productive Lake Mariut, Egypt. Westlake (1966) suggested that annual net primary productivity is roughly 50% of gross productivity. A value for respiration of 40% of gross productivity may represent a reasonable general approximation, and I have used this value.

When ^{14}C data alone are used, gross productivity is often assumed to be 125% of net productivity (see e.g., Brylinsky and Mann, 1973). However, it is often difficult to interpret published ^{14}C data in terms of net primary productivity because of differences in techniques used and trophic conditions of ecosystems. I assumed that ^{14}C measurements approximated net productivity (although the values probably lie between net and gross). For convenience I assumed 50% carbon in the dry weight of aquatic plants. Westlake (1966) indicates that macrophytes range between 40% and 48% carbon. Thus, this summary (Table 9–3) represents some increased resolution over previous studies, but the estimates are tentative. However, some important points emerge.

The extremes of the ranges represent relative oligotrophic and eutrophic (infertile and fertile) conditions. However, the most productive lakes in one category may be less productive than the least productive lakes in another category (Table 9–3). This contrast demonstrates that terms such as eutrophic and oligotrophic are relative, and that their contexts of usage should be carefully defined, particularly now, when they may have emotional connotations as well. Some of the characteristics relative to productivity (phytoplankton) and the trophic status of freshwater lakes are given in Table 9–4. Vollenweider (1968) has presented a detailed evaluation of some of these relationships.

Rivers range down to a minimum of < 1 mg carbon fixed per meter per day because heavily shaded rivers and streams may have very little autochthonous production (although some have substantial allochthonous inputs, for example,

Table 9–4 Some general characteristics of lakes of various trophic status[a]

Trophic status[b]	Primary productivity[b] (mg C/m²/day)	Total organic carbon (mg/liter)	Phytoplankton density (cm³/m³)	Phytoplankton biomass (mg C/m³)	Chlorophyll-a (mg/m³)	Dominant phytoplankton	Light extinction (η/m)[c]	Total P (µg/liter)	Total N (µg/liter)	Total inorganic solids (mg/liter)
Ultraoligo-trophic	< 50	—	< 1	< 50	0.01–0.5		0.03–0.8	< 1–5	< 1–250	2–15
OLIGOTROPHIC	50–300	< 1–3	—	20–100	0.3–3	Chrysophyceae, Cryptophyceae, Dinophyceae, Bacillario-phyceae	0.05–1.0	—	—	—
Oligomeso-trophic	—	—	1–3	—	—		—	5–10	250–600	10–200
MESOTROPHIC Meso-eutrophic	250–1,000	< 1–5	— 3–5	100–300	2–15		0.1–2.0	— 10–30	— 500–1,100	— 100–500
EUTROPHIC	600–8,000	5–30	—	> 300	10–500	Bacillario-phyceae, Cyanophyta, Chlorophyta, Euglenophyta	0.5–4.0	—	—	—
Hyper-eutrophic	—	—	> 10	—	—		—	30–> 5,000	500–> 15,000	400–60,000
DYSTROPHIC (humic)	< 50–500	3–25	—	< 50–200	0.1–10		1.0–4.0	< 1–10	< 1–500	5–200

[a] Based on Birge and Juday (1934), Sakamoto (1966), Vollenweider (1968), Schindler and Holmgren (1971), Powers *et al.* (1972), Talling *et al.* (1973), and others.

[b] As measured by the ^{14}C method.

[c] $I_z = I_o e^{\eta}$, where I_z = light intensity at some depth z, I_o = surface light intensity, and η = extinction coefficient.

Hynes, 1963; Fisher and Likens, 1973). River ecosystems apparently can be very productive, however, and they may be among our most productive systems when heavily fertilized (see Odum, 1956).

Arctic, antarctic, and alpine lakes by and large are less productive on an annual basis than others, primarily because of limitation in growing season (covered by ice and snow during part of the year) and because of the seasonal delay or limitation in recharge of nutrients from the drainage basin. Of particular interest in this regard are some findings by Hobbie (1964), which indicate that in two Alaskan lakes some of the highest daily productivity values and between 41% and 83% of the annual production occurred beneath an ice cover. Lake Vanda in Antarctica has a permanent ice cover some 4 m thick, yet the euphotic zone extends to a depth of 60 m in summer. This large lake is one of the least productive lakes in the world (summer photosynthetic ^{14}C fixation of 14 mg C/m^2/day; Goldman et al., 1967). Temperature was suggested as the major limiting factor. Oligotrophic Lake Tahoe in the Sierra Nevada Mountains of California–Nevada has a euphotic zone about 100 m thick in summer, but ^{14}C productivity averaged only about 99 mg C/m^2/day during 1959–1962 (Goldman, 1967). In contrast, the lower basin of Lake Constance, second largest of the alpine lakes, is considered to be eutrophic; primary productivities as high as 450 mg C/m^2/day have been reported during October (Findenegg, 1966).

Within the temperate zone, Lake Ashtabula in North Dakota is a fertile reservoir on the Cheyenne River and has an annual net primary productivity of about 340 g C/m^2 (Peterka and Reid, 1968). Lake Werowrap, a saline lake in Australia, apparently has the highest known annual productivity for an undisturbed lake in the temperate zone (435 g C/m^2 based upon measurements throughout the year with ^{14}C; Walker, 1973). Macrophytes were not present in Lake Werowrap and about 91% of the total annual production for the lake occurred in the top 50 cm of the water column (Walker, 1973). Pederborgsø in Denmark is heavily enriched with nutrients and has an annual productivity of 943 g C/m^2 based on the ^{14}C method (original value multiplied by 1.45, cf. Mathiesen, 1963; Jónasson and Kristiansen, 1967). Søllerod Sø in Denmark is also heavily enriched with sewage and has an annual gross productivity of 520 g C (Steemann Nielsen, 1955, or a net productivity of 312 g C/m^2/year assuming respiration is 40% of gross productivity). Rodhe (1969) summarizes data for various Danish lakes and one Swedish lake all fertilized by domestic or industrial effluents, and gives a maximum daily rate for gross primary productivity of 6000 mg C/m^2. Shallow and highly eutrophic Clear Lake in California has a maximum daily rate of about 2500 mg C/m^2 based on the ^{14}C method (Goldman, 1968). The largest volume of freshwater in the world, Lake Baikal, has an average phytoplankton productivity of about 310 mg C/m^2/day (Kozhov, 1963), an annual net productivity of 122.5 g C/m^2 (Moskalenko, 1972), and is probably mesotrophic. This massive reservoir of water is currently being subjected to major cultural influences. Some estimates of annual net phytoplankton productivity have been provided for the Laurentian Great Lakes by Vollenweider et al. (1974). Lake Erie is the most productive (240–

250 g C/m²/year), followed by Lake Ontario (180–190 g C/m²/year), Lake Michigan (140–150 g C/m²/year), Lake Huron (80–90 g C/m²/year), and Lake Superior (40–80 g C/m²/year).

As might be expected, tropical lakes may be exceedingly productive if nutrients are plentiful. It has been known for a long time that very high productivities may be obtained in shallow tropical ponds and rice paddy cultures. The question of what the upper limit for aquatic primary productivity in natural ecosystems might be is of interest here. Wetzel (1966) suggested that only in highly enriched situations are productivities likely to exceed 5 g C/m²/day, whereas Lund (1967) believed that the upper limit would be about 10–13 g C/m²/day. In a recent detailed study of Lake Lanao, Phillipines, Lewis (1974) reported a maximum daily net productivity of 5 g C/m²/day and an annual value of 620 g C/m². In Lake Nakuru, Africa, with its large populations of flamingoes, Melack and Kilham (1971) reported a maximum daily gross productivity of 34 g O_2/m² (∼ 12.8 g C/m²). This is one of the highest values ever reported for an inland water body and undoubtedly is related to high nutrient levels maintained by the flamingoes (Table 9–5). Talling *et al.* (1973) also considered the problem of maximum phytoplankton production based upon literature values and studies of two productive saline lakes in Ethiopia, and stated that for mass cultures under natural illumination, ". . . net yield of 10 g C/m²/day or more are well-attested, whereas with natural phytoplankton there are very few detailed records of net or gross yields in excess of ∼ 4 g C/m²/day. . . ." They propose that a rate of gross photosynthesis per unit biomass at light saturation greater than 30 mg O_2/mg chl–a/hour would be rare. Chlorophyll a values in the euphotic zone of lakes could reach maximum levels on theoretical grounds at about 200–300 mg/m² (Talling *et al.*, 1973). Thus, gross productivity values of 43 and 57 g O_2/m²/day (Talling *et al.*, 1973) for the two Ethiopian lakes, L. Aranquadi and L. Kilotes, are near the theoretic limit. However, a larger value (17.5 g C/m²/day) has been measured (if ¹⁴C results truly indicate net productivity) in Red Rock Tarn (Table 9–5), a shallow saline lake in Australia (Hammer, 1970; Hammer *et al.*, 1973).

For world primary production estimates only two general categories seem useful at this time: (1) lakes and rivers, and (2) swamps and marshes. The areal extent of all lakes and rivers is about 2 × 10⁶ km². Mean productivity values for these ecosystems are very difficult to estimate. Based largely on the major freshwater reservoirs (L. Baikal and The Laurentian Great Lakes—31% of the Earth's surface freshwaters), and considering macrophytes and periphyton as well as phytoplankton productivity, a mean net productivity for lakes and rivers is assumed to be 200 g C/m²/year, which gives an annual total net production of about 0.4 × 10⁹ t of carbon or 0.8 × 10⁹ t of dry matter. For freshwater swamps and marshes the area is about 2 × 10⁶ km², while the mean net productivity may be taken as 1500 g C/m²/year, giving a total annual value of 3 × 10⁹ t C or 6 × 10⁹ t dry matter. The estimate for lakes and rivers is lower, and for swamps and marshes is higher than given earlier (Whittaker and Likens, 1969, 1973).

Cultural Eutrophication

A variety of environmental factors may regulate or limit net primary productivity in aquatic ecosystems. Brylinsky and Mann (1973) have suggested from a "large-scale" correlation analysis (43 lakes and 12 reservoirs scattered from the tropics to the arctic) that aquatic production in freshwater ecosystems is primarily controlled by input of solar radiation as this is determined by latitude. Latitude as an independent variable explained about 56% of the variance in productivity. No one would deny that solar radiation is vital to photosynthesis. There are, however, reasons to consider Brylinsky and Mann's interpretation of their correlations to be misleading, and that nutrient input and availability is a much more important variable in determining productivity of freshwater ecosystems, particularly on a regional basis. There are many examples of lakes that are subject to the same solar input but that differ widely in productivity because of the availability of nutrients, particularly inorganic nitrogen and phosphorus. It is interesting to note that the most productive lake thus far observed apparently is not in the tropics (Table 9–5). Also, from data in Table 9–3 it is apparent that some polar and some alpine lakes have productivities higher than some tropical lakes. Moreover, as Brylinsky and Mann point out, the use of ambient concentrations of inorganic phosphorus and nitrogen as independent variables is not a valid assessment of nutrient effects on productivity. These vital nutrients cycle so rapidly among living particulate, dead particulate, and dissolved forms (e.g., Whittaker, 1961; Lean, 1973) that concentrations in the water are of little diagnostic value. Furthermore, luxury uptake of inorganic phosphorus by phytoplankton may mask any relationship between nutrient availability and productivity (cf. Lehman *et al.*, 1975). Therefore, an input–

Table 9–5   Maximum daily gross productivity values for some saline lakes at different latitudes

Lake	Latitude	Total Inorganic dissolved solids (mEq/liter)	Maximum gross productivity (g C/m^2/day)	Reference
Nakuru	0°S	78–168	12.8[b]	Melach and Kilham (1971)
Aranguadi	9°N	70–80	21.4[b]	Talling *et al.* (1973)
Mariut	31°N	200[a]	10.8[c]	Aleem and Samaan (1969)
Red Rock Tarn	38°S	300[a]	29.2[d]	Hammer *et al.* (1973)

[a] Estimated.

[b] Dissolved oxygen production × 0.375.

[c] Carbon-14 value increased by a correction factor equivalent to 10% of the rate of photosynthesis at light saturation.

[d] Carbon-14 value increased by 167%.

output flux for nutrients as well as the rate of internal cycling must be known to evaluate fully the role of nutrients in governing primary productivity in aquatic ecosystems.

Nutrient availability and solar radiation are certainly two of the important variables, but temperature, available substrate, and grazing by zooplankton and benthic invertebrates also may play a role. These subjects are dealt with extensively elsewhere and are not considered here (see Goldman, 1966; Russell-Hunter, 1970; Likens, 1972b). Something should, however, be said about the effects of man on lakes, particularly with regard to cultural eutrophication.

Rodhe (1969) considers that lakes with a gross productivity of > 75 g $C/m^2/year$ are naturally eutrophic and those > 350 g $C/m^2/year$ are polluted. The term "polluted" has taken on ambiguous and sometimes emotional meaning; Rodhe uses the term in reference to excessive cultural eutrophication, in which bodies of water are fouled by domestic or industrial wastes. Eutrophication refers to the increased biologic response that results from nutrient enrichment (well fed), whereas pollution carries the connotation of biologic damage resulting from excessive inputs of nutrients (overfed), or of poisons.

If we conservatively consider values > 250 g $C/m^2/year$ for gross productivity (150 g $C/m^2/year$ net productivity) to characterize culturally eutrophic lakes, then there are many thousands of such lakes in the United States. No one knows exactly how many lakes there are in the United States, but of the more than 100,000 estimated, possibly one-third are showing signs of cultural eutrophication (Hasler and Ingersoll, 1968). Ketelle and Uttormark (1971) have compiled the names of 425 lakes in the United States (by no means representing a complete list or even a random sampling) that "have deteriorated to the extent that protective action is no longer sufficient and rehabilitation is required if satisfactory quality is to be re-established." Of this compilation, 340 of the lakes have a surface area larger than 40.5 ha (100 acres). The problem is not localized by any means, and the rate of eutrophication, although highly variable in different lakes, reflects the rate of human population increase and use.

Of the five Laurentian Great Lakes, Superior and Huron and to a lesser extent Michigan are still relatively unproductive, whereas Ontario and Erie, particularly Erie's western basin, are eutrophic (Beeton, 1969; Beeton and Edmondson, 1972; Ragotzkie, 1974). However, Beeton and Edmondson (1972) have pointed out that in these very large lakes inshore areas are affected first by the nutrients and toxins that drain from the land. Thus, inshore areas of Erie, Michigan and Ontario and particularly Lake Huron's Saginaw Bay and Lake Michigan's Green Bay are much more eutrophic than are offshore waters. Standing crops of phytoplankton tell something of the rate of change with cultural eutrophication, although the relationship between biomass and productivity is poorly quantified. Between 1927 and 1964 in Lake Erie the average number of algal cells in the water at a given time increased by 44.3 cells/ml/year. A slower but significant increase of 13 cells/ml/year occurred in Lake Michigan between 1926 and 1958 (Davis 1966). Lake Superior has the largest area of any freshwater lake (82,103 km^2; Ragotzkie, 1974) and represents about 8% of the Earth's surface covered by freshwaters. If the phytoplanktonic primary productivity averaged about 50 g C/m²/year (Vollenweider *et al.*, 1974), the annual production for the lake would be 4.1 $\times$ 10⁶ t C, or more than 1% of the Earth's total for lakes and rivers. In contrast, daily productivity in Lake Erie may be as much as 45 times greater than that of Lake Superior (Beeton, 1969). Because of its vast size, a culturally eutrophic Lake Superior could alter significantly the estimate for productivity of the Earth's lakes and rivers.

Because a principal result of eutrophication is increased biologic productivity, the effects on man are mixed. Increased nutrient inputs may result in a very high rate of primary productivity. If the algal increase is readily grazed by consumers, and more and larger "desirable" fish species can be sustained, man may welcome the eutrophication. If, however, with increased fertilization the algal succession produces species not readily eaten by consumers, or the algae foul the water by their presence (floating scums) or by their metabolism, man is not pleased. Some lakes are considered eutrophic or "polluted" because of blooms of particular nuisance algae, often blue-green species, which adversely affect man's use through odor, taste, or appearance (see Edmondson, 1969).

Conclusion

Although the areal extent of inland aquatic ecosystems on the Earth is relatively small, these ecosystems are some of the most productive. A variety of environmental factors, including availability of nutrients; heat and light; water-body morphology; morphometry and substrate; rate of grazing or harvest; and drainage-area size, use, and geology, may regulate or limit net primary productivity in these diverse and complex aquatic ecosystems. Carbon may be fixed in them by phytoplankton, macrophytes, periphyton, or chemosynthetic microorganisms, or it may be input, as dissolved or particulate organic matter, from the drainage basin. Evaluation of all these inputs and their regulating factors is important for predicting eutrophication trends in aquatic ecosystems.

Through cultural changes man accelerates the inputs of nutrients into lakes—from agricultural fertilizer, sewage, detergents, industrial wastes, atmospheric pollution—and greatly increases their biologic productivity. This productivity may be in forms less welcome to us. In some cases industrial or agricultural toxins can either reduce productivity or prevent human consumption of fish contaminated with pesticides or other toxins. This is our dilemma. The growth of man's population has created an increasing need for high-protein food from water bodies, but the growth of industry and population provides increasing pressures on water bodies that may reduce either their productivity, or the suitability of what they produce for food. It is these culturally induced changes in lakes that are observed to be spreading rapidly throughout the United States and elsewhere. It is heartening to see that cultural eutrophication may be alleviated or reversed locally (Lake Washington: Edmondson, 1972; Green Lake: Oglesby, 1969), but such reversal is usually very expensive and politically difficult. Unfortunately, if population and industry continue to grow, consequences of that growth may progressively restrict man's use of lakes and rivers for food and recreation.

Acknowledgments

I thank B. Peterson, J. Meyer, and R. Wetzel for comments and suggestions on the manuscript.

References

Aleem, A. A., and A. A. Samaan. 1969. Productivity of Lake Mariut, Egypt. II. Primary production. *Int. Rev. Ges. Hydrobiol.* 54:491–527.

Allen, H. 1971. Primary productivity, chemoorganotrophy, and nutritional interactions of epiphytic algae and bacteria on macrophytes in the littoral of a lake. *Ecol. Monogr.* 41:97–127.

Beeton, A. M. 1969. Changes in the environment and biota of the Great Lakes. In *Eutrophication: Causes, Consequences, and Correctives*, pp. 150–187. Washington, D.C.: National Academy of Sciences.

Beeton, A. M. and W. T. Edmondson. 1972. The eutrophication problem. *J. Fish. Res. Bd. Can.* 29:673–682.

Birge, E. A., and C. Juday. 1934. Particulate and dissolved organic matter in inland lakes. *Ecol. Monogr.* 4:440–474.

Brylinsky, M., and K. H. Mann. 1973. An analysis of factors governing productivity in lakes and reservoirs. *Limnol. Oceanogr.* 18:1–14.

Culver, D. A., and G. J. Brunskill. 1969. Fayetteville Green Lake. V. Studies of primary production and zooplankton in a meromictic marl lake. *Limnol. Oceanogr.* 14:862–873.

Davis, C. C. 1966. Plankton studies in the largest great lakes of the world, Special Rep. 14, Center for Great Lakes Studies, pp. 1–36. Ann Arbor, Michigan: Univ. of Michigan.

Edmondson, W. T. 1969. Eutrophication in North America. In *Eutrophication: Causes, Consequences, and Correctives*, pp. 124–149. Washington, D.C.: National Academy of Sciences.

————. 1972. Nutrients and phytoplankton in Lake Washington. In *Nutrients and Eutrophication*, G. E. Likens, ed., Special Symp. Vol. 1, pp. 172–193. Lawrence, Kansas: American Society of Limnology and Oceanography.

Findenegg, I. 1964. Bestimmung des Trophiegrades von Seen nach der Radiocarbonmethode. *Naturwissenschaften* 51(15):1–2.

————. 1966. Phytoplankton und Primäproduktion einiger ostschweizerischer Seen und des Bodensees. *Schweiz. Z. Hydrol.* 28:148–171.

Fisher, S. G., and G. E. Likens. 1973. Energy flow in Bear Brook, New Hampshire: An integrated approach to stream ecosystem metabolism. *Ecol. Monogr.* 43:421–439.

Fogg, G. E. 1969. Oxygen—versus ^{14}C-methodology. In *A Manual on Methods for Measuring Primary Production in Aquatic Environments*, R. A. Vollenweider, ed., *IBP Handbook No. 12*, pp. 76–78. Philadelphia, Pennsylvania: Davis.

Goldman, C. R. (ed.). 1966. *Primary Productivity in Aquatic Environments*. [*Mem. Ist. Ital. Idrobiol. 18* (Suppl.)] Berkeley, California: Univ. of California Press.

————. 1967. Integration of field and laboratory experiments in productivity studies. In *Estuaries*, G. H. Lauff, ed. Washington, D. C.: American Association for the Advancement of Science Publ. 83:346–352.

————. 1968. Aquatic primary production. *Amer. Zool.* 8:31–42.

————, D. T. Mason, and J. E. Hobbie. 1967. Two Antarctic desert lakes. *Limnol. Oceanogr.* 12:295–310.

Hammer, U. T. 1970. Primary production in saline lakes. *Austral. Soc. Limnol. Bull.* 3:20.

————, K. F. Walker, and W. D. Williams. 1973. Derivation of daily phytoplankton production estimates from short-term experiments in some shallow eutrophic Australian saline lakes. *Austral. J. Marine Freshwater Res.* 24:259–266.

Hasler, A. D., and B. Ingersoll. 1968. Dwindling lakes. *Natural Hist.* 77(9):1–6.

Hobbie, J. E. 1964. Carbon 14 measurements of primary production in two arctic Alaskan lakes. *Verhandl. Int. Ver. Limnol.* 15:360–364.

Hogetsu, K., and S. Ichimura. 1954. Studies on the biological production of Lake Suwa. VI. The ecological studies on the production of phytoplankton. *Jap. J. Bot.* 14:280–303.

Hutchinson, G. E. 1957. *A Treatise on Limnology* Vol. 1: *Geography, Physics, and Chemistry*, 1015 pp. New York: Wiley.

————. 1970. The biosphere. *Sci. Amer.* 223(3):44–53.

————. 1973. Eutrophication. *Amer. Sci.* 61:269–279.

Hynes, H. B. N. 1963. Imported organic matter and secondary productivity in streams. *Proc. 16th Int. Congr. Zool.* Washington, D.C., 1963, 4:324–329.

Jónasson, P. M., and J. Kristiansen. 1967. Primary and secondary production in Lake Esrom. Growth of *Chironomus anthracinus* in relation to seasonal cycles of phytoplankton and dissolved oxygen. *Int. Rev. Ges. Hydrobiol.* 52:163–217.

Jordan, M., and G. E. Likens. 1975. An organic carbon budget for an oligotrophic lake in New Hampshire, U.S.A. *Verhandl. Int. Ver. Limnol.* 19. (In Press)

Ketelle, M. J., and P. D. Uttormark. 1971. Problem lakes in the United States. Tech. Rep. 16010 EHR 12/71, 282 pp. Madison, Wisconsin: Univ. of Wisconsin, Water Resources Center.

Kozhov, M. 1963. *Lake Baikal and Its Life*, 344 pp. The Hague: Junk.

Lean, D. R. S. 1973. Phosphorus dynamics in lake water. *Science* 179:678–680.

Lehman, J. T., D. B. Botkin, and G. E. Likens. 1975. The assumptions and rationales of a computer model of phytoplankton population dynamics. *Limnol. Oceanogr.* (In Press)

Lewis, W. M. Jr. 1974. Primary production in the plankton community of a tropical lake. *Ecol. Monogr.* 44:377–409.

Likens, G. E. 1972a. Eutrophication and aquatic ecosystems. In *Nutrients and Eutrophication*, G. E. Likens, ed., Special Symp. Vol. 1, pp. 3–14. Lawrence, Kansas: American Society of Limnology and Oceanography.

———— (ed.). 1972b. *Nutrients and Eutrophication*. Special Symp. Vol. 1, 328 pp. Lawrence, Kansas: American Society of Limnology and Oceanography.

Lund, J. W. G. 1967. Planktonic algae and the ecology of lakes. *Sci. Progr. London* 55:401–419.

————, F. J. H. Mackereth, and C. H. Mortimer. 1963. Changes in depth and time of certain chemical and physical conditions and of the standing crop of *Asterionella formosa* Hass. in the North Basin of Windermere in 1947. *Phil. Trans. Roy. Soc. B* 246:255–290.

Mathiesen, H. 1963. Om planteplanktonets produktion of organisk stof nogle naeringsrige søer på Sjaelland. Ferskvandsfiskeribladet 1963 (1, 2). 8 pp.

Melack, J. M., and P. Kilham. 1971. Primary production by phytoplankton in East African alkaline lakes. *Bull. Ecol. Soc. Amer.* 52:45.

Merilänen, J. 1970. On the limnology of the meromictic lake Valkiajärvi, in the Finnish Lake District. *Ann. Bot. Fenn.* 7:29–51.

Moskalenko, B. K. 1972. Biological productive system of Lake Baikal. *Verhandl. Int. Ver. Limnol.* 18:568–573.

Nace, R. L. 1960. Water management, agriculture, and ground-water supplies. *U.S. Geol. Surv. Circ.* 415:1–11.

Odum, H. T. 1956. Primary production in flowing waters. *Limnol. Oceanogr.* 1:102–117.

Oglesby, R. T. 1969. Effects of controlled nutrient dilution on the eutrophication of a lake. In *Eutrophication: Causes, Consequences, and Correctives,* pp. 483–493. Washington, D.C.: National Academy of Sciences.

Penman, H. L. 1970. The water cycle. *Sci. Amer.* 223(3):99–108.

Peterka, J. J., and L. A. Reid. 1968. Primary production and chemical and physical characteristics of Lake Ashtabula Reservoir, North Dakota. *Proc. N. Dak. Acad. Sci.* 22:138–156.

Powers, C. F., D. W. Schultz, K. W. Malueg, R. M. Brice, and M. D. Schuldt. 1972. Algal responses to nutrient additions in natural waters. II. Field experiments. In *Nutrients and Eutrophication,* G. E. Likens, ed., Special Symp. Vol. 1, pp. 141–156. Lawrence, Kansas: American Society of Limnology and Oceanography.

Ragotzkie, R. A. 1974. The Great Lakes rediscovered. *Amer. Sci.* 62:454–464.

Rodhe, W. 1958. The primary production in lakes: Some results and restrictions of the ^{14}C method. *Rapp. Proc. Verb. Cons. Int. Explor.* Mer 144:122–128.

―――. 1969. Crystallization of eutrophication concepts in northern Europe. In *Eutrophication: Causes, Consequences, and Correctives,* pp. 50–64. Washington, D.C.: National Academy of Sciences.

Russell-Hunter, W. D. 1970. *Aquatic Productivity.* 306 pp. New York: Macmillan.

Sakamoto, M. 1966. The chlorophyll content in the euphotic zone in some Japanese lakes and its significance in the photosynthesis production of phytoplankton communities. *Bot. Mag. Tokyo* 79:77–88.

Schindler, D. W., and S. K. Holmgren. 1971. Primary production and phytoplankton in the Experimental Lakes Area, Northwestern Ontario, and other low-carbonate waters, and a liquid scintillation method for determining ^{14}C activity in photosynthesis. *J. Fish. Res. Bd. Can.* 28:189–201.

Sorokin, Y. I. 1966. On the trophic role of chemosynthesis and bacterial biosynthesis in water bodies. In *Primary Productivity in Aquatic Environments,* C. R. Goldman, ed. [*Mem. Ist. Ital. Idrobiol. 18* (Suppl.)], 181–205. Berkeley, California: Univ. of California Press.

Steemann Nielsen, E. 1955. The production of organic matter by the phytoplankton in a Danish lake receiving extraordinary great amounts of nutrient salts. *Hydrobiology* 7:68–74.

Talling, J. F., R. B. Wood, M. V. Proper, and R. M. Baxter. 1973. The upper limit of photosynthetic productivity by phytoplankton: Evidence from Ethiopian lakes. *Freshwater Biol.* 3:53–76.

Vollenweider, R. A. 1968. *Scientific fundamentals of the eutrophication of lakes and flowing waters, with particular reference to nitrogen and phosphorus as factors in eutrophication.* Paris: Organization for Economic Cooperation and Development (DAS/CSI/68.27), 159 pp.

——— (ed.). 1969. A Manual on Methods for Measuring Primary Production in Aquatic Environments, *International Biological Programme Handbook* No. 12:1–213. Philadelphia, Pennsylvania: Davis.

———, M. Munawar, and P. Stadelmann. 1974. A comparative review of phytoplankton and primary production in the Laurentian Great Lakes. *J. Fish. Res. Bd. Can.* 31:739–762.

Walker, K. 1973. Studies on a saline lake ecosystem. *Austral. J. Mar. Freshwater Res.* 24:21–71.

Westlake, D. F. 1963. Comparisons of plant productivity. *Biol. Rev.* 38:385–425.

———. 1966. Some basic data for investigations of the productivity of aquatic macrophytes. In *Primary Productivity in Aquatic Environments*, C. R. Goldman, ed. [*Mem. Ist. Ital. Idrobiol.*, 18 (Suppl.)], 229–248. Berkeley, California: Univ. of California Press.

Wetzel, R. G. 1966. Variations in productivity of Goose and hypereutrophic Sylvan Lakes, Indiana. *Invest. Indiana Lakes and Streams* 7:147–184.

———, and P. H. Rich. 1973. Carbon in freshwater systems. In *Carbon and the Biosphere*, G. M. Woodwell and E. V. Pecan, eds., pp. 241–263. Springfield, Virginia: Tech. Inf. Center, U.S. Atomic Energy Commission (CONF-720510). (*Brookhaven Symp. Biol.* 24:241–263.)

———, P. H. Rich, M. C. Miller, and H. L. Allen. 1972. Metabolism of dissolved and particulate detrital carbon in a temperate hard-water lake. *Mem. Ist. Ital. Idrobiol.* 29 (Suppl.):185–243.

Whittaker, R. H. 1961. Experiments with radiophosphorus tracer in aquarium microcosms. *Ecol. Monogr.* 31:157–188.

———. 1970. *Communities and Ecosystems*, 162 pp. New York: Macmillan.

———, and G. E. Likens. 1969. Net primary production and plant biomass for major ecosystems and for the Earth's surface. [Table presented at the *Brussels Symp. Productivity of Forest Ecosystems (1969)* and published by Whittaker (1970) and Whittaker and Woodwell (1971).]

———, and G. E. Likens. 1973. Carbon in the biota. In *Carbon and the biosphere,* G. M. Woodwell and E. V. Pecan, eds., pp. 281–302. Springfield, Virginia: Tech. Inf. Center and Atomic Energy Comm. (CONF–720510). (*Brookhaven Symp. Biol.* 24:281–302.)

———, and G. M. Woodwell. 1971. Measurement of net primary production of forests. (French summ.) In *Productivity of Forest Ecosystems: Proc. Brussels Symp. 1969*, P. Duvigneaud, ed. *Ecology and Conservation*, Vol. 4, 159–175. Paris: UNESCO.

10

Primary Production of
the Major Vegetation Units
of the World

Helmut Lieth

The primary productivity of the world is of paramount importance for man. Primary productivity captures that portion of solar energy that supports the life of all components of the biosphere. The largest portion of human food is provided by the productivity of plant life on land. From land production also comes our greatest single substance for construction and fabrication—wood—and a host of other products. The productivity of vegetation is one major aspect (the accumulation of toxic materials in the environment and potential psychologic effects are others) of the carrying capacity of the earth for man—its ability to support human populations on a long-term basis. Fossil fuels are accumulated profits from past primary production. The mantle of vegetation protects the Earth's surface against destructive erosion; and it provides an important part of the environmental context in which man and his societies have developed and in which man himself feels most at home. It is by primary productivity and the growth of plants by the creation of organic matter through photosynthesis that the life of the vegetational mantle and thereby of man is maintained.

Such thoughts were responsible for the creation of the International Biological Program (IBP) in the early nineteen sixties. It is to the credit of this program that we can present production figures today from regions that were not studied previously. We present here a new and independent appraisal of total world production in addition to a breakdown into production levels for major kinds of communities. Increased knowledge of energy values for biologic material makes it possible to express the production values both in tons of dry matter and in calories.

KEYWORDS: Primary production; terrestrial vegetation types; geoecology; world overview; global pattern.

Primary Productivity of the Biosphere, edited by Helmut Lieth and Robert H. Whittaker.
© 1975 by Springer-Verlag New York Inc.

Dry-Matter Production

The major vegetation units of the world are rather stable over long periods if there is no human interference. In contrast their classification in the literature of the last 20 years has been unstable; there are inescapable reasons why the classification of plant communities remains subjective to some degree. In order to employ a reasonably standard classification system, we have chosen the UNESCO scheme (Ellenberg and Müller-Dombois, 1967) in its modified form (Olson, 1970). For land vegetation this system outlines eight formation classes comprising 28 subclasses and a large number of further subunits. The formation subclasses of this system coincide best with the "biome types" as they are understood by the IBP and with the "formation types" of many authors (e.g., Whittaker 1962, 1970). Table 10–1 summarizes production estimates for about 20 such vegetation units. The production values are calculated separately for two different categories: annual dry-matter production and annual energy fixation. Carbon content can be estimated as 45% of the dry matter, and CO_2 is estimated as 1.6 times the dry matter.

Table 10–1 shows the vegetation unit (column 1), the area covered by this vegetation type (column 2), the rate of primary productivity (columns 3 and 4), and the total annual dry-matter production for the vegetation type (column 5). The sum total for the earth amounts to 155.2×10^9 t—55×10^9 t for the oceans and 100.2×10^9 t for the continental areas.[1] The values given in Table 10–1 coincide reasonably well with the recent estimates of Whittaker and Likens (Whittaker, 1970; Whittaker and Woodwell, 1971) of 164×10^9 t dry matter for the world, and of Golley (1972) of 143.8×10^9 t dry matter. Among other recent estimates, those of Basilevich *et al.* (1970) differ most from our figures with a sum total of 225×10^9 t: 55×10^9 t for the oceans (no difference) and 170×10^9 t dry matter for land. A comparison of the more recent global productivity estimates is provided in Chapter 13.

Energy Fixation

A separate appraisal for the annual energy fixation of the same vegetation units is given in columns 6–8 of Table 10–1. Column 6 gives the mean figure for the combustion value of the vegetation type, considering actual compositions of vegetation samples as described by Lieth and Pflanz (1968). This is converted into calories fixed per square meter in column 7 by multiplying the figure of column 6 with the figure of column 4 in the first half of the table. Column 8 represents the total estimate of energy fixation for the entire vegetation unit. The total for the land surface is 426×10^{18} cal/year. Estimating marine primary productivity as 55×10^9 t/year, with the caloric equivalents

[1] A new assessment of world net primary productivity was made after this book manuscript had gone to press. In light of new IBP data and those reported and evaluated in Chapters 11, 12, 13, and 15 of this book, a new estimate of 121.7×10^9 t of dry matter for 149×10^6 km^2 of land area was reported (Lieth, 1975). This figure comes even closer to the estimates of the other authors discussed in this paragraph.

Table 10–1   Net primary production and energy fixation for the world (ca. 1950)[a]

1	2	3	4	5	6	7	8
		Net primary productivity			Mean combustion value (kcal/g)	Annual energy fixation	
Vegetation unit	Area (10⁶ km²)	Range (g/m²/year)	Approx. mean	Total production (10⁹ t)		Mean (10⁶ cal/m²)	Total (10¹⁸ cal)
Forest	50.0						277.0
Tropical rain forest	17.0	1000–3500	2000	34.0	4.1	8.2	139.4
Raingreen forest	7.5	600–3500	1500	11.3	4.2	6.3	47.2
Summergreen forest	7.0	400–2500	1000	7.0	4.6	4.6	32.2
Chaparral	1.5	250–1500	800	1.2	4.9	3.9	5.9
Warm temperate mixed forest	5.0	600–2500	1000	5.0	4.7	4.7	23.5
Boreal forest	12.0	200–1500	500	6.0	4.8	2.4	28.8
Woodland	7.0	200–1000	600	4.2	4.6	2.8	19.6
Dwarf and open scrub	26.0		90	2.4			10.2
Tundra	8.0	100–400	140	1.1	4.5	0.6	4.8
Desert scrub	18.0	10–250	70	1.3	4.5	0.3	5.4
Grassland	24.0		600	15.0			60.0
Tropical grassland	15.0	200–2000	700	10.5	4.0	2.8	42.0
Temperate grassland	9.0	100–1500	500	4.5	4.0	2.0	18.0
Desert (extreme)	24.0		1	—			0.1
Dry desert	8.5	0–10	3	—	4.5	—	0.1
Ice desert	15.5	0–1	0	—		—	—
Cultivated Land	14.0	100–4000	650	9.1	4.1	2.7	37.8
Freshwater	4.0		1250	5.0			21.4
Swamp and marsh	2.0	800–4000	2000	4.0	4.2	8.4	16.8
Lake and stream	2.0	100–1500	500	1.0	4.5	2.3	4.6
Total for continents:	149.0		669	100.2			426.1

[a] *Columns 1–8:* [1] Subdivisions are named according to Ellenberg and Müller-Dombois (1967) and Olson (1970). [2] Basically result of the effort of three consecutive groups of geobotany students at the University of North Carolina, Chapel Hill. Adjustments and compromises were made in some cases. [3] Values were deduced from our own compilations of productivity data, with results very similar to those of Whittaker and Woodwell (1971). [4] Original, cf. Whittaker and Likens (1971), Whittaker (1970), and Odum (1971). [5] Product of the positions in columns 2 and 4. All values were rounded off to one decimal point. [6] Original, cf. Jordan (1971a) and Odum (1971). Values of 4.5 computed for reefs, estuaries, and inshore waters; 4.9 for open ocean and upwelling areas. [7] Product of columns 4 and 6. [8] Product of columns 2 and 7.

given in the footnote to Table 10–1, we obtain 261×10^{18} cal/year for the oceans and a total for the world of 687×10^{18} cal/year fixed in net primary productivity. If we consider 610×10^{18} kcal for the total annual solar radiation (full spectrum, at the earth's surface), the total energy fixation averages 0.11% based on 0.06% for the ocean and 0.24% for the land surfaces.

Our world total of 687×10^{18} cal/year coincides well with Golley's (1972) figure of 652×10^{18} cal/year. The two assessments reinforce each other, as our estimates are based for the most part on a different data pool, with overlap in the two calculations occurring only in the tropical regions. Golley relied heavily on a compilation by Cummins and Wuycheck (1971), which was then available only in mimeographed form, and which was published after we had assembled our data for Table 10–1. Our own listing relies heavily on European data already available and on several hundred self-checks during the years 1962–1966 [a thesis by Pflanz (1964) and reports by Velemis, Powell, and Vaasma, mostly unpublished, with the exception of Lieth (1965a) and Lieth and Pflanz (1968)—because the author of this chapter changed continents].

Comparison of the energy figures in Table 10–1 leads to an observation about the adaptations of different vegetation types. Among the forest types, caloric contents are correlated with climate and taxonomic groups. Caloric values are generally higher in temperate than in tropical forests and are higher in gymnosperms than in angiosperms. At the extremes of this range, the combustion values in (angiosperm) tropical rainforests are 20–25% lower than in (gymnosperm) boreal forests. This points to a hypothesis on the success of the angiosperms over the gymnosperms during the last 60 million years (Lieth, 1972; cf. Jordan, 1971b). It is notable that the gymnosperms have, in most temperate areas, been pushed to environments that are marginal (because of aridity, or cold, or infertility) for tree growth and into early successional stages, whereas they have been essentially wiped out of the lowland tropics. Perhaps a key adaptive advantage of the angiosperms is their ability to construct wood with much less expenditure of energy per unit weight.

Other Community Properties

In many papers productivity is considered also in relationship to biomass, assimilatory surface, and chlorophyll content. Ranges of values for these factors have been compiled from the literature as shown in Table 10–2 to support the data on dry-matter productivity and energy binding, and to provide a basis for the biosphere characterization in Chapter 15.

Knowledge of biomass (dry organic matter of organisms present at a given time, sometimes referred to as standing crop, as distinguished from productivity as a rate value) is essential in understanding nutrient pools in organisms as part of the study of nutrient cycling and biogeochemistry. The higher the productivity of a community, the greater the amount that is likely to accumulate as biomass. The correlation is loose, however, and is not widely useful for the estimation of productivity itself. Biomass is much affected by ages of the dominant plants, and these ages differ much in successional communities. Grasslands

10. Primary Production of the Major Vegetation Units of the World

Table 10–2   Biomass of mature stands, leaf area indices, and chlorophyll contents of vegetation units[a,b]

1	2	3	4
Vegetation unit	Mature biomass (kg/m^2)	Leaf area index or assimilating surface (m^2/m^2)	Total chlorophyll (g/m^2)
Tropical rain forest	45 (*1, 9*) 75? (*2*)	6–10–12–16.6 (*1,14*)	3–9 (*14*)
Raingreen forest	42 (*1,9*)	6–7–10 (*1, 11*)	2–? (*11*)
Summergreen forest	42–46 (*4*)	3–12 (*4,14*)	2–6 (*6,14*)
Chaparral	26 (*8*)	4–7–12 (*3,8*)	?
Warm temperate mixed forest	24 (*1*)	5–14 (*1,14*)	3–8 (*14*)
Boreal forest	20–52 (*1*)	7–15 (*1,2,5*)	1.4 (*5*)
Woodland	2–20 (*1,7*)	4.2 (*14*)	?–2 (*11*)
Tundra	0.1–3 (*7*)	0.5–1–1.3 (*12,13*)	0.4–0.6 (*12*)
Desert scrub	0.1–4 (*7*)	?	?
Tropical grassland	?–5	1–5 (*12,14*)	1.7–5 (*14*)
Temperate grassland	?–3	?–5–9–16 (*5,6,17*)	0.6–5 (*5,6*)
Dry desert	0	0	0
Ice desert	0	0	0
Cultivated land (annual crops)	3.5	4–12 (*6,15,16*)	1–5 (*5,6,15,16*)
Swamp and marsh	2.5–? (*10*)	?–11–23.3 (*6,14*)	0.3–4.3 (*5,14*)
Lake and stream	?–0.1 (*7*)	?	0.005–0.12–1.3 (*14*)
Algal mass culture (10-cm layer)			(summer) 10–20 (*14*)
Reefs and estuaries	0.04–4 (*7*)	-	0.1–1.3–?
Continental shelf	0.001–0.04 (*7*)	-	0.02–1.33 (*14*)
Open ocean	?–0.005 (*7*)	-	0.03–0.045 (*7,14*)
Upwelling zones	0.005–0.1? (*7*)	-	0.05–? (14)

[a] *Columns 1–4:* [1] Correspond to Table 10–2. [2] Dry matter, values close to maxima for mature communities of a given type, cf. ranges given by Whittaker and Likens (Whittaker, 1970). [3] Ranges of leaf surface area (m^2/m^2) of ground surface. [4] Ranges of chlorophyll content (g/m^2) of ground surface.

[b] *Sources (indicated in parentheses):* (*1*) Art and Marks (1971); (*2*) Rodin and Bazilevich (1967); (*3*) Martens (1964); (*4*) Lieth (1962); Lieth *et al.* (1965); (*5*) Bray, in Lieth (1962); (*6*) Medina and Lieth (1963, 1964); Medina and San Jose (1970); (*7*) Whittaker (1970); (*8*) Lossaint and Rapp (1971); (*9*) Kira and Ogawa (1971); (*10*) Reader (1971); (*11*) Bandhu (1971); (*12*) Dennis and Tieszen (1971); (*13*) Vareschi (1953); (*14*) Aruga and Monsi (1963); (*15*) Kreh (1965); (*16*) Schultz (1962); (*17*) Geyger (1964). (Secondary literature has been cited whenever possible because of the very large number of primary sources.)

and other fire-susceptible communities, even those that are highly productive, tend to have low biomass compared with other communities. Biomass ranges for terrestrial communities are generally from 0.1 to 5 kg/m^2 in many grasslands, desert scrubs, and tundra communities; 5–20 kg/m^2 in many woodlands (of small trees), shrublands (e.g., chaparral), and young forests; and 20–60 kg/m^2 for many mature forests (see Whittaker, 1966, 1970). Additional world

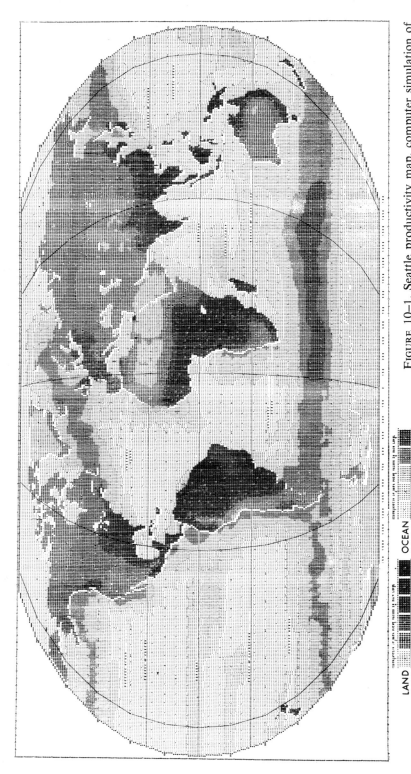

LAND 〈legend swatches〉 OCEAN 〈legend swatches〉

A   B   C   D   E   F   1   2   3   4   5
> 50 50 250 500 1000 2000 <   50 $\frac{100}{200}$ $\frac{200}{400}$ 400
     250 500 1000 2000

Productivity ranges (g/m²/year, dry matter)

FIGURE 10–1. Seattle productivity map, computer simulation of Lieth's (1964) primary productivity map, first computerized cartographic representation of total world primary productivity.

biomass data are given by Bowen (1966), Rodin and Bazilevich (1967), Whittaker (1970), Olson (1970), and Bazilevich and Rodin (1971).

Leaf surface is generally expressed as the "leaf area index," in square meters of leaf surface area over 1 m^2 of ground surface. This is clearly an important dimension related to production, for it defines the leaf area through which the gaseous exchange of photosynthesis must occur. Leaf-area indices are correlated with productivity, but only roughly so, and in a way that does not permit effective prediction of production from the indices. Evergreen communities generally have higher indices than deciduous ones of similar productivity. Most gymnosperm forests have high indices (even after they are divided by two for comparison with broadleaf forests, because surface areas are computed for the whole surface of gymnosperm needles but only for one side of broad leaves). Whereas biomass ranges tend to be more in contrast among different communities than productivity ranges, leaf-area ranges tend to be convergent. A wide variety of communities have leaf-area indices of 3–6 if they are deciduous, or up to 8 (or, for conifer needles, 16) if they are evergreen. Lower values, of course, occur in dry grasslands, desert scrubs, and tundra; and higher values are reported for special systems.

Chlorophyll content may appear to be the community property most directly relevant to the prediction of productivity. The correlation is, again, loose and the use for prediction insecure. Efficiency of energy capture by chlorophyll differs widely within and between communities. Our experience shows that chlorophyll content is rarely at the minimum level (see also Gabrielsen, 1960); chlorophyll may even serve, in some cases, as a shading pigment that prevents other leaves from being overirradiated. Ranges of chlorophyll content are, like leaf-area indices, convergent among communities; the span of 2–4 or 6 g chlorophyll/m^2 of ground surface should include a wide range of more productive communities, and 0.4–2 should include most others except those of extreme environments. Unlike the leaf-area index, chlorophyll can be compared between aquatic and terrestrial communities. Chlorophyll contents of plankton communities are very low (about 1.3 down to 0.05 and even 0.005 g/m^2) compared with terrestrial communities.

Some other community properties have been considered on a worldwide scale; these include gross primary productivity (Golley, 1972), nutrient pools (Rodin and Bazilevich, 1967; Young, 1968; Bazilevich and Rodin, 1971), litter accumulation (Bray and Gorham, 1964) and decomposition rates (Lieth, 1963; Olson, 1963), and albedo (Bray, 1962).

Productivity Mapping

The information summarized in Table 10–1 can be utilized for mapping the productivity of the world. In the first such effort (see Chapter 4), the primary productivity data available at the time were supplemented with agricultural and forest yield data using estimated corrections to community productivity, to produce a world productivity map (Lieth, 1964; published also in Duvigneaud,

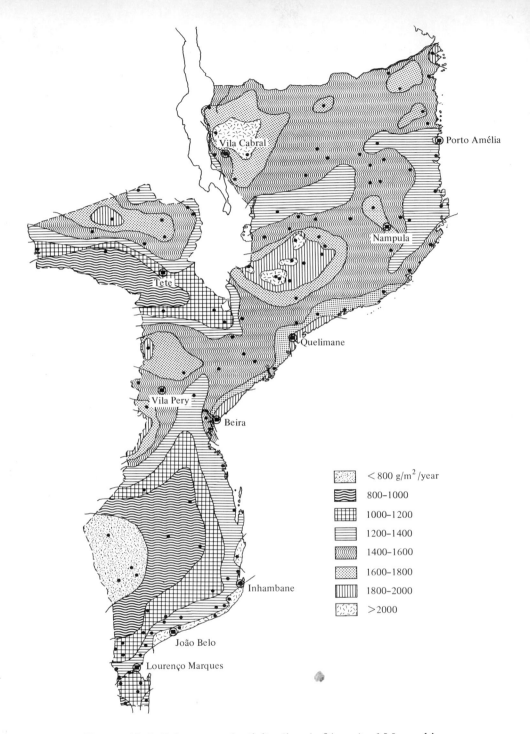

FIGURE 10–2. Primary productivity (in g/m²/year) of Mozambique calculated from the Miami Model (see Chapter 12) (Barreto and Soares, 1972). This map is essentially a converted precipitation map as precipitation is limiting factor at almost all stations. (Scale: approx. 1:10 million.)

1967 and Reichle, 1970). Of maps produced since that time, the most recent world productivity map of Basilevich *et al.* (1970), comes much closer to the total value of our 1964 map than did their original map (Rodin and Bazilevich, 1967). The 1964 map, which gives a world production figure in fair agreement with that of Table 10–1, was used to calculate the carbon exchange between atmosphere and biosphere (Junge and Czeplak, 1968).

A first effort such as this map invites improvement. Not only are more (and in some cases better) productivity measurements available now than in 1964, but also the use of a computer may be most helpful in summarizing, correlating, and interpolating data and in printing out the map itself. Two kinds of productivity maps are feasible: one made on the basis of actual productivity measurements, and the other based on predictions of productivity from environmental data. The map produced along the first of these lines was essentially an updated version of the terrestrial part of the 1964 productivity map, produced as a student project primarily by T. Zaehringer and B. Berryhill. The result is called the Innsbruck Productivity Map, because it was first shown at a 1971 productivity symposium in Innsbruck, Austria (Lieth, 1972) see Figure 12–13, this volume. Since then we have completed with the help of two students, E. Hsiao and P. Van Wyck, a computer simulation of the ocean portions of the 1964 productivity map (Fig. 8–1). The combined ocean productivity and Innsbruck Productivity Map make up the first computer-simulated map of global productivity, the Seattle Productivity Map (presented at the 5th General Assembly of the IBP, August 1972, Seattle) shown in Figure 10–1.

The construction of the Seattle Productivity Map was accomplished by juxtaposing the ocean map and the matching Innsbruck Productivity Map on one sheet. Such overprinting can be accomplished by means of a short program developed by E. Box (see Chapter 13 for details) and R. Lewyckyj. Perhaps a revision of the Seattle Productivity Map will be the basis for validating future predictive models of ocean productivity, to complement the present discussion of the terrestrial models.

In any landscape, primary productivity varies over short distances. Such differences in productivity are inferred from differences in topography and water availability, soil quality, and successional stages. One consequence is that statistical methods must be used to gain regional averages of primary productivity. The first assessment of this nature was made by Filzer (1951) for the agricultural productivity of Central Europe. A second consequence is the need for productivity maps on different scales—from the global and continental, to the regional, down to the local on a scale that may be useful in land management. Recently we have prepared a preliminary assessment of regional productivity in North Carolina (Whigham *et al.*, 1971; see also Chapter 6, this volume). Chapter 6 shows the details of the procedure and some resulting maps for North Carolina. Maps of this kind are now available for Wisconsin (Stearns *et al.*, 1971), Tennessee (DeSelm *et al.*, 1971), and New York and Massachusetts (Art *et al.*, 1971) as discussed in Chapter 7.

The local map that treats the productivity of particular tracts of land as a guide to their use remains to be developed. It cannot be stated that either these or the regional maps will, in the short term, exert much influence over land use in the United States where many still regard land as abundant.

All the world estimates I have given are representative for ca. 1950. The accelerating rate at which the world is being transformed and the biosphere is being affected by man hardly needs emphasis; the final chapter of this book considers further the prospects for the relationship of man and the biosphere. Suffice it to say that by 1980 or 1990 it will hardly be worthwhile to categorize vegetation types in the manner used for the present; the primary production of the world surely will have been altered. Approaches through environmental correlation and modeling like that of the Miami Model map discussed in Chapter 12 may become more appropriate in representing the potential productivity of great areas from which natural vegetation has been or will be displaced. As an example of this approach, Figure 10–2 shows the map of Mozambique (Barreto and Soares, 1972). Other examples are shown in Chapter 7 by the evapotranspiration models used to construct the productivity pattern of the United States. I conclude with the hope that the disparity of actual and potential production may not be so great as to threaten both man and the biosphere.

Acknowledgments

This experimental work on primary productivity and seasonal modeling is sponsored by the Deciduous Forest Biome Project, International Biological Program, funded by the National Science Foundation under Interagency Agreement AG–199, 40–193–69 with the Atomic Energy Commission–Oak Ridge National Laboratory. We acknowledge the technical help of both Diana Watley and Cynthia Grossman in preparing this manuscript; Ms. Watley assisted in the preparation of the "Miami Model."

References

Art, H. W., P. L. Marks, and J. T. Scott. 1971. Productivity profile of New York, 30 pp. Eastern Deciduous Forest Biome Memo Rep. 71-12.

———, and P. L. Marks. 1971. A summary table of biomass and net annual primary production in forest ecosystems of the world, In *Forest Biomass Studies*, H. E. Young, ed., pp. 1–34. (*Int. Union of Forest Res. Organizations Conf., Sect. 25, Gainesville, Florida.*) Orono, Maine: Univ. of Maine, Life Sciences and Agriculture Experiment Station.

Aruga, Y., and M. Monsi. 1963. Chlorophyll amount as an indicator of matter productivity in bio-communities. *Plant Cell Physiol.* 4:29–39.

Bandhu, D. 1971. A study of the productive structure of tropical dry deciduous forest at Varanasi, 111 pp. Ph.D. Thesis. Varanasi: Banares Hindu University.

Barreto, L. S., and F. A. Soares. 1972. Carta provisória da produtividade primária liquida dos ecossistemas terrestres de Moçambique. *Rev. Cienc. Agron. Lourenço Marques* 5:11–18.

Bazilevich, N. I., L. E. Rodin, and N. N. Rozov. 1970. Untersuchungen der biologischen Produktivität in geographischer Sicht. (In Russian.) *5te Tagung Geogr. Ges. USSR, Leningrad.*

Bazilevich, N. I., and L. Ye. Rodin. 1971. Geographical regularities in productivity and the circulation of chemical elements in the earth's main vegetation types. *Sov. Geogr. Rev. Transl.* 12:24–53.

Bowen, H. J. M. 1966. *Trace Elements in Biochemistry*, 241 pp. New York: Academic Press.

Bray, J. R. 1962. The primary productivity of vegetation in central Minnesota, USA and its relationships to chlorophyll content and albedo (German summ.) In *Die Stoffproduktion der Pflanzendecke*, H. Lieth, ed., pp. 102–109. Stuttgart: Fischer.

————, and E. Gorham. 1964. Litter production in forests of the world. *Advan. Ecol. Res.* 2:101–157.

Cummins, K. W., and J. C. Wuycheck. 1971. Caloric equivalents for investigations in ecological energetics. *Mitt. Int. Ver. Limnol.* 18:1–158.

Dennis, J. G., and L. L. Tieszen. 1971. Primary production and nutrient dynamics of tundra vegetation at Barrow, Alaska. In Preliminary report of Project 3111, S. Bowen, ed., 1971 Progr. Rep. Tundra Biome US-IBP. 1:35–37.

DeSelm, H. R., D. Sharpe, P. Baxter, R. Sayres, M. Miller, D. Natella, and R. Umber. 1971. Tennessee productivity profiles. Eastern Deciduous Forest, 182 pp. (mimeogr.) Biome Memo Rep. 71–13.

Duvigneaud, P. (ed.) 1967. *L'écologie, science moderne de synthèse*. In *Écosystèmes et biosphère*, Vol. 2, 135 pp. Brussels, Ministère de l'Éducation Nationale et Culturelle.

Ellenberg, H., and D. Mueller-Dombois. 1967. Tentative physiognomic-ecological classification of plant formations of the earth. *Ber. Geobot. Inst. ETH Stiftg. Rübel, Zürich* 37:21–55.

Filzer, P. 1951. *Die natürlichen Grundlagen des Pflanzenertrages in Mitteleuropa*, 198 pp. Stuttgart: Schweizerbart.

Gabrielsen, E. K. 1960. Chlorophyllkonzentration und Photosynthese. *Handb. Pflanzenphysiol.* 2:156–167. Berlin and Heidelberg: Springer.

Geyger, E. 1964. Methodische Untersuchungen zur Erfassung der assimilierenden Gesamtoberflächen von Wiesen. *Ber. Geobot. Inst. ETH Stiftg. Rübel, Zürich* 35:41–112.

Golley, F. B. 1972. Energy flux in ecosystems. In *Ecosystem Structure and Function*, J. A. Wiens, ed. Corvallis, Oregon: Oregon State Univ. *Ann. Biol. Colloq.* 31:69–90.

Jordan, D. F. 1971a. Productivity of a tropical forest and its relation to a world pattern of energy storage. *J. Ecol.* 59:127–143.

————. 1971b. A world pattern of plant energetics. *Amer. Sci.* 59:425–433.

Junge, C. E., and G. Czeplak. 1968. Some aspects of the seasonal variation of carbon dioxide and ozone. *Tellus* 20:422–434.

Kira, T., and H. Ogawa. 1971. Assessment of primary production in tropical and equatorial forests. In *Productivity of Forest Ecosystems: Proc. Brussels Symp., 1969*, P. Duvigneaud, ed., *Ecology and Conservation*, Vol. 4:309–322. Paris: UNESCO.

Kreh, R. 1965. Untersuchungen über den Aufbau und die Stoffproduktion eines Sonnenblumenbestandes. 71 pp. with table volume. Doctoral thesis, Stuttgart-Hohenheim: Landwirtschftl. Hochschule.

Lieth, H. 1962. *Die Stoffproduktion der Pflanzendecke*, 156 pp. Stuttgart: Fischer.

214

Part 3: Global Productivity Patterns

———. 1963. The role of vegetation in the carbon dioxide content of the atmosphere. *J. Geophys. Res.* 68:3887–3898.

———. 1964. Versuch einer kartographischen Darstellung der Produktivität der Pflanzendecke auf der Erde. *Geographisches Taschenbuch*, 1964/65, pp. 72–80. Wiesbaden: Steiner-Verlag.

———. 1965. Ökologische Fragestellungen bei der Untersuchung der biologischen Stoffproduktion. *Qual. Plant. Mater. Veg.* 12:241–261.

———. 1975. The primary productivity in ecosystems. Comparative analysis of global patterns. In *Unifying Concepts in Ecology*, W. H. Van Dobben and R. H. Lowe-McConnel, eds. The Hague: Junk. (*In press.*)

———. 1972. Über die Primärproduktion der Pflanzendecke der Erde. *Z. Angew. Bot.* 46: 1–37.

Lieth, H., D. Osswald, and H. Martens. 1965. Stoffproduktion, Spross/Wurzel-Verhältnis, Chlorophyllgehalt und Blattfläche von Jungpappeln. *Mitt. Ver. Forstl. Standortsk. Forstpflanz.* 1965:70–74.

———, and B. Pflanz. 1968. The measurement of calorific values of biological material and the determination of ecological efficiency (French summ.). In *Functioning of Terrestrial Ecosystems at the Primary Production Level: Proc. Copenhagen Symp. 1965*, F. E. Eckardt, ed. *Natural Resources Res.* Vol. 5:233–242. Paris: UNESCO.

Lossaint, P., and M. Rapp. 1969. Repartition de la matière organique et cycle du carbone dans les groupements forestiers et arbustifs méditerranéens sempervirents. In *Productivity of Forest Ecosystems: Proc. Brussels Symp. 1969*, P. Duvigneaud, ed. *Ecology and Conservation*, Vol. 4:597–614. Paris: UNESCO.

Martens, H. J. 1964. Untersuchungen über den Blattflächenindex und die Methoden zu seiner Messung. 51 pp. Thesis, Stuttgart: Univ. of Stuttgart.

Medina, E., and H. Lieth. 1963. Contenido de clorofila de algunas asociaciones vegetales de Europa Central y su relacion con la productividad. *Qual. Plant Mater. Veg.* 9:219–229.

———, and H. Lieth. 1964. Die Beziehungen zwischen Chlorophyllgehalt, assimilierender Fläche und Trockensubstanzproduktion in einigen Pflanzengemeinschaften. *Beitr. Biol. Pflanz.* 40:451–494.

———, and J. J. San Jose. 1970. Analisis de la productividad de caña de azucar. II. *Turrialba* 20:149–152.

Odum, E. P. 1971. *Fundamentals of Ecology.* 3rd. ed., 574 pp. Philadelphia: Saunders.

Olson, J. S. 1963. Energy storage and the balance of producers and decomposers in ecological systems. *Ecology* 44:322–331.

———. 1970. Geographic index of world ecosystems. In *Analysis of Temperate Forest Ecosystems*, D. Reichle, ed., *Ecological Studies* 1:297–304. New York: Springer-Verlag.

Pflanz, B. 1964. Der Energiegehalt und die ökologische Energieausbeute verschiedener Pflanzen und Pflanzenbestände. Thesis, 42 pp. Stuttgart: Univ. of Stuttgart.

Reader, R. 1971. Net primary productivity and peat accumulation in Southeastern Manitoba. Master's thesis, 220 pp. Winnipeg: Univ. of Manitoba.

Reichle, D., ed. 1970. *Analysis of Temperate Forest Ecosystems. Ecological Studies* 1, 304 pp. New York: Springer.

Rodin, L. E., and N. I. Bazilevich. 1967. *Production and Mineral Cycling in Terrestrial Vegetation*, 288 pp. Edinburgh: Oliver and Boyd.

Schultz, G. 1962. Blattfläche und Assimilationsleistung in Beziehung zur Stoffproduktion. Untersuchungen an Zuckerrüben. *Ber. Deut. Bot. Ges.* 75:261–267.

Stearns, F., N. Kobriger, G. Cottam, and E. Howell. 1971. Productivity profile of Wisconsin, 35 pp. (with appendices, mimeogr.) Eastern Deciduous Forest Biome Memo Rep. 71-14.

Vareschi, V. 1953. Sobre las superficies de asimilacion de sociedades vegetales de cordilleras tropicales y estratropicales. *Bol. Soc. Venez. Cienc. Nat.* 14:121–173.

Whigham, D., H. Lieth, R. Noggle, and D. Gross. 1971. Productivity profile of North Carolina: Preliminary results, 42 pp. (with appendixes.) Eastern Deciduous Forest Biome Memo Rep. 71-9.

Whittaker, R. H. 1962. Classification of natural communities. *Bot. Rev.* 28:1–239.

————. 1966. Forest dimensions and production in the Great Smoky Mountains. *Ecology* 47:103–121.

————. 1970. *Communities and Ecosystems*, 162 pp. New York: Macmillan.

————, and G. E. Likens. 1969. World productivity estimate, Brussels Symp. 1969, published in Whittaker (1970) and Whittaker and Woodwell (1971).

————, and G. M. Woodwell. 1971. Measurement of net primary production of forests. (French summ.) In *Productivity of Forest Ecosystems: Proc. Brussels Symp. 1969*, P. Duvigneaud, ed. *Ecology and Conservation*, Vol. 4:159–175. Paris: UNESCO.

Young, H. E., ed. 1968. *Symposium on Primary Productivity and Mineral Cycling in Natural Ecosystems*, 245 pp. Orono, Maine: Univ. of Maine.

11

Net Primary Productivity in Tropical Terrestral Ecosystems

Peter G. Murphy

Ranging from lowland evergreen rain forest to alpine tundra, the variety of terrestrial ecosystems lying within tropical latitudes exceeds that of any other region on earth. Our knowledge of net primary productivity (NPP) rates in tropical ecosystems must be described as fragmentary. The relatively few available data pertain to a diverse assortment of samples subject to different levels of precipitation and disturbance.

The published data on organic productivity are dispersed widely in the literature; recent efforts to review and summarize this information have, therefore, been welcome. Notable among the treatments of productivity on a worldwide basis are reviews by Odum and Odum (1959), Pearsall (1959), Lieth (1962), Westlake (1963), Rodin and Bazilevich (1967), Art and Marks (1971), Jordan (1971a), and Lieth (1972, 1973). Productivity in tropical ecosystems has been reviewed by Golley (1972), Golley and Lieth (1972), and Golley and Misra (1972). Other papers concern specific areas within the tropics: India (Misra, 1972), Nigeria (Hopkins, 1962), and the western Pacific region (Kira and Shidei, 1967). A paper by Bourlière and Hadley (1970) on the ecology of savannas reviews the productivity data for that important category of tropical ecosystem.

The objective of this chapter is to present and summarize the available data relating to annual NPP in tropical terrestrial ecosystems, including data too recently collected to have been included in earlier reviews.

KEYWORDS: Net primary productivity; terrestrial vegetation types; tropical region; geoecology.

Primary Productivity of the Biosphere, edited by Helmut Lieth and Robert H. Whittaker.

Available Data

Table 11–1 contains data relative to NPP in a variety of tropical ecosystems. It should be emphasized that each category of tropical ecosystem included in the table is composed of a large variety of subtypes. Tropical grassland, for example, varies from short, sparse herbaceous communities in which bare soil is clearly visible, to tall and dense communities, depending upon local conditions. Because of variation within the categories of ecosystems, and in order to allow a more accurate interpretation of the data, each value of NPP is accompanied by information on the site from which it was obtained. The table includes geographic location in addition to annual rainfall and approximate length of growing season as defined by rainfall pattern when the data were available or could be estimated.

The estimates of total NPP in Table 11–1 are based upon a variety of methods of measurement. In many instances total NPP (aboveground + belowground) had to be estimated from information on some component of the total, such as aboveground NPP in grasslands and leaf-litter production in forests. The factors used in adjusting the original data to obtain total NPP are specified in the footnotes to Table 11–1.

Grassland

The NPP of grasslands varies widely depending on the total annual rainfall and its distribution by seasons. Walter (1954) demonstrated a direct relationship between water availability and aboveground productivity for arid and semiarid desert and grassland in southwest Africa where annual rainfall ranges from 100 to 600 mm. In certain geographic areas, India for example, a prolonged dry season of up to 9 months duration greatly restricts the growing season and consequently the total annual NPP.

Figure 11–1 shows the relationship between total annual rainfall and total annual NPP for tropical grasslands in India, Australia, and Africa. Most of the published reports of productivity in tropical grasslands are based upon the periodic harvesting of aboveground replicated samples and do not include data on belowground parts. Varshney (1972), however, reported that belowground parts accounted for $\sim 40\%$ of total NPP in grassland near Varanasi, India. For lack of more extensive information, Varshney's value is assumed to be representative for tropical grassland; data on aboveground NPP were adjusted accordingly for inclusion in Table 11–1 and Figure 11–1. It is apparent from Fig. 11–1 that grassland productivity on sites that receive less than 700 mm of annual rainfall is low. The lowest value reported is 40 g/m²/year for grassland at Jodhpur, India, for a dry year in which rainfall totaled only 92.7 mm (Gupta *et al.*, 1972). On sites that receive between 700 and 1000 mm of rain annually, total annual NPP ranged from 650 to 3810 g/m²/year. The large variations in NPP within this relatively small range of rainfall may be related to any one or a combination of factors including periodicity of rainfall, rate of evapotranspiration, soil permeability and fertility, species characteristics, and grazing pressure. Of the published data for unirrigated grassland, the maximum site value is

Table 11-1 Total annual net primary productivity (oven-dry basis) in tropical terrestrial ecosystems[a,b]

Ecosystem	Location	Approximate growing season (days)	Annual rainfall (mm)	Total Annual NPP (g/m²/year)	Reference
Desert	Southwest Africa	—	155	200*	Walter (1954)
Grassland, *Dichanthium annulatum*-dominated	Varanasi, India 25°18'N, 83°1'E	120	725	1420	Ambasht et al. (1972)
Grassland, *Dichanthium annulatum*-dominated (protected 1 year from grazing)	Varanasi, India 25°18'N, 83°1'E	120	725	1060	Choudhary (1972)
Grassland, *Dichanthium annulatum*-dominated (protected 3 years from grazing)	Varanasi, India 25°18'N, 83°1'E	120	725	650**	Choudhary (1972)
Grassland	Varanasi, India 25°18'N, 83°1'E	120	725	790*	Singh (1968)
Grassland, *Heteropogon contortus*-dominated	Chakia, Varanasi, India 25°18'N, 83°1'E	150	1000	3810	Ambasht et al. (1972)
Grassland, *Panicum miliare*-dominant, forbs important	Kurukshetra, India 29°58'N, 76°50'E	120	770	2980	Singh and Yadava (1972)
Grassland	Jodhpur, India 26°15'N, 73°3'E	90	289	180*	Gupta et al. (1972)

[a] Adjustments have been made in the original data as indicated by the following symbols:
* Includes an estimated 40% for unmeasured belowground parts (based on the data of Varshney, 1972).
** Includes an estimated 52% for unmeasured belowground parts (based on data of Choudhary, 1972).
† Includes an estimated 30% for unmeasured belowground parts.
‡ Based all or in part on rates of CO_2 exchange.
§ Estimated by multiplying annual leaf litter production or annual leaf litter respiration by three (based on data presented by Bray and Gorham, 1964).
Some of these estimates may be high because in some cases litter samples included small branches.
[b] OD, Oven-dry weight estimated as 50% of fresh weight.

Table 11-1   continued

Ecosystem	Location	Approximate growing season (days)	Annual rainfall (mm)	Total Annual NPP (g/m²/year)	Reference
Grassland, *Heteropogon contortus*-dominated	Delhi, India 28°54'N, 77°13'E	120	800	1330*	Varshney (1972)
Grassland	Udaipur, India 24°32'N, 73°25'E	90	627	300*	Vyas et al. (1972)
Grassland	South West Africa	—	360	520*	Walter (1954)
Grassland, *Andropogon* sp.-dominated	Pretoria, South Africa 25°43'S, 28°16'E	—	607	150*	Bourlière and Hadley (1970)
Grassland	Springbok Flats, South Africa 29°35'S, 17°55'E	—	162	170*	Louw (1968)
Grassland, (native pasture)	Katherine, Australia 14°15'S, 132°20'E	110	660	250*	Norman (1963)
Savanna, *Prosopis* sp.-dominated	Jodhpur, India 26°15'N, 73°3'E	90	361	1450	Bazilevich and Rodin (1966)
Savanna	Rajastan, India 31°20'N, 72°00'	—	610	730	Rodin and Bazilevich (1966)
Savanna, *Dichanthium* sp.-dominated	Gir Forest, India	120	820	680*	Bourlière and Hadley (1970) (data of Hodd)
Savanna, *Themeda* sp.-dominated	Gir Forest, India	120	820	480*	Bourlière and Hadley (1970) (data of Hodd)
Savanna (derived)	Olokemeji, Nigeria 8°57'N, 6°30'E	270	1168	1130*	Hopkins (1965, 1968)
Savanna	Shika, Nigeria 8°57'N, 6°30'E	200	1118	570*	Rains (1963)
Savanna (derived)	Eruja, Ghana 8°N, 2°W (approx)	—	1500	1450*	Nye and Greenland (1960)

Savanna/forest mosaic	Lamto, Ivory Coast 7°43′N, 6°30′W	270	1370	1660*,OD	Roland (1967)
Savanna, *Cenchrus* sp.- and *Chloris* sp.-dominated	Richard-Toll, Senegal 14°53′N, 14°58′W	60	300	70*	Morel and Bourlière (1962)
Savanna, *Aristida papposa*-dominated	Lidney, Chad 17°N, 19°E (approx)	40	320	70*	Gillet (1967)
Savanna, *Cenchrus biflorus*-dominated	Mohi, Chad 17°N, 19°E (approx)	40	320	200*	Gillet (1967)
Savanna, *Brachiaria deflexa*-dominated	Rimé, Chad 17°N, 19°E (approx)	40	320	230*	Gillet (1961)
Savanna, *Cassia tora*-dominated	Tebede, Chad 17°N, 19°E (approx)	40	320	530*	Gillet (1967)
Savanna *Themeda* sp.- and *Heteropogon* sp.-dominated	Matopos, Rhodesia 17°50′S, 29°30′E	—	650	230*	Bourlière and Hadley (1970) (data of West)
Savanna (tall grass)	Serengeti, Tanzania 6°48′S, 33°58′E	—	700	870*	Bourlière and Hadley (1970) (data of Verschuren)
Savanna, *Themeda* sp.- and *Heteropogon* sp.-dominated	Kivu, Albert Park	—	860	800*	Bourlière and Hadley (1970) (data of Verschuren)
Savanna, *Imperata* sp.-dominated	Kivu, Albert Park	—	860	2920*	Bourlière and Hadley (1970) (data of Verschuren)
Savanna, *Trachypogon* sp.-dominated	Llanos of Venezuela	130	1300	530*	Blydenstein (1962)
Savanna (derived)	Costa Rica 10°30′N, 84°30′W	—	2044	2320*	Daubenmire (1972)

[a] Adjustments have been made in the original data as indicated by the following symbols:

* Includes an estimated 40% for unmeasured belowground parts (based on the data of Varshney, 1972).

** Includes an estimated 52% for unmeasured belowground parts (based on data of Choudhary, 1972).

† Includes an estimated 30% for unmeasured belowground parts.

‡ Based all or in part on rates of CO_2 exchange.

§ Estimated by multiplying annual leaf litter production or annual leaf litter respiration by three (based on data presented by Bray and Gorham, 1964).

Some of these estimates may be high because in some cases litter samples included small branches.

[b] OD, Oven-dry weight estimated as 50% of fresh weight.

Table 11-1 *continued*

Ecosystem	Location	Approxi-mate growing season (days)	Annual rainfall (mm)	Total Annual NPP (g/m²/year)	Reference
Savanna (irrigated), *Heteropogon contortus*-dominated	Jhansi, India 25°29′N, 78°32′E	—	—	3400	Ann. Report Indian Forest and Grassland Res. Inst. (1969)
Savanna (irrigated) *Schima* sp.-dominated	Jhansi, India 25°29′N, 78°32′E	—	—	4900	Ann. Report Indian Forest and Grassland Res. Inst. (1969)
Dry deciduous forest	Varanasi, India 25°18′N, 83°1′E	120	1040	1550	Bandhu (1971)
Seasonal forest	Ivory Coast 5°N, 5°W (approx)	—	1500	1340	Müller and Nielsen (1965)
Mixed dry forest	Ibadan, Nigeria 7°26′N, 3°48′E	—	1230	1140§	Madge (1965)
Deciduous forest in Savanna	Calabozo Plains, Venezuela 8°48′N, 67°27′W	—	1200	2460§	Medina and Zelwer (1972)
Rain forest	Thailand 7°35′N, 99°00′E	365	> 2000	2860	Kira *et al.* (1967)
Montane (1460 m) rain forest	Tjibodas, Java 7°S, 107°E (approx)	365	> 2000	2430§	Wanner (1970)
Lowland rain forest	Sarawak 3°N, 112°E (approx)	365	3800	3210§	Wanner (1970)
Rain forest	Ivory Coast 5°N, 5°W (approx)	365	—	2460§	Bernhard-Reversat *et al.* (1972)
Rain forest	Ghana 8°N, 2°W (approx)	365	—	2430	Nye (1961)
Rain forest	Congo 3°S, 13°48′E	365	—	3150	Bartholomew *et al.* (1953)

Rain forest	Manaus, Brazil 3°01'S, 60°00'W	365	1800	1680§	Klinge (1968)
Evergreen Cloud Forest (1000 m)	Rancho Grande, Venezuela 5°N, 65°W (approx)	365	1750	2340§	Medina and Zelwer (1972)
Montane (500 m) rain forest, *Dacryodes excelsa*-dominated	El Verde, Puerto Rico 18°N, 66°W (approx)	365	3800	1030	Jordan (1971b)
Montane (500 m) rain forest, *Dacryodes excelsa*-dominated	El Verde, Puerto Rico 18°N, 66°W (approx)	365	3800	1230‡	H. T. Odum and Jordan (1970) and Odum (1970)
Montane Successional Rain Forest (500 m)	El Verde, Puerto Rico 18°N, 66°W (approx)	365	3800	540	Jordan (1971b)
Montane rain forest, *Goethalsia meiantha*-dominated	Costa Rica 10°30'N, 84°30'W	365	—	350‡	(This is net *ecosystem* production.) Lemon *et al.* (1970)
Bamboo Brake in monsoon forest, *Oxytenanthera albociliata*-dominated	Burma 23°N, 95°E (approx)	—	—	2780†	Rozanov and Rozanov (1964)
Bamboo Brake in rain forest, *Dendrocalamus brandisii*-dominated	Burma 23°N, 95°E (approx)	—	—	2300†	Rozanov and Rozanov (1964)
Bamboo Brake (arid area) *Dendrocalamus strictus*-dominated	Burma 23°N, 95°E (approx)	—	—	1530†	**Rozanov and Rozanov** (1964)
Mangrove, *Rhizophora mangle*-dominated	Puerto Rico 18°N, 67°W (approx)	365	—	930‡	Golley *et al.* (1962)

[a] Adjustments have been made in the original data as indicated by the following symbols:

    * Includes an estimated 40% for unmeasured belowground parts (based on the data of Varshney, 1972).

    ** Includes an estimated 52% for unmeasured belowground parts (based on the data of Choudhary, 1972).

    † Includes an estimated 30% for unmeasured belowground parts.

    ‡ Based all or in part on rates of CO_2 exchange.

    § Estimated by multiplying annual leaf litter production or annual leaf litter respiration by three (based on data presented by Bray and Gorham, 1964).

    Some of these estimates may be high because in some cases litter samples included small branches.

[b] OD, Oven-dry weight estimated as 50% of fresh weight.

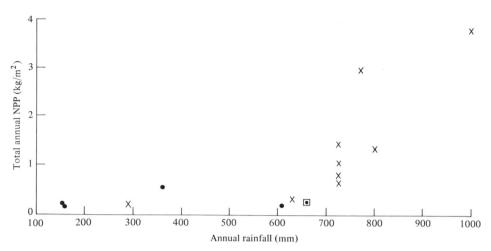

FIGURE 11–1. Relationship between annual rainfall and total annual NPP in dry matter for tropical grasslands. ●, Africa; ✕, India; and □, Australia.

3810 g/m²/year measured in a successional *Heteropogon contortus*-dominated grassland near Varanasi, India; it is estimated to receive in excess of 1000 mm of rainfall annually, most of which is distributed over a 3-month period (Ambasht *et al.*, 1972). This exceptionally productive grassland becomes very dense and tall (> 1.5 m) when protected from grazing (Ambasht, *personal communication*). In many areas of India where annual rainfall is high, woodland or forest is the ultimate end point of succession, but grassland is maintained by the pressures of grazing and other disturbances. Such successional grasslands appear to be the most productive. Based on 11 representative samples, the average total annual NPP for tropical grassland is estimated to be 1080 g/m²/year.

Savanna

Bourlière and Hadley (1970) define savanna as ". . . a tropical formation where the grass stratum is continuous and important but is interrupted by trees and shrubs; the [grass] stratum is burnt from time to time, and the main growth patterns are closely associated with alternating wet and dry seasons." Figure 11–2 plots total annual NPP as a function of annual rainfall for savannas in India, Venezuela, Costa Rica, and Africa. As indicated in Fig. 11–2, savanna may exist in areas that receive an annual amount of rainfall as low as that received in some grassland areas. The dramatic effects of irrigation on two savanna areas in Jhansi, India, is apparent in Figure 11–2. The irrigated savannas in India are three to four times as productive as unirrigated savannas in that country.

As in the case of grassland, total NPP was estimated from data on above-ground productivity. Because most of the productivity estimates for savanna were based on measurements of peak standing crop (trees excluded), the esti-

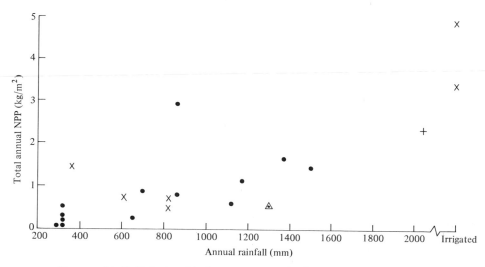

FIGURE 11–2. Relationship between annual rainfall and total annual NPP in Savanna. ●, Africa; ×, India; +, Costa Rica; and △, Venezuela.

mates of total NPP are assumed to be low. Mathews and Westlake (1969) demonstrated that actual NPP may exceed peak standing crop by a factor of 1.5–3.5 in communities with high rates of turnover. The data in Table 11–1 and Fig. 11–2 were not adjusted for this error.

The maximum site value for tropical savanna, excluding irrigated areas, is 2920 g/m²/year for a sample in the Congo where annual rainfall averages 860 mm (Bourlière and Hadley, 1970). The minimum site value is 70 g/m²/year for savannas in Chad (Gillet, 1967) and Senegal (Morel and Bourlière, 1962) that receive only ~ 300 mm of rainfall annually. On the basis of 19 representative samples, the average annual NPP for unirrigated savanna is estimated to be 890 g/m²/year.

Seasonal forest (Raingreen forest)

Forests displaying conspicuous seasonal properties, such as leaf fall and temporary cessation of growth, are found throughout the tropics wherever seasonal drought alternates with relatively wet periods. All or just a few of the tree species of a given stand may show seasonal properties, depending upon the duration of the dry period.

Forest productivity has been estimated allometrically and from measurements of annual litter fall, annual leaf fall, and rates of litter respiration. Based upon data presented by Bray and Gorham (1964) annual leaf fall is considered to represent one-third of total annual NPP. The minimum value of NPP for seasonal forest is 1140 g/m²/year, estimated from data on leaf-litter production for a forest at Ibadan, Nigeria (Madge, 1965). Based on litter-production measurements by Medina and Zelwer (1972) in deciduous forest patches in the savanna of the Calabozo Plains of Venezuela, the maximum site value is 2460 $g/m^2/year$. Average NPP, based on only four samples, is 1620 $g/m^2/year$.

Evergreen rain forest

Forests receiving abundant year-round rainfall and lacking distinct seasonality in leaf fall are, on the average, the most productive of any of the tropical terrestrial ecosystems measured to date. The forests grouped in this category range from lowland types to montane types. Total annual NPP, based on nine representative samples, averages 2400 $g/m^2/year$. The maximum site value is 3210 $g/m^2/year$ for lowland forest in Sarawak, estimated from rates of litter respiration measured by Wanner (1970). From this value productivities range far downward to the value of 540 $g/m^2/year$ for a successional montane rain forest.

Productivity of tropical rain forest has been of special interest, and some very high estimates have been published. It should be kept in mind that extensive areas on tropical podzol soils are apparently of low productivity (Janzen, 1974) and that in some montane forests productivity may be limited by high precipitation and humidity, with intense soil leaching and restricted transpiration. The mean of 2400 $g/m^2/year$ seems a reasonable value, but even this could be revised downward. If the 3:1 ratio of total to litter productivity used for some values in Table 11–1 is too high (as may well be the case in these forests), a ratio of 2.5:1 reduces the mean for the same nine samples to 2170 $g/m^2/year$. Using a 2.0:1 ratio reduces the mean to 1960 $g/m^2/year$; adding two more montane samples to the set gives an 11-sample mean of 2120 $g/m^2/year$, with the high ratio of 3:1, or of 1930 $g/m^2/year$ with the ratio of 2.5:1. Brünig (1974) estimates 2100 $g/m^2/year$, aboveground. A more accurate mean will depend not only on more reliable measurements, but on a weighting of different kinds of rain forests with different productivities by their relative areas.

Two studies have attempted to measure the integrated metabolism of tropical rain forest. H. T. Odum and Jordan (1970) measured rates of CO_2 exchange in a lower montane rain forest in Puerto Rico. They estimated total daily respiration to be 16.4 g C/m^2. The ecosystem was considered to be near steady state, and total gross photosynthesis was therefore assumed to be equal to total respiration. NPP was estimated as 1230 $g/m^2/year$ by subtracting autotrophic respiration from gross photosynthesis. This value agrees reasonably with the allometrically derived value of 1030 $g/m^2/year$ of Jordan (1971b) for the same site. A 3-year-old successional rain forest in the same area of Puerto Rico was found to have a total annual NPP of 540 $g/m^2/year$ (Jordan, 1971b). Lemon et al. (1970) measured rates of CO_2 exchange in a 50-year-old rain forest in Costa Rica and found that net *ecosystem* production equaled 350 $g/m^2/year$, indicating that the ratio of gross photosynthesis to total respiration in that particular ecosystem was greater than unity.

Bamboo brake

Total annual NPP in three bamboo brakes occurring in forest openings in Burma was high, ranging from 1530 to 2780 $g/m^2/year$ and averaging 2200 $g/m^2/year$ (based on the data of Rozanov and Rozanov, 1964). In arriving at these estimates it was assumed that the aboveground productivity, which was the only portion measured, accounted for 70% of total NPP.

Mangrove

Total NPP in a *Rhizophora mangle*-dominated ecosystem in southeastern Puerto Rico was estimated to be 930 g/m²/year. The estimate is based on rates of CO_2 exchange reported by Golley *et al.* (1962).

Summary

Figure 11–3 summarizes rates of NPP in tropical terrestrial ecosystems. Mean levels of productivity range widely, from 200 g/m²/year in desert to 2000–2400 g/m²/year in evergreen rain forest. The wide range in values of NPP exhibited within several categories of ecosystems, particularly savanna, grassland, and evergreen rain forest, represent, in part, a probable result of differences in methods used in measuring NPP. But the variation also reflects the pooling of widely varying communities into very general categories. As more data become available these general categories should be partitioned into more meaningful subtypes. Nevertheless, the available data are sufficient to give an idea of the order of magnitude of NPP in a variety of tropical terrestrial ecosystems.

Comparisons of NPP between ecosystems of different climatic areas are difficult, but it does appear that tropical ecosystems are more productive on an annual basis than their temperate counterparts. Tropical forests as a whole, with a mean annual NPP of 2160 g/m²/year, exceed temperate forests, averaging 1300 g/m²/year, by a factor of 1.7; boreal forests, averaging only 800 g/m²/year, by a factor of 2.7. The annual NPP of tropical grassland (much of it in

FIGURE 11–3. Average annual net primary productivity (dry matter) in various tropical ecosystems. ●, Mean (or individual value); —, range; and ☐, standard error of the mean.

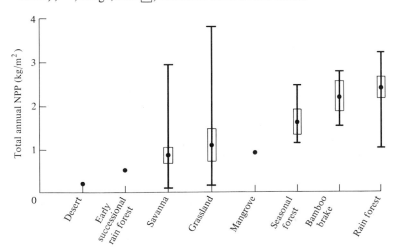

forest climates), averaging 1080 g/m²/year, exceeds that of temperate grass-land, averaging 500 g/m²/year, by a factor of 2.2. The mean NPP of tropical savannas (890 g/m²/year), which may be more directly comparable with temperate grasslands, is 1.8 times that of the latter. The estimates of temperate rates of NPP are those of Whittaker and Likens in Whittaker and Woodwell (1971). Further discussions of correlations between NPP and environmental parameters follow in Chapter 12.

References

Ambasht, R. S., A. N. Maurya, and U. N. Singh. 1972. Primary production and turn-over in certain protected grasslands of Varanasi, India, In *Tropical Ecology with an Emphasis on Organic Production*, P. M. Golley and F. B. Golley, eds., pp. 43–50. Athens, Georgia: Univ. of Georgia.

Annual report of the Indian Forest and Grassland Research Institute at Jhansi. 1969. (*Unpublished data.*)

Art, H. W., and P. L. Marks. 1971. A summary of biomass and net annual primary production in forest ecosystems of the world. In *Forest Biomass Studies*, H. E. Young, ed., pp. 1–34. (*Internat. Union Forest Research Organizations Conf., Sect. 25. Gainesville, Florida.*) Orono, Maine: Univ. of Maine.

Bandhu, D. 1971. A study of the productive structure of tropical dry deciduous forest at Varanasi. Ph.D. thesis. Banaras: Banaras Hindu Univ.

Bartholomew, M. V., J. Meyer, and H. Laudelot. 1953. Mineral nutrient immobilization under forest and grass fallow in the Yangambi (Belgian Congo) region. *Publ. INEAC Ser. Sci.* 57:27 p.

Bazilevich, N. I., and L. E. Rodin. 1966. The biological cycle of nitrogen and ash elements in plant communities of the tropical and sub-tropical zones. *Forestry Abstr.* 27:357–368.

Bernhard-Reversat, F., C. Huttel, and G. Lemee. 1972. Some aspects of the seasonal ecologic periodicity and plant activity in an evergreen rain forest of the Ivory Coast. (In French with English summ.), In *Tropical Ecology with an Emphasis on Organic Production*, P. M. Golley and F. B. Golley, eds., pp. 217–218; 219–234. Athens, Georgia: Univ. of Georgia.

Blydenstein, J. 1962. The *Trachypogon* savanna in the high plains. Ecological study of the area surrounding Calatgo, Guarico State, Venezuela. *Bol. Soc. Venez. Cienc. Nat.* 23:139–206.

Bourlière, F., and M. Hadley. 1970. The ecology of tropical savannas. In *Ann. Rev. Ecol. Syst.* Vol. 1, pp. 125–152.

Bray, R., and E. Gorham. 1964. Litter production in forests of the world. In *Advan. Ecol. Res.*, J. B. Cragg, ed., Vol. 2, pp. 101–157. New York: Academic Press.

Brünig, E. F. 1974. Ökosysteme in den Tropen. *Umschau* 74(13):405–410.

Choudhary, V. B. 1972. Seasonal variation in standing crop and net above-ground production in *Dichanthium annulatum* grassland at Varanasi. In *Tropical Ecology with an Emphasis on Organic Production*, P. M. Golley and F. B. Golley, eds., pp. 51–57. Athens, Georgia: Univ. of Georgia.

Daubenmire, R. 1972. Standing crops and primary production in savanna derived from semideciduous forest in Costa Rica. *Bot. Gaz.* 133:395–401.

Gillet, H. 1961. Pâturages sahéliens de Ranch de l'ouadi rimé. *J. Agr. Trop. Bot. Appl.* 8:465–536.

————. 1967. Essai d'évaluation de la biomasse végétale en zone sahélienne (végétation annuelle). *J. Agr. Trop. Bot. Appl.* 14:123–258.

Golley, F. B. 1972. Summary. In *Tropical Ecology with an Emphasis on Organic Production*, P. M. Golley and F. B. Golley, eds., pp. 407–413. Athens, Georgia: Univ. of Georgia.

————, and H. Lieth. 1972. Bases of organic production in the tropics. In *Tropical Ecology with an Emphasis on Organic Production*, P. M. Golley and F. B. Golley, eds., pp. 1–26. Athens, Georgia: Univ. of Georgia.

————, and R. Misra. 1972. Organic production in tropical ecosystems. *BioScience* 22:735–736.

Golley, F., H. T. Odum, and K. Wilson. 1962. The structure and metabolism of a Puerto Rican red mangrove forest in May. *Ecology* 43:9–19.

Gupta, R. K., S. K. Saxena, and S. K. Sharm. 1972. Above-ground productivity of grasslands at Jodhpur, India. In *Tropical Ecology with an Emphasis on Organic Production*, P. M. Golley and F. B. Golley, eds., pp. 75–93. Athens, Georgia: Univ. of Georgia.

Hopkins, B. 1962. Biological productivity in Nigeria. *Sci. Assoc. Nigeria Proc.* 1/3: 20–28.

————. 1965. Observations on savanna burning in the Olokemeji Forest Reserve, Nigeria. *J. Appl. Ecol.* 2:367–381.

————. 1968. Vegetation of the Olokemeji Forest Reserve, Nigeria. V. The vegetation of the savanna site with special reference to its seasonal changes. *J. Ecol.* 56:97–115.

Janzen, D. H. 1974. Tropical blackwater rivers, animals, and mast fruiting by the Dipterocarpaceae. *Biotropica* 6:69–103.

Jordan, C. F. 1971a. A world pattern in plant energetics. *Amer. Sci.* 59:425–433.

————. 1971b. Productivity of a tropical rain forest and its relation to a world pattern of energy storage. *J. Ecol.* 59:127–142.

Kira, T., and T. Shidei. 1967. Primary production and turnover of organic matter in different forest ecosystems of western Pacific. *Jap. J. Ecol.* 17:70–87.

————, H. Ogawa, K. Yoda, and K. Ogino. 1967. Comparative ecological studies in three main types of forest vegetation in Thailand. IV. Dry-matter production, with special reference to the Khao Chong rain forest. *Nature Life SE Asia* 5:149–174.

Klinge, H. 1968. Litter production in an area of Amazonian terra firma forest. I. Litter-fall, organic carbon and total nitrogen contents of litter. *Amazonia* 1(4): 287–301.

Lemon, E. R., L. H. Allen, and L. Muller. 1970. Carbon dioxide exchange of a tropical rain forest. II. *BioScience* 20:1054–1059.

Lieth, H. 1962. *Die Stoffproduktion der Pflanzendecke*. 156 p. Stuttgart: Fischer.

————. 1972. Über die Primärproduktion der Pflanzendecke der Erde. *Zeit. Angew. Bot.* 46:1–37.

————. 1973. Primary production: Terrestrial ecosystems. *Human Ecol.* 1:303–332.

Louw, A. J. 1968. Fertilizing natural veld on red loam soil of the Springbok flats. 2. Effect of sulphate of ammonia and superphosphate on air-dry yield and mineral content. *S. Afr. J. Agr. Sci.* 11:629–636.

Madge, D. S. 1965. Leaf fall and litter disappearance in a tropical forest. *Pedobiologia* 5:273–288.

Mathews, C. P., and D. F. Westlake. 1969. Estimation of production by populations of higher plants subject to high mortality. *Oikos* 20:156–160.

Medina, E., and M. Zelwer. 1972. Soil respiration in tropical plant communities. In *Tropical Ecology with an Emphasis on Organic Production*, P. M. Golley and F. B. Golley, eds., pp. 245–267. Athens, Georgia: Univ. of Georgia.

Misra, R. 1972. A comparative study of net primary productivity of dry deciduous forest and grassland of Varanasi, India. In *Tropical Ecology with an Emphasis on Organic Production*, P. M. Golley and F. B. Golley, eds., pp. 279–293. Athens, Georgia: Univ. of Georgia.

Morel, G., and F. Bourlière. 1962. Relations écologiques des avifaunes sédentaires et migratices dans une savane sahélienne du bas Sénégal. *Terre et Vie* 16:371–393.

Müller, D., and J. Nielsen. 1965. Production brute, pertes par respiration et production nette dans la forêt ombrophile tropicale. *Forstl. Forsøgsv. Danm.* 29:69–160.

Norman, M. J. T. 1963. The pattern of dry matter and nutrient content changes in native pastures at Katherine, N.T. *Austral. J. Exp. Agr. Animal Husbandry* 3:119–124.

Nye, P. H. 1961. Organic matter and nutrient cycles under moist tropical forest. *Plant Soil* 8:333–346.

———, and D. J. Greenland. 1960. The soil under shifting cultivation. *Tech. Commonw. Bur. Soils* 51:1–156.

Odum, E. P., and H. T. Odum. 1959. *Fundamentals of Ecology*, Chapter 3, 546 pp. Philadelphia, Pennsylvania: Saunders.

Odum, H. T. 1970. Summary: An emerging view of the ecological system at El Verde. In *A Tropical Rain Forest*, H. T. Odum and R. F. Pigeon, eds., Vol. I, pp. 191–289. Washington, D.C.: U.S. Atomic Energy Commission, Div. of Technical Information.

———, and C. F. Jordan. 1970. Metabolism and evapotranspiration of the lower forest in a giant plastic cylinder. In *A Tropical Rain Forest*, H. T. Odum and R. F. Pigeon, eds., Vol. I, pp. 165–189. Washington, D.C.: U.S. Atomic Energy Commission, Div. of Technical Information.

Pearsall, W. H. 1959. Production ecology. *Sci. Progr.* 47:106–111.

Rains, A. B. 1963. Grassland research in northern Nigeria 1952–1962. *Misc. Pap. Inst. Agr. Res. Samaru* 1:1–67.

Rodin, L. E., and N. I. Bazilevich. 1967. *Production and Mineral Cycling in Terrestrial Vegetation*, 288 pp. London: Oliver and Boyd.

Roland, J.-C. 1967. Recherches écologiques dans la savane de Lamto (Côte d'Ivoire): Données préliminaires sur le cycle annuel de la végétation herbacée. *Terre et Vie* 21:228–248.

Rozanov, B. G., and I. M. Rozanova. 1964. The biological cycle of nutrient elements of bamboo in the tropical forests of Burma. *Bot. Zh.* 49(3):348–357.

Singh, J. S. 1968. Net above-ground community productivity in the grasslands at Varanasi. In *Proc. Symp. Rec. Advan. Trop. Ecol. 2*, R. Misra and B. Gopal, eds., pp. 631–653. Varanasi: International Society of Tropical Ecology.

———, and P. S. Yadava. 1972. Biomass structure and net primary productivity in the grassland ecosystem at Kurukshetra. In *Tropical Ecology with an Emphasis on Organic Production*, P. M. Golley and F. B. Golley, eds., pp. 59–74. Athens: Georgia: Univ. of Georgia.

Varshney, C. K. 1972. Productivity of Delhi grasslands. In *Tropical Ecology with an Emphasis on Organic Production*, P. M. Golley and F. B. Golley, eds., pp. 27–42. Athens: Georgia: Univ. of Georgia.

Vyas, L. N., R. K. Garg, and S. K. Agarwal. 1972. Net above-ground production in the monsoon vegetation at Udaipur. In *Tropical Ecology with an Emphasis on Organic Production*, P. M. Golley and F. B. Golley, eds., pp. 95–99. Athens, Georgia: Univ. of Georgia.

Walter, H. 1954. Le facteur eau dans les régions arides et sa signification pour l'organisation de la végétation dans les contrées sous-tropicales. In *Colloques Internationaux des Centre National de la Recherche Scientifique*. Vol. 59, pp. 27–39, Paris: Centre National dela Recherche Scientifique. Les Divisions Ecologiques du Monde.

Wanner, H. 1970. Soil respiration, litter fall and productivity of tropical rain forest. *J. Ecol.* 58:543–547.

Westlake, D. F. 1963. Comparisons of plant productivity. *Biol. Rev.* 38:385–425.

Whittaker, R. H., and G. M. Woodwell. 1971. Measurement of net primary production of forests. (French summ.) In *Productivity of Forest Ecosystems: Proc. Brussels Symp. 1969*, P. Duvigneaud, ed. *Ecology and Conservation*, Vol. 4:159–175. Paris: UNESCO.

Part 4

Utilizing the Knowledge of
Primary Productivity

The incentives for measuring the productivity pattern on earth vary among the scientists involved. Providing a base level for ecosystems research was one, uncovering the relationship between productivity and environmental factors a second, and investigating the use of the produced matter by man still another. This section consists of four chapters that illustrate the use of productivity data. Our selection of such examples is narrow, and is guided by our own interests.

Chapters 12 and 13 deal with the modeling and computerized evaluation of the productivity pattern on earth. This type of model is based on first correlating productivity with environmental parameters, then using environmental measurements to predict local productivities, and after that displaying these in the productivity pattern as a map and integrating from the map the production of geographic areas and the world. The first models of this type by Lieth have received a wide distribution (see literature list of Chapter 12); we hope its illustration here will encourage others to use this effective approach.

Chapter 14 deals with the transfer of assimilates to chemical and structural categories and its use for evolution research and theoretical considerations about ecosystems. This chapter should be regarded as a projection toward future research. It is perhaps in this area that future research will uncover maximization and optimization principles for ecosystems that are unrecognized at present.

Chapter 15 summarizes the global productivity pattern and considers man's use in relationship to it. No question is more important to man than the adequacy of the productivity that feeds him. Although it is true that man's population is, as a whole, poorly fed, the mere ratio of food calories available to human needs inadequately states the problem. The composition of food, particularly protein content, can be crucial. Moreover, the essential question is not the present ratio of food to need, but the trend of that ratio, and the more profound question of the stability of civilized populations. If man can achieve a stable population, wise management of the biosphere will be necessary to provide its food. Understanding of the biosphere will be the necessary basis of that management. We should like to hope, despite discouraging signs, that a management of the earth's living mantle that is both wise and conserving will one day come about, and that such knowledge of that mantle's productivity as we summarize here may contribute to it.

12 Modeling the Primary Productivity of the World

Helmut Lieth

The many problems of energy and nutrient flow and their relationship to the structure of communities and potential for harvest make primary productivity interesting. The correlation between the productivity and character of vegetation cover, and the potential for agriculture and the environmental aspects of cultural development, have created additional interest. This volume emphasizes the fact that assessment of primary productivity is a time-consuming and expensive procedure. In some cases, it is even logistically impossible to measure the current productivity rate directly. Under such circumstances, one is inclined to look for indirect ways to estimate the productive capacity of any given region. The most feasible approach to the task is the elaboration of models that predict productivity from environmental parameters that have been measured in a reasonably dense network over the world.

We present and discuss three of the various attempts to build such predictive models. The first model predicts productivity from annual precipitation and temperature averages. This model was first presented at the 1971 Miami symposium and was publicized in summary form as the Miami Model (Lieth, 1972b, 1973). The development of this model is described in detail in this chapter.

The second model predicts productivity from actual annual evapotranspiration. This model was developed for the C. W. Thornthwaite Memorial Symposium at the 22nd International Geographical Congress in Montreal in 1972. Several maps were produced using various data pools. This set of models and maps was called the Montreal Models, and the second of these was named the

KEYWORDS: Primary productivity; terrestrial ecosystems; global modeling; environmental correlation models; precipitation; temperature; evapotranspiration; vegetation period; computer model; global pattern.

Primary Productivity of the Biosphere, edited by Helmut Lieth and Robert H. Whittaker.

Part 4: Utilizing the Knowledge of Primary Productivity

C. W. Thornthwaite Memorial Model. The models appear in Lieth and Box (1972).

In the third approach, we used the correlation of vegetation period to productivity previously suggested by Gessner (1959) and Lieth (1962, 1965b).

The Miami Model: Primary Productivity Predicted from Annual Precipitation and Temperature Averages

As we have said, primary productivity is dependent on a number of environmental conditions, and the first in order among these on land are temperature and available water. If one can establish valid correlation models between these factors and the primary productivity, a worldwide data pool of meteorologic records can be utilized. The relationships among precipitation, temperature, and vegetation types were demonstrated several years ago (Lieth, 1956; Lieth and Zauner, 1957). Major vegetation types show different relationships to these climatic variables (Fig. 12–1). The types overlap with regard to annual precipitation and temperature because of the effects of other factors (nutrients and other soil characteristics, fire, continentality, floristic history) that may also affect productivity. Other studies (Lieth, 1961–1968) have established effects of rainfall and temperature on primary productivity, and a logical approach is to start modeling with these as principal factors. In order to model the terrestrial productivity on a global scale, we need the following data sets and models.

1. A set of representative productivity data paired with the environmental parameters of interest
2. A model to convert the numerical values of each environmental parameter into productivity values
3. A computer map with an adequate and even distribution of environmental datum points
4. An information system to combine the model output from each environmental parameter for each datum point on the map

The fourth step yields a map model that predicts the primary productivity of the world from the environmental parameters we utilized. Such a map may then be compared with a productivity map constructed from other information.

The set of productivity data

Our modeling exercises are based on the compilation shown in Table 12–1, with selected productivity data paired with average annual temperature values and average annual precipitation totals from nearby meteorologic stations. The productivity data were derived from recent publications, most of which we had not used for earlier assessments. The meteorologic data were taken from our *Climate Diagram World Atlas* (Walter and Lieth, 1960–1967). We grouped the data from the northern hemisphere into four transects, each covering tundra to tropics: (1) North and South America, (2) Europe and Africa, (3) Russia

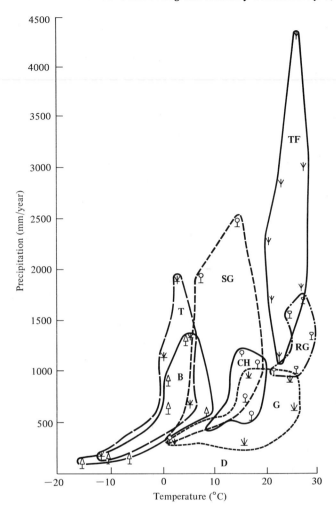

FIGURE 12–1. Graphic model of relationships among temperature, precipitation, and vegetation formation classes, adapted from Lieth (1956). Abscissa, mean annual temperature in degrees centigrade; ordinate, mean annual precipitation in millimeters. Curves circumscribe meteorologic stations corresponding to given vegetation type. (—) Enclose evergreen types: B, Boreal forest; CH, chaparral; TF, tropical rain forest. (- - -) Enclose seasonally green vegetation types: T, tundra; SG, summergreen (deciduous) forest; RG, raingreen forest; G, grassland; D, desert and semidesert.

and Asia Minor, and (4) East and Southeast Asia. A fifth region was added, from Southwest Africa, to test previous assumptions. The geographic distribution of the datum points is presented in Figure 12–2. From Table 12–1 we attempted a correlation model that could be used to predict the primary productivity from precipitation and temperature values (see Figs. 12–3 and 12–4).

Table 12–1   Selected productivity data for modeling relations to climate[a]

1 Station name	2 Climate-diagram code number		3 Mean annual temp. (C°)	4 Mean annual precipitation (mm)	5 Annual total dry matter productivity[b] (g/m²)	6 Source
Region I						
1 Barrow	Supplement sheet		−12.2	104	100–450	Lieth (1962)
2 Kehora	401	030	2.3	641	710	Reader (1971)
3 Kehora	401	030	2.3	641	990	Reader (1971)
4 Kehora	401	030	2.3	641	1629	Reader (1971)
5 Knoxville	401	305	15.2	1156	2408	Whittaker and Woodwell (1970)
6 New Bern	401	312	17.5	1409	1280	Nemeth (1971)
7 Raleigh	401	313	15.5	1145	1900	Wells and Lieth (1970)
8 Coweeta/N.C.	—		12.2	1800	1203	Whittaker and Woodwell (1970)
9 Iuncos/P.R.	403	007	24.8	1697	1033*	Jordan (1971)
10 Osa/C.R.	—		25.0	4500	1067*	Ewel (1971a)
11 Darien/R.P.	—		27.0	2000	2106*	Ewel (1971b)
12 Calabozo/V.[a]	500	075	27.1	1334	1100	Medina (1970)
Region II						
1 Abisco	107	497	−1.0	267	450*	Lieth (1962)
2 Lund	105	088	7.3	616	1560	Art and Marks (1971)
3 Søborg	105	102	7.5	585	1350	Lieth (1962)
4 Gembloers	103	317	9.2	816	1440	Art and Marks (1971)
5 Heilbronn	106	100	9.7	675	1350	Lieth (1962)
6 Heilbronn	106	100	9.7	675	1270	Lieth (1962)
7 Murrhardt	106	305	8.2	951	2300	Lieth et al. (1965)
8 Lorch	106	475	7.4	823	880	Lieth (1962)
9 Nördlingen	106	325	7.9	634	830	Lieth (1962)
10 Nördlingen	106	325	7.9	634	1050	Lieth (1962)
11 Wielicka	106	385	7.8	686	1017	Walter (1968)
12 Luzern	107	225	8.6	1121	980	Art and Marks (1971)
13 Bondaye	303	510	26.5	1633	1340	Walter (1968)

[a] *Columns 1–6:* [1] The name of the climate record station. Calabozo value used differs from that in the climate diagram. [2] The climate-diagram code number from Walter–Lieth Climate Diagram World Atlas (1960–1967). Dashes indicate that such a diagram was not available, and the figures used in columns 3 and 4 were taken either from the authors of column 5, or from the U. S. Weather Service. [3] Annual mean temperature as given in the diagram. [4] Annual sum of precipitation as given in the diagram. [5] Productivity figures as given by the authors [6].

[b] Asterisk (*) indicates that further calculations were necessary.

1	2		3	4	5	6
Station name	Climate-diagram code number		Mean annual temp. (C°)	Mean annual precipitation (mm)	Annual total dry matter productivity[b] (g/m²)	Source

Region III

Station name			Mean annual temp.	Mean annual precip.	Productivity	Source
1 Archangelsk	110	218	0.4	466	560	Drozdov (1971)
2 Onega	110	217	0.9	497	600	Drozdov (1971)
3 Vologda	110	239	2.4	288	600	Drozdov (1971)
4 Velsk	110	255	−1.5	519	790	Drozdov (1971)
5 Porezkoje	110	178	3.4	508	900	Drozdov (1971)
6 Briansk	110	082	4.7	469	1100	Drozdov (1971)
7 Kiev	110	039	6.8	528	840	Drozdov (1971)
8 Voronesh	110	083	5.6	480	720	Drozdov (1971)
9 Kursk	110	081	5.2	564	810	Drozdov (1971)
10 Namangan	111	037	13.4	188	1040	Drozdov (1971)
11 Roshdestvens-koje	111	078	3.6	135	870	Drozdov (1971)
12 Kokpetky	111	077	1.5	272	1030	Drozdov (1971)
13 Vechnbask-untschak	111	051	7.7	254	380	Drozdov (1971)
14 Termez	111	012	17.3	183	430	Drozdov (1971)
15 Aralskoje More	111	058	6.6	102	120	Drozdov (1971)
16 Selemiya	201	086	16.7	346	240	Drozdov (1971)
17 Palmyra	201	065	19.1	131	70	Drozdov (1971)

Region IV

Station name			Mean annual temp.	Mean annual precip.	Productivity	Source
1 Kigiljaka Mys	111	403	−14.2	94	100	Drozdov (1971)
2 Markovo	111	350	−9.4	200	250	Drozdov (1971)
3 Kumagaya	206	154	13.3	1335	1540	Art and Marks (1971)
4 Kumagaya	206	154	13.3	1335	1075	Art and Marks (1971)
5 Kumagaya	206	154	13.3	1335	3100	Art and Marks (1971)
6 Kyoto	206	132	13.8	1600	1500	Art and Marks (1971)
7 Kyoto	206	132	13.8	1600	3530	Art and Marks (1971)
8 Kyoto	206	132	13.8	1600	2500	Art and Marks (1971)
9 Chantoburi	205	055	27.2	3235	2850	Kira and Ogawa (1969)
10 Nakorn Sawan	205	116	28.2	1222	2860	Kira and Ogawa (1969)
11 Buitenzorg	207	005	25.0	4117	3275*	Lieth (1962)

Part 4: Utilizing the Knowledge of Primary Productivity

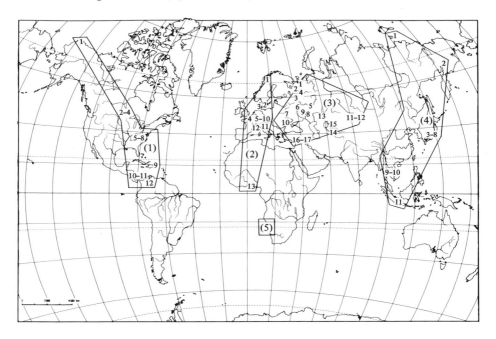

FIGURE 12–2. Location of 52 datum points used to construct our productivity models. Numbers of regions and locations refer to data set listed in Table 12–1.

Correlation models of temperature and precipitation versus productivity

A first pair of models were developed in collaboration with T. Wolaver (Lieth, 1972). They were refined, by the exclusion of extreme values, with E. Box, to give the results shown in Figures 12–3 and 12–4.

Figures 12–3a and 12–3b show the relationship between mean annual temperature and productivity. Figure 12–3a shows the data and the curve yielded by calculating the least squares for this data set. The two curves and the representative data from each of the five transects are shown in Figure 12–3b. The upper curve in Figure 12–3b is an optimum curve drawn by Nyquist analysis, giving a temperature delineation that encompasses the extreme points. The lower curve is an arbitrarily chosen precipitation exclusion curve; we eliminated from temperature considerations all values below the exclusion curve from $-10°C$ and 0 mm ppt, and 25°C and 1000 mm ppt. The Nyquist curve indicates (for the existing data) that maximum productivity occurs at about 25°C. However, in Figure 12–3, the maximum field value appears at $\sim 13°C$. This value represents a young, vigorously growing Japanese plantation, and may not be indicative of the normal growth of climax vegetation in that area. The optimum temperature for productivity, in the range of 15°–25°C, agrees with the optimum temperature range for photosynthesis. More data from the humid tropics are needed to define the optimum temperature for primary productivity.

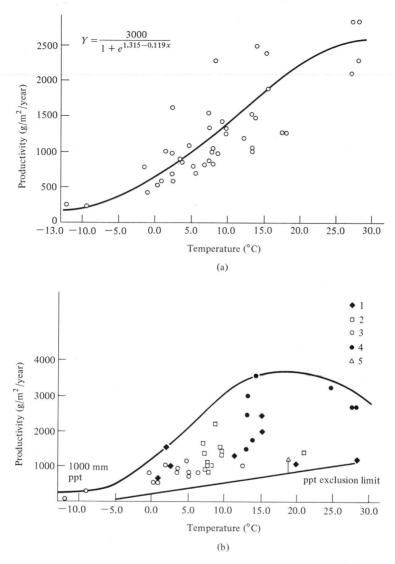

FIGURE 12–3. Net primary productivity versus mean annual temperature. (a) Equation calculated from data in Table 12–1 after excluding extreme values. (b) Distribution pattern of datum points from each region, as listed in Table 12–1. Abscissa, temperature in degrees centigrade; ordinate, grams net dry matter produced per square meter per year. Further details in text.

The data enclosed by the two curves in Figure 12–3b were used to calculate the relationship between temperature and productivity, assuming that large-area average productivity does not exceed 3 kg/m²/year, and that the curve has a sigmoid shape. The first assumption is derived from our collection of productivity values (see Lieth, 1962–1974b). The second argument is deduced from

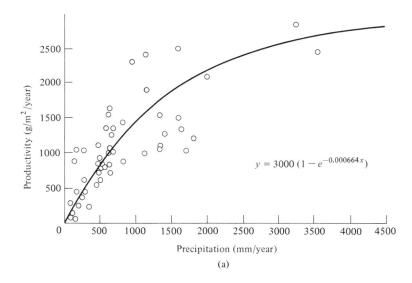

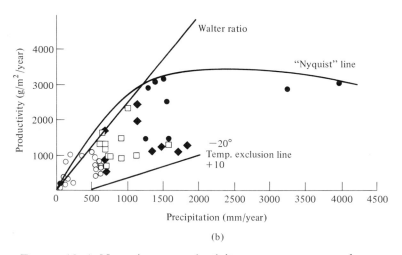

FIGURE 12–4. Net primary productivity versus mean annual precipitation. (a) Equation, (b) graphic analysis. Datum-point symbols are same as those in Fig. 12–3. Abscissa, millimeters annual precipitation; ordinate, dry matter net productivity in grams per square meter per year. In (b) lower curve indicates exclusion level for low temperature values; upper curve is "Walter ratio" and assumes production level of 2 g dry matter for each millimeter of annual precipitation; curve over maximum values is Nyquist line suggesting saturation curve.

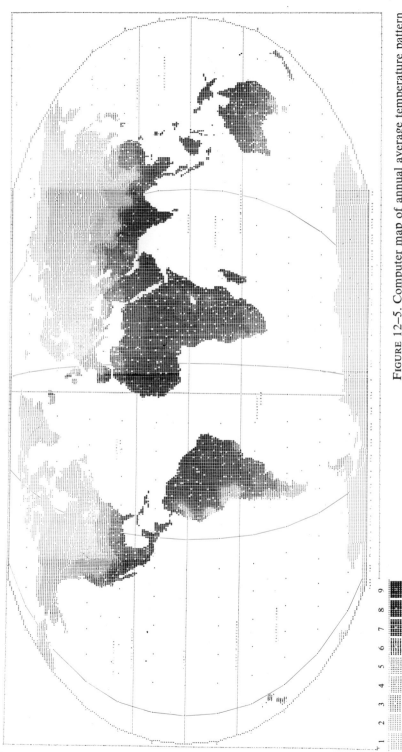

FIGURE 12–5. Computer map of annual average temperature pattern of world constructed from data set used for Miami Model (Fig. 12–8). Project was executed mainly by T. Tyndall. Shade levels indicate temperature signatures in degrees centigrade.

the shape of the Nyquist curve. The predictive formula we derived for Figure 12–3a is

$$y = \frac{3000}{1 + e^{1.315 - 0.119x}} \qquad (12\text{-}1)$$

where y is the productivity level ($g/m^2/year$), x is the mean annual temperature (°C), and e is the natural log base. Over all, the relation suggests the van't Hoff rule, with productivity doubling every 10°C between the temperatures of $-10°$ and 20°C. This relation has much to do with lengths of growing seasons and rates of processes other than photosynthesis itself; the doubling relation does not appear to apply to plankton (see Chapter 8) or temperate forests (Whittaker, 1966; Whittaker and Woodwell, 1971).

The same approach was taken for precipitation versus productivity (Fig. 12–4). The lower portion of Figure 12–4b summarizes the general considerations. The upper straight slope represents "Walter's ratio," which evaluates the aboveground productivity data collected by Walter (1939, 1964, p. 275, cf. Lieth, 1962) in South West Africa, multiplied by two for total productivity values. This ratio predicts that in arid climates 2 g dry matter per square meter are produced for each millimeter of precipitation. This relation cannot be extended to humid climates, as Figure 12–4 indicates. The upper Nyquist curve over the maximum datum points suggests that the precipitation versus productivity equation follows the usual assumption for yield factors, the saturation curve (Mitscherlich, 1954). The lower curve marks the exclusion threshold for data pairs where the productivity is clearly limited by low temperatures. This curve was set arbitrarily from 0°C and 500 mm ppt, to 20°C and 1500 mm ppt. All data used to calculate the least-squares formula are shown in the upper portion of Figure 12–5. Assuming that the maximum productivity is 3 kg/m²/ year and saturation-curve form, the relation was calculated as

$$y = 3000 \, (1 - e^{-0.000664x}) \qquad (12\text{-}2)$$

where y is the productivity level ($g/m^2/year$), x is the precipitation (mm), and e is the natural log base.

Computer maps for data input

To convert the correlation models into spatial models (maps) it is necessary to compile a network of locations with the required environmental measurements. The environmental data from each station are then converted into a productivity level for which the productivity values can be interpolated on a regional base and arbitrary delineations of productivity levels introduced to demonstrate possible patterns.

This procedure is usually done manually by cartographers and is called surface mapping. Today the development of computer mapping routines enables us to combine the individual objectives mentioned previously—taking an outline map of the world, inserting datum points of environmental variables, converting the environmental values into productivity values, averaging the productivity values of neighboring datum points, calculating regional slopes where necessary, break-

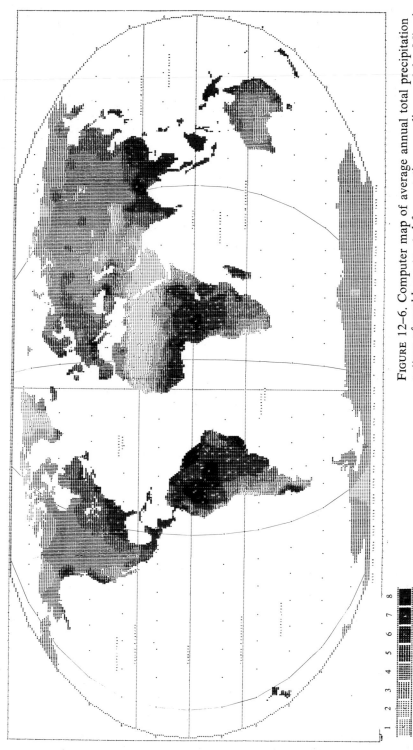

FIGURE 12-6. Computer map of average annual total precipitation pattern of world constructed from station sampling used for Miami Model (Fig. 12-8). Project was executed mainly by D. Martin. Shade levels indicate precipitation signatures (mm/yr).

	1	2	3	4	5	6	7	8
	0	50	100	250	500	1000	2000	above
	50	100	250	500	1000	2000	4000	4000

Precipitation ranges (mm/year)

ing the slopes into level ranges—with the production of a final product: the predictive productivity map.

The computer mapping routine used for our purpose is SYMAP. Its subroutines and procedures are adequately described in Reader *et al.* (1972) and Dudnik (1972). We developed for this mapping routine a world outline map that simulates an existing outline map called Robinson's projection. With this outline map we simulated our first primary productivity map (Lieth, 1964) in order to see how to manipulate most efficiently existing and newly generated data pools. The basic computer outline was gradually improved and utilized for all projects described in this volume under the name UNC Biosphere model (University of North Carolina Biosphere model). Three compatible outlines exist: (1) total globe, (2) land only, and (3) oceans only. The compatibility of outlines 2 and 3 is demonstrated in Chapter 10, Figure 10–1. The compatibility of 1 and 2 is shown in the construction of the Thornthwaite Memorial Model. The individual steps of constructing the spatial models from the two environmental variables, annual temperature and annual total precipitation require a brief demonstration.

In order to predict the primary productivity from the environmental parameters we selected from our climate diagram world atlas (Walter–Lieth, 1960–1967) about 1000 stations so spaced over the continental areas as to adequately cover the computer world map. Figures 12–5 and 12–6 show the maps for annual average temperature and annual total precipitation. The maps were produced by University of North Carolina graduate students as class projects and were checked against existing maps in order to compare adequacy and limitations of the data set used.

Looking over the two maps it is at once apparent that the severest limitations are in mountainous areas and in coastal areas of the extreme north and south. Such limitations can be resolved in principle, but the size of the computer-base maps needs to be increased substantially as does the number of datum points to be entered on such a map. This is basically possible with the available computer routines, but the present accuracy of our correlation models does not warrant the high cost of such a project.

Further properties and limitations of the computer maps are discussed in Chapter 13.

Converting the environmental maps into productivity maps

Each individual environmental map can be converted into a productivity map if the correlation model is known and the transformation to production estimates is valid. (To use, for example, the annual temperature alone would be invalid because the large areas of hot deserts would appear as highly productive.) The utilization of precipitation alone seems more reasonable, for precipitation has an intrinsic dependent relationship to mean annual temperature. The map of productivity predicted from precipitation only is presented therefore in Figure 12–7. This map may be compared with the other maps in this book for areas quite similar (or very different) productivity levels.

The construction of a productivity map by using more than one environmental

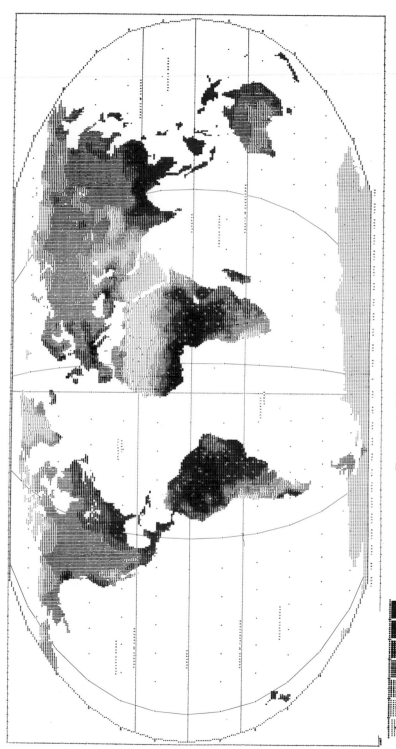

FIGURE 12-7. Net primary productivity in dry matter g/m²/year, as predicted from annual precipitation values. Equation (12-2), represented in Fig. 12-4 was used to convert precipitation map (Fig. 12-6) into primary productivity map. Comparison of this map with Miami Model (Fig. 12-8) indicates which areas of world, according to Eq. (2), have precipitation as minimum factor, and which areas have temperature as limiting factor for primary productivity.

< 110–250– 500–1000–1500–2000 <

Productivity ranges (g/m²/year, dry matter)

parameter requires the development of an information system with which we can select the most likely productivity level for each combination of environmental variables. This is necessary because the normal variability of each correlation model may yield a different productivity level for the precipitation–temperature combination of any individual datum point in our computer map. The logic we applied for the map presented in Figure 12–8 involves application of Liebig's law: the minimum factor controls the productivity level. Therefore, the computer program selected for each station the lower of the two productivity values predicted from the environmental variables, and this value then contributed to the calculation of productivity level patterns in the final map.

This map (Fig. 12–8) was first presented at the Miami Symposium from which this book originates (see also Lieth 1973). In the many summaries already printed (see reference list), this map is referred to as the Miami Model.

The Montreal Models: Primary Productivity Predicted from Evapotranspiration

During the 22nd International Geographical Congress in Montreal, a new map model was presented by Lieth and Box (1972) predicting the primary productivity from actual evapotranspiration. As the paper is fully documented in Lieth and Box (1972) we include only a brief summary.

In order to model the terrestrial primary productivity from evapotranspiration on a global scale, we need the following data sets and models.

1. A computer map of the actual evapotranspiration from the land areas of the world
2. A model to convert evapotranspiration data into productivity values
3. A combination of 2 with 1, for example, a map of terrestrial primary productivity predicted from actual evapotranspiration

The computer map model of terrestrial actual evapotranspiration of the world

This map was constructed by Elgene Box as a simulation of Geiger's, (1965) map. Initially, the map was simulated using about 850 datum points, that is, the midpoints of each 10° × 10° quadrat over the entire globe (684 points, owing to 15° projection overlap), supplemented by additional points for quadrats containing both land and sea portions. The chosen datum-points were located geographically, and their evapotranspiration values were estimated according to their location within Geiger's contour intervals. These values were then checked against precipitation values for nearby weather stations, taken from the Climate-Diagram World Atlas (Walter and Lieth, 1960–1967). Generally, evapotranspiration values were not allowed to exceed 50% of the precipitation in drier areas and 75% of the precipitation in wetter areas. The contour intervals used for the computer simulation are the same as those of Geiger's map, namely intervals of 250 mm of evapotranspiration, ranging from 0 to 2000 mm, with the following modifications. We have

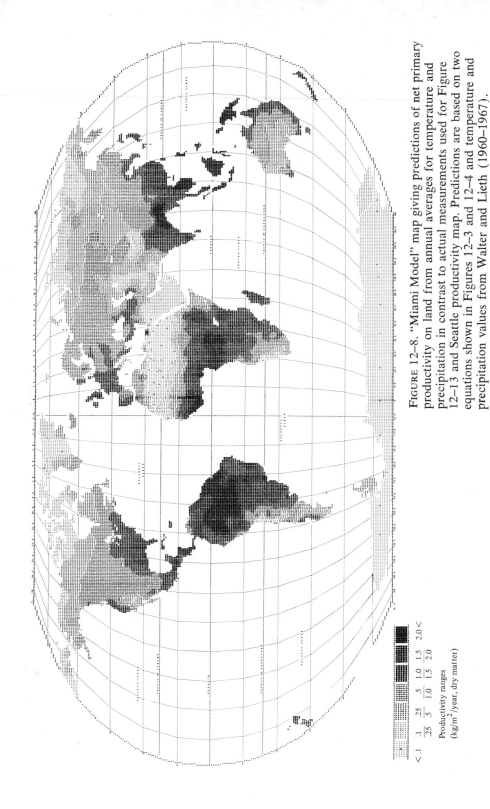

FIGURE 12–8. "Miami Model" map giving predictions of net primary productivity on land from annual averages for temperature and precipitation in contrast to actual measurements used for Figure 12–13 and Seattle productivity map. Predictions are based on two equations shown in Figures 12–3 and 12–4 and temperature and precipitation values from Walter and Lieth (1960–1967).

<.1   .1   .25   .5   1.0   1.5   2.0 <
      ---   ---   ---   ---   ---
      .25   .5   1.0   1.5   2.0

Productivity ranges
(kg/m²/year, dry matter)

1. Divided Geiger's lowest interval (0–250 mm) into two intervals of 0–125 mm and 125–250 mm, based on precipitation measurements
2. Combined intervals 1000–1250 mm and 1250–1500 mm into one interval of 1000–1500 mm
3. Combined intervals 1500–1750 mm and 1750–2000 mm into one interval of 1500–2000 mm

In order to improve the simulation of the more complex land areas, the initial 300 land datum points were supplemented by an additional 300 datum points, with more points proportionally being added in regions of more topographic complexity, such as areas cut by mountain ranges. In order to improve further the fit with Geiger's contours, datum points were also moved geographically, but only within the appropriate contour interval, and only after the position of the new datum point had been checked against Geiger's evapotranspiration contours and against the available precipitation data. The movement of datum points was an absolute necessity because the perfectly latticelike pattern of quadrat midpoints results in unrealistic, rectilinear contours. The resulting map of global actual evapotranspiration is based on 1125 datum points (Fig. 12–9).

The SYMAP world map module can produce a map of almost any size (F-MAP elective 1 of the SYMAP program), but all our world maps measure 65 × 130 cm. A map of this size contains ~ 90,000 print positions for the entire globe (land and ocean areas). From the set of about 1120 datum points used for Figure 12–9, we isolated the land areas with the same land-mass outline (A-OUTLINE of the SYMAP program) that we have used for other world maps, such as the Innsbruck and the Miami maps. Although not printed the ocean datum points are used to construct the map. This prevents contour discrepancies owing to meaningless linear interpolation across large water bodies. The procedure amounts to cutting out the land areas from the global evapotranspiration map with a pair of scissors (Fig. 12–9) and results in ~ 25,700 print positions. Figure 12–10 shows the map of actual evapotranspiration from the land areas.

The model to convert evapotranspiration data into productivity values

For this task we used the data set gathered for the Miami Model (see Table 12–1 and Fig. 12–2). The evapotranspiration value for each location in the data set was picked from the Geiger map, with checks made against available precipitation data (Walter and Lieth 1960–1967). We recognize the shortcomings of the procedure, but this was the only way to pair these two important parameters, evapotranspiration and NPP. Figure 12–11 shows the scattering of the datum points and the curve of the averages for evapotranspiration classes, given in increments of 125 mm. The class averages become erratic below 200 and above 750 mm. This reflects our ignorance of tropical regions and the need for intensive research in this area. Generally the curve of the class averages tends toward a saturation curve of a form similar to the one we used for the precipitation-versus-production equation of the Miami Model (Fig. 12–4). The least-squares model that we constructed from our data has the form

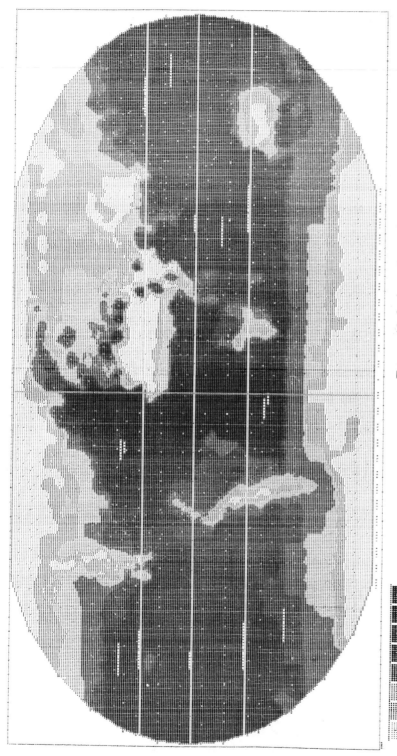

FIGURE 12–9. Annual actual evapotranspiration pattern of globe, computer simulation of Geiger's wall map, made with same dimensions of all maps in UNC Biosphere Model series. These maps are based on Robinson's projection, which shows continents as areas of reduced evapotranspiration. Map was executed by E. Box.

<125 250 500 750 1000 1500 2000 < mm

Actual evapotranspiration ranges
(mm/year)

$$P = 3000 \left[1 - e^{-0.0009695(E-20)}\right] \qquad (12\text{-}3)$$

where P is the annual net primary productivity (g/m^2), E is the annual actual evapotranspiration (mm), and e is the base of natural logarithms. The curve is shown in Figure 12–11 as a dashed line.

Conversion of the evapotranspiration map into a productivity map

Taking the same datum points used for Figure 12–10, we converted each evapotranspiration value into a productivity value using equation (12–3). The SYMAP output yielded from the data sets specified above is the second of the two Montreal Models we constructed and was entitled the C. W. Thornthwaite Memorial Model. The map is shown in Figure 12–12.

Correlation of Primary Productivity to the Length of the Vegetation Period

The assumption that primary productivity and length of vegetation period are highly correlated stems from the agricultural experience that areas with extended warm summers may provide for either two crops or for one high-yielding crop requiring a long vegetation period. In areas of comparable soil fertility, lengths of frost-free period show positive correlations with the level of yield.

The US–IBP Eastern Deciduous Forest Biome, through its Biome Wide Studies group, has investigated this correlation throughout the Eastern United States. Two properties were needed for the evaluation: (1) the primary productivity of large regions, and (2) the length of the vegetation period, determined as accurately as possible in a way that has biologic meaning.

The first property is discussed in detail in Chapters 6 and 7. We use data from these chapters with the reservations expressed there.

The second property, the length of vegetation period throughout this region, was elaborated by Reader (1973). Reader used the flowering times of dogwood, redbud, and lilac (see Reader *et al.*, 1974, for details) as an indicator for the beginning of the vegetation period, and the leaf coloring of yellow poplar, red maple, and dogwood as an indicator for the ending of the vegetation period.

The two parameters employed simultaneously for each region, taking a county as the point unit, enabled Reader to construct a regression between primary productivity and length of vegetation period. The scattergram is presented in Figure 12–13. The regression line follows the equation

$$P = -157 + 5.17S \qquad (12\text{-}4)$$

where P is the primary productivity ($g/m^2/year$) and S is the photosynthetic season (days).

For the purpose of global modeling, the key points of this equation can be evaluated for a 1-year period. If the equation were correct for the entire world, it would mean that about a 1-month vegetation period is necessary for a

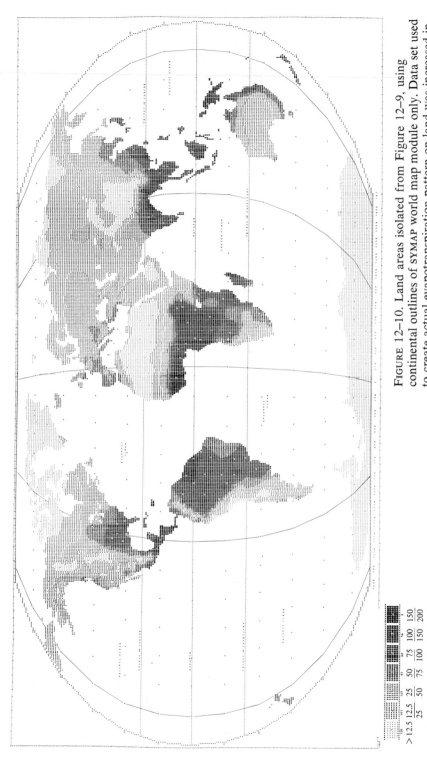

FIGURE 12–10. Land areas isolated from Figure 12–9, using continental outlines of SYMAP world map module only. Data set used to create actual evapotranspiration pattern on land was increased in density (over density used for Fig. 12–9) to improve accuracy of map.

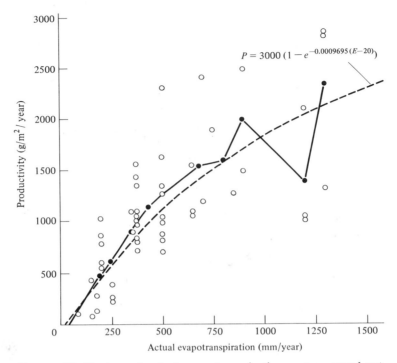

FIGURE 12–11. Annual actual evapotranspiration versus annual net primary productivity. The figure demonstrates development of equation (12–3). Abscissa, actual evapotranspiration (mm); ordinate, annual primary productivity (g/m²). (○) Datum points pairing productivity values from locations shown in Figure 12–2, and actual evapotranspiration values picked from Figure 12–10. (—●—) Class averages for 250- and 125-mm evapotranspiration classes. Graphic display of Eq. (12–3) derived from least squares of curve of class averages. Form of equation is shown on graph. Equation was used to convert actual evapotranspiration map shown in Figure 12–10 into primary productivity map, C. W. Thornthwaite Memorial Model (Fig. 12–12).

threshold production, and that the entire year as a period of active plant growth would give us an average ceiling of 1730 g/m²/year primary productivity. This seems low compared to the productivity ceiling reached with the other models based on biologic productivity measurements. There is, however, a large variation inherent in the data set for southern counties, and the top counties taken as reference points would come close to a ceiling of 2500 g/m²/year. A more important result of this model is that it suggests that our present over-all land-use practices are far from the optimum that can be attained by natural and seminatural ecosystems. We expect that further refinement of this model will yield a sigmoid curve similar to that of Eq. (12-1).

Under the assumption, again, that Eq. (12-4) is valid for the entire earth, we can use it to construct the global productivity pattern provided we have

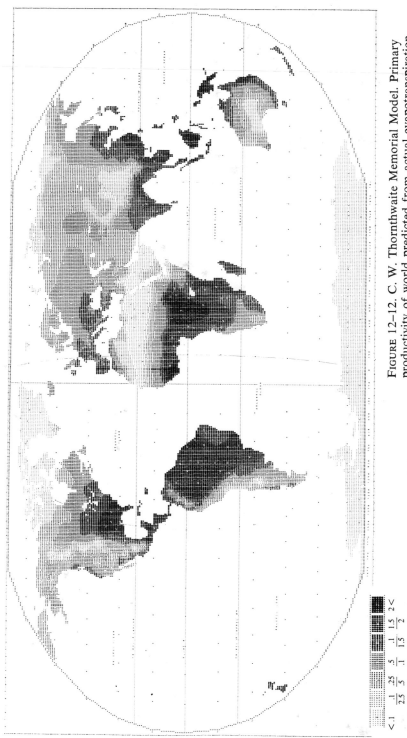

FIGURE 12–12. C. W. Thornthwaite Memorial Model. Primary productivity of world predicted from actual evapotranspiration, using Box–Lieth model to convert Geiger's evapotranspiration map (Fig. 12–8) into productivity map. Map comes closest to agreement with present knowledge of productivity pattern of land areas.

Productivity ranges
(kg/m² /year, dry matter)

$<.1$ $\frac{.1}{2.5}$ $\frac{.25}{.5}$ $\frac{.5}{1}$ $\frac{.1}{1.5}$ $\frac{1.5}{2}$ $2<$

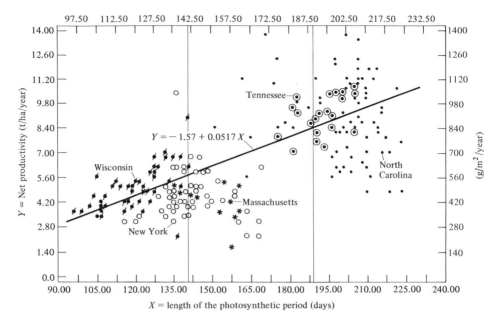

FIGURE 12–13. Correlation between length of vegetation period as independent variable and primary productivity as dependent variable in Eastern United States. Although no good correlation may be established for single states, the positive trend is visible over the North–South profile from Wisconsin to North Carolina. Abscissa, photosynthetic period in days; left ordinate, primary productivity (t/ha/year); right ordinate, primary productivity (g/m²/year). Regression equation is given in figure for metric tons per hectare, whereas, in Eq. (12-4) it is given for grams per square meter. (X = length of photosynthetic period (days).

a network of stations with recorded length of vegetation period. Such a network was constructed in a student project by Wyatt and Sharp (Lieth, 1974a) with the climate-diagram world atlas of Walter and Lieth (1961–1967) as the data base. The length of vegetation period was calculated in months for about 600 stations evenly distributed over the world (see Lieth, 1974a, for details). With this network, Wyatt and Sharp constructed a computer map using the UNC Biosphere Model. This map can be converted directly into a productivity map because Eq. (12-4) describes a straight line. The map is shown in Figure 12–14. The graduation of productivity levels here is such that the productivity of 1 month equals 160 g/m². The map is similar to the Miami Model and the Thornthwaite Memorial Model, as we expected but yields a much lower global productivity estimate, about 73.5 × 10⁹ t/year (Lieth, 1975). It appears much coarser in pattern because of the lower station density used. The UNC Biosphere Model is still too superficial to be used for global productivity calculation, but in principle it could be used as were the two previous maps in Chapter 13.

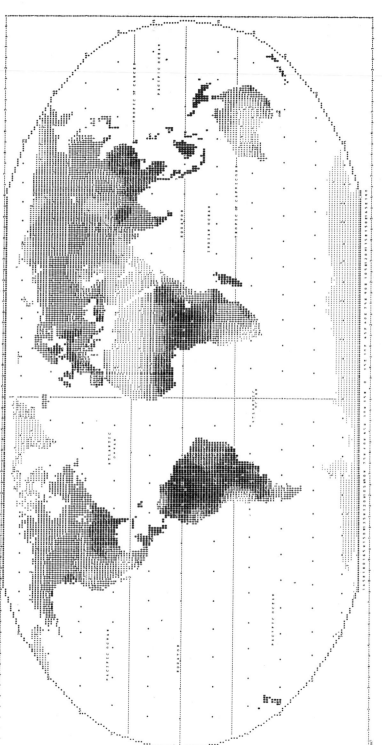

Figure 12–14. Global productivity pattern predicted from length of vegetation period using regression equation (12–4) and Figure 12–13. Resulting map has similar patterns as the maps in Figures 12–8, 12–12, and 12–15; but it yields a much lower total terrestrial productivity value.

| $<.4$ | $\dfrac{.4}{2.4}$ | $\dfrac{2.4}{4.4}$ | $\dfrac{4.4}{6.4}$ | $\dfrac{6.4}{8.4}$ | $\dfrac{8.4}{10.4}$ | $10.4<$ |

Growing season (months)

| <75 | $\dfrac{75}{350}$ | $\dfrac{350}{560}$ | $\dfrac{560}{880}$ | $\dfrac{880}{1200}$ | $\dfrac{1200}{1520}$ | $1520<$ |

Productivity ranges
(g/m^2/year, dry matter)

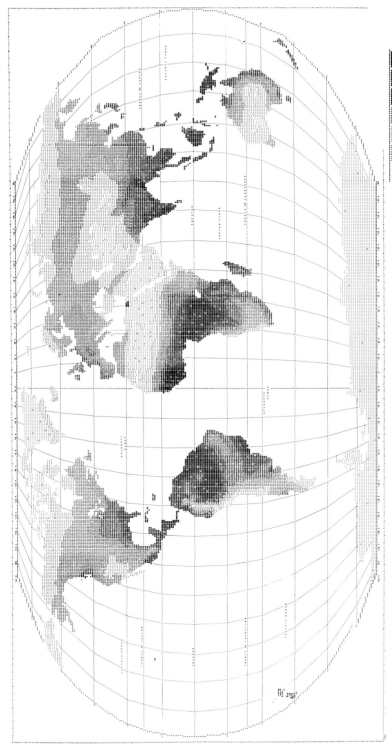

FIGURE 12–15. Innsbruck Productivity Map, computer simulation of Lieth's (1964) primary productivity map. Values shown are in approximate agreement with production rates given in Table 10–1 for individual vegetation types.

Comparison of Function-Generated Maps with the Traditional Maps Based on Data Extrapolation

The model-derived maps described so far can be accepted as first-level and approximate, predictive models for the primary productivity of the world. In the following section, maps derived from Eq. (12-1)–(12-3) are discussed and compared with the Innsbruck and Seattle maps (see Figs. 12–15 and 10–1).

There are three main requirements for a predictive model of primary productivity based on biologic considerations:

1. It must originate close to the zero point for productivity and it must satisfy the experimentally observed values for the independent variable (parameter).
2. It must approach but generally not exceed a ceiling value, a fact that is well established by Mitscherlich's yield law.
3. The form of the curve itself should reflect the best fit for an available data set.

Equation (12-3) (dashed curve in Figure 12–11) was calculated from the class-average curve (filled circles in Fig. 12–11). This curve fulfills all requirements stated above and also allows for the logically required 30–40% runoff from the precipitation measured at any location. Because both curves [Eqs. (12-4) and (12-11)] were derived from entirely different sources for the independent variable, the similarity of the results reinforces each model.

Comparing the Thornthwaite Memorial Model with the Miami Model or with the Innsbruck Productivity Map produces the same conclusion as with a comparison between the Innsbruck Productivity Map and the Miami Model. In general, we find agreement among all three maps, but clearly gaps exist between many individual regions. Further refinements of both the Innsbruck-type maps and the predictive-model maps are required before a satisfying picture of the global primary productivity and the controlling environmental parameters can be drawn.

References

Art, H. W., and P. L. Marks. 1971. A summary table of biomass and net annual primary production in forest ecosystems of the world. In *Forest Biomass Studies*, H. E. Young, ed., pp. 3–34. (*Int. Union of Forest Res. Organizations Conf. Sect. 25, Gainesville, Florida.*) Orono, Maine: Univ. of Maine, Life Sciences and Agriculture Experiment Station.

Drozdov, A. V. 1971. The productivity of zonal terrestrial plant communities and the moisture and heat parameters of an area. *Soviet Geogr. Rev. Transl.* 12: 54–60.

Dudnik, E. E. 1972. SYMAP *Manual*. Cambridge, Massachusetts: Laboratory of Computer Graphics and Spatial Analysis, Graduate School of Design, Harvard Univ.

Ewel, T. 1971a. Biomass changes in early tropical forest succession. *Turrialba* 21: 110–112.

———. 1971b. Ph.D. thesis, Chapel Hill, North Carolina: Univ. of North Carolina.

Geiger, R. 1965. World Atmosphere Series of Maps, Map No. WA6: *Annual Effective Evapotranspiration*. Darmstadt: Justus Perthes.

Gessner, F. 1959. *Hydrobotanik*, Vol. 2. Berlin: Deutscher Verlag der Wissenschaften.

Jordan, C. F. 1971. Productivity of a tropical forest and its relation to a world pattern of energy storage. *J. Ecol.* 59:127–143.

Kira, T., and H. Ogawa. 1971. Assessment of primary production in tropical and equatorial forests. (French summ.) In *Productivity of Forest Ecosystems: Proc. Brussels Symp. 1969,* P. Duvigneaud, ed. *Ecology and Conservation,* Vol. 4: 309–321. Paris: UNESCO.

Lieth, H. 1956. Ein Beitrag zur Frage der Korrelation zwischen mittleren Klimawerten und Vegetationsformationen. *Ber. Deut. Bot. Ges.* 69:169–176.

————, and F. Zauner. 1957. Vegetationsformationen und mittlere Klimadaten. *Flora* 144:290–296.

————. 1961. La produccion de sustancia organica por la capa vegetal terrestre y sus problemas. *Acta Cient. Venez.* 12:107–114.

————. 1962. *Die Stoffproduktion der Pflanzendecke.* 156 pp. Stuttgart: Fischer.

————. 1963. The role of vegetation in the carbon dioxide content of the atmosphere. *J. Geophys. Res.* 68:3887–3898.

————. 1964. Versuch einer kartographischen Darstellung der Produktivität der Pflanzendecke auf der Erde. In *Geographisches Taschenbuch 1964/65,* pp. 72–80. Wiesbaden: Steiner.

————, D. Osswald, and H. Martens. 1965. Stoffproduktion, Spross/Wurzel-Verhältnis, Chlorophyllgehalt und Blattfläche von Jungpappeln. *Mitt. Ver. Forstl. Standortsk. Forstpflanz.* 1965:70–74.

————. 1965a. Ökologische Fragestellungen bei der Untersuchung der biologischen Stoffproduktion. *Qual. Plant. Mater. Veg.* 12:241–261.

————. 1965b. Indirect methods of measurement of dry matter production. *Methodology of Plant Ecophysiology: Proc. Montpellier Symp.,* F. E. Eckardt, ed. *Arid Zone Research,* Vol. 25:513–518. Paris: UNESCO.

————. 1968. The determination of plant dry matter production with special emphasis on the underground parts. (French summ.) In *Functioning of Terrestrial Ecosystems at the Primary Production Level: Proc. Copenhagen Symp. 1965,* F. E. Eckardt, ed. *Natural Resources Res.,* Vol. 5:179–186. Paris: UNESCO.

————. 1970. Predicted annual fixation of carbon for the land-masses and oceans of the world. [Map on front end paper.] In *Analysis of Temperate Forest Ecosystems,* D. Reichle, ed. *Ecological Studies,* Vol. 1. Heidelberg: Springer-Verlag.

————. 1971. Mathematical modeling for ecosystems analyses. (French summ.) In *Productivity of Forest Ecosystems: Proc. Brussels Symp. 1969,* P. Duvigneaud, ed. *Ecology and Conservation,* Vol. 4:567–575. Paris: UNESCO.

————. 1972a. Über die Primärproduktion der Pflanzendecke der Erde. (*Symp. German Botanical Soc., Innsbruck, Austria, Sept., 1971.*) *Z. Angew. Bot.* 46: 1–37.

————. 1972b. Modeling the primary productivity of the world, 10 pp. (offset) Deciduous Forest Biome Memo Rep. 72-9, March 1972.

————. 1973. Primary production: Terrestrial ecosystems. *Human Ecol.* 1:303–332.

————, ed. 1974a *Phenology and Seasonality Modeling. Ecological Studies,* Vol. 8, 444 pp. New York: Springer-Verlag.

————, 1974b. Primary productivity of successional stages. In *Vegetation Dynamics,* R. Knapp, ed. *Handbook of Vegetation Science,* Vol. 8:187–193. The Hague: W. Junk.

————. 1975. The primary productivity in ecosystems. Comparative analysis of global patterns. In *Unifying Concepts in Ecology,* W. H. Dobben and R. H. Lowe-McConnel, eds. The Hague: W. Junk. (*In press.*)

————, and E. Box. 1972. Evapotranspiration and primary productivity; C. W. Thornthwaite Memorial Model. *Publications in Climatology,* Vol. 25(2):37–46. Centerton/Elmer, New Jersey: C. W. Thornthwaite Associates.

Medina, E. 1970. Estudios eco-fisiologicos de la vegetation tropical. *Bol. Soc. Venez. Cinc. Nat.* 29:63–88.

Mitscherlich, E. A. 1954. Bodenkunde für Landwirte. In *Forstwirte und Gärtner,* 7th ed. Berlin and Hamburg: Parey.

Nemeth, J. 1971. Doctoral thesis, Raleigh, North Carolina: North Carolina State Univ.

Reader, R. 1971. Net primary productivity and peat accumulation in Southeastern Manitoba, 220 pp. Master's thesis. Winnipeg: Univ. of Manitoba.

————. 1973. Leaf Emergence, Leaf Coloration, and Photosynthetic Period-productivity Models for the Eastern Deciduous Forest Biome, 167 pp. and appendix. Doctoral thesis. Chapel Hill, North Carolina: Univ. of North Carolina, Ecology Curriculum.

————, and UNC Applied Programming Group. 1972. SYMAP *(Version 5.16A) Instruction Manual,* 65 pp. Document No. LSR-139-0. Research Triangle Park, North Carolina: Triangle Universities Computation Center.

————, J. S. Radford, and H. Lieth. 1974. Modeling important phytophenological events in Eastern North America. In *Phenology and Seasonality Modeling,* H. Lieth, ed. *Ecological Studies,* Vol. 8:329–348. New York: Springer-Verlag.

Walter, H. 1939. Grasland, Savanne und Busch der ariden Teile Afrikas in ihrer ökologischen Bedingtheit. *Jahrb. Wiss. Bot.* 87:850–860.

————. 1964. *Die Vegetation der Erde in öko-physiologischer Betrachtung.* Vol. 1: *Die tropischen und subtropischen Zonen,* 538 pp. Jena: Fischer.

————. 1968. *Die Vegetation der Erde in öko-physiologischer Betrachtung,* Vol. 2: *Die gemässigten und arktischen Zonen,* 1001 pp. Jena: Fischer.

————, and H. Lieth. 1960–1967. *Climate Diagram World Atlas.* Jena: Fischer.

Wells, C., and H. Lieth. 1970. Preliminary assessment of the productivity of a *Pinus taeda* plantation in the Piedmont of North Carolina, 41 pp. (mimeogr.) Report to the Deciduous Forest Biome Headquarters.

Whittaker, R. H. 1966. Forest dimensions and production in the Great Smoky Mountains. *Ecology* 47:103–121.

————, and G. M. Woodwell. 1971. Measurement of net primary production of forests. (French summ.) In *Productivity of Forest Ecosystems: Proc. Brussels Symp. 1969,* P. Duvigneaud, ed. *Ecology and Conservation,* Vol. 4:159–174. Paris: UNESCO.

13

Quantitative Evaluation of Global Primary Productivity Models Generated by Computers

Elgene Box

That net plant production plays a key role in ecologic, environmental, and planning considerations requires little supporting discussion. In addition to its implication for the upper limit to the Earth's sustainable human population, knowledge of the Earth's production and its spatial distribution permits us to estimate such characteristics of our planet as

1. Geographic distribution of the Earth's potential food resources
2. Sizes and geographic distribution of the various reservoirs in the Earth's carbon and oxygen cycles
3. Limits toward which we might be able to increase regional productivity levels artificially, assuming that such grandiose projects were environmentally desirable
4. Effects of the destruction of major vegetation formations, such as the Amazon rain forest, on the Earth's atmospheric composition and climate
5. Potential and actual productivity levels of individual countries, hence the maximum carrying capacities of their national and regional ecosystems

For basically rural, agricultural economic patterns, roughly at the level of eighteenth-to-nineteenth century Europe, the world is already overpopulated beyond its carrying capacity. Consequently, we have forced ourselves into our present pattern of urban population centers in order to preserve the productivity of the remaining landscape. However, the urban pattern has made possible the growth that is now the main destroyer of the productive landscape—in the industrial countries through direct urban–industrial expansion, and in the non-

KEYWORDS: Primary productivity; computer model; quantitative evaluation; global pattern; geoecology.

Primary Productivity of the Biosphere, edited by Helmut Lieth and Robert H. Whittaker.

industrial countries by unprecedented acceleration of population growth. Estimates of the maximum population that the Earth can sustain have often been between the figures of 6 and 15 billion. We shall exceed 6 billion before the end of the twentieth century. The resources of the world are already hard-pressed by 4 billion, and a sizable fraction of those 4 billion is fed by using fossil fuel to increase agricultural productivity (Chapter 15). Further growth to and beyond 6 billion is likely to imply, at the worst, breakdown of the world order, at the least (I will not say at best) drastic changes in economic systems, with increasing central planning and less personal freedom.

In his attempt to evaluate the total production of the Earth, both cumulatively and geographically, Lieth (Chapter 12) has directed the preparation of four computerized global productivity models, the results of which are graphically displayed as statistical maps by the SYMAP (synagraphic mapping) computer program (Harvard Graduate School of Design), as discussed below. A fifth map was obtained by combining the simulations of actual land and sea productivities into one map. As a statistical map data values are provided at arbitrary geographic locations, and the values at the intermediate positions, including the resulting contour lines and zonation, are interpolated based on a specified number of points. Thus the thematic mapping of a continuous surface is performed by approximation based on discrete increments, using arbitrarily distributed statistical data. Lieth's five maps are listed below. Most of these are identified by the cities in which they were first presented.

1. *Innsbruck Productivity Map* (Fig. 12–15): Terrestrial productivity was simulated from average productivity values of the various terrestrial vegetation formations (biomes) (Lieth, 1972); see page 260.

2. *Miami Model Productivity Map* (Fig. 12–8): Terrestrial productivity was predicted from two least-squares functional correlations between productivity and climatic variables (mean annual temperature and total annual precipitation) using in each case the smaller of the predicted productivity figures (Lieth, 1971, 1973); see page 251.

3. Montreal Model Productivity Map (*C. W. Thornthwaite Memorial Model* (Fig. 12–12): Terrestrial productivity was predicted from a least-squares functional correlation between productivity and annual actual evapotranspiration (Lieth and Box, 1972); see page 257.

4. *Ocean Productivity Map* (Fig. 8–1): Marine productivity was simulated from average-productivity values of the marine ecosystems (Lieth and Box, 1972); see page 174.

5. *Seattle Productivity Map* (Fig. 10–1): An overprinting of the Innsbruck and Ocean Productivity maps, showing the earth's entire surface within one map (Lieth and Box, 1973); see page 210.

The Innsbruck and Ocean Productivity maps are SYMAP-produced statistical simulations, based on data sets of 930 and 1621 datum points, respectively, of an earlier world map of actual productivity (Lieth, 1965). The two maps were based on average productivity measurements of 20 vegetation formations, as shown in Table 13–2, and on existing maps of existing vegetation covers. Lieth then mechanically overprinted the two complementary maps within a single frame to produce the Seattle Productivity Map of the plant productivity of the entire earth. The Miami and Montreal Models are functional models the mathematical results of which are displayed as SYMAP maps. In both maps the primary productivity is predicted from climatic parameters: the Miami Model from temperature and precipitation measurements from the *Climate-Diagram World Atlas* (Walter and Lieth, 1960–1967), and the Montreal Model from a SYMAP statistical simulation (by Box) based on a world map of actual evapotranspiration (by Geiger, 1965).

Statistical Mapping and Map Evaluation

The results of the Lieth maps are proof that the SYMAP program can be a very useful tool for producing maps of various phenomena—both simple statistical simulations of existing maps (spatial models) and statistical displays of the results of computer-generated mathematical models (functional models). SYMAP is by no means restricted to the topic of productivity, but is a general mapping program that can be used to compute and display distributions of almost any parameter for which an areal distribution is meaningful. The production of such computer maps has definite advantages over hand-drawn maps, in addition to the obvious advantages of production speed and functional modeling. Maps that are processed and printed by computer can be analyzed and interpreted by computer also. It is important, then, to have additional programs for the reading and further processing of such preproduced and prestored maps. For geoecology and for the earth and ecologic sciences in general, the following analytic and operational capabilities appear most needed:

1. Interconversion between latitude/longitude and SYMAP (projection) coordinates (FLEXROB, MAPCOUNT)
2. Determination of zonal areas (MAPCOUNT)
3. Digitization of zonal outlines (MAPZONES)
4. Character-by-character comparison of maps (MAPMATH)
5. Superimposition of maps of identical dimensions within a single frame (MAPMERGE)
6. User-specifiable character-by-character mathematical operations on one or more maps to produce a new map (as opposed to operations performed before the spatial interpolation on the data points themselves, which may be located quite differently on different maps) (MAPMATH)

The author has written several map-processing programs, which are indicated above in parentheses.[1] For example, the overprinting of Lieth's land and sea productivity maps to produce the Seattle map now can be accomplished computationally using the MAPMERGE program, and the different land productivity maps can be compared character-for-character against each other by the MAP-

---

[1] Complete documentation including listings of the programs and explanations, with examples of the use of the various options, is in publication as a special report of the Kernforschungsanlage Jülich (Box, 1975). Preliminary documentation may be obtained from the author.

MATH program in order to locate areas of discrepancy. This chapter describes what is considered to be the most useful of these map-processing programs, the MAPCOUNT program, which combines the first two operations in the above list. MAPCOUNT is then applied to the Lieth productivity maps in an attempt to quantify the world's total annual primary productivity.

SYMAP Earth Models and Resulting Maps

Before evaluating the Lieth maps, it is necessary to have some understanding of how the maps were produced and of some of the mathematical properties and problems of such areal models. In order to produce any map, SYMAP requires the provision of a map outline, for example, the land areas or ocean areas; the locations in SYMAP (row/column) coordinates and the values for the datum points; and certain specifications for the production of the desired map. Legends such as parallels and meridians, the tropics, and the equator, and labels may be included optionally. In order for the resulting map to be readily analyzable by MAPCOUNT in terms of areas, the model must approximate real areas, as rendered by some planar mathematical projection of the Earth's spherical surface. The mathematical formulation of this projection must be provided to MAPCOUNT in the form of a computational algorithm by means of a subroutine DEFORM of standard format. All Lieth's world productivity maps are produced using the Robinson (1974) projection of which the basic characteristics of the computer version were partially described by Lieth (1972). The Robinson projection is an uninterrupted, pseudocylindrical projection that has no prime properties, but is an optimal compromise between the properties of equivalence (preservation of areal relationships) and conformality (preservation of angles). As a result, it is very good for visual displays. Vertical distance from the equator is directly proportional to latitude (i.e., parallels are equally spaced) up to about 45°; then the spacing decreases so that areas of larger polar regions such as Greenland are distorted by a factor of no more than slightly over two.

The outlines of the Earth's land and ocean areas are represented by closed sequences of points (ordered row-column pairs) within the SYMAP grid. These point sequences define the perimeters of the desired regions, which are referred to as SYMAP islands. The outlines were traced by Lieth's students from a large wall map. The problem of representing areas by such outlines is rendered more complicated, however, by the fact that printed rows and columns cannot be used to represent boundaries because print positions have areas themselves but boundaries do not. SYMAP circumvents this problem, which would otherwise render each island too large, by shifting the specified boundaries down and to the left by half of a row/column, so that boundaries fall between print positions. Lieth's land outline was produced first, and the outline for the oceans was traced from a printed land outline in such a way that the SYMAP shift results in a border of one blank row/column between the land and oceans at the top and the left sides of land masses. Consequently, the ocean outline is too small in total area, even in its conception.

The actual appearance and content of a map is dependent, as well, upon a number of parameters and modes of operation, the values of which are specified in the SYMAP F-MAP package. In the SYMAP program, the interpolation is performed according to a formula that assumes that the effect of one point upon its neighbors decreases with the square of the distance between them. The number of points used for the interpolations may be variable or fixed. For search radius and number of points for interpolation, we generally have taken the SYMAP default values (fixed search radius and from four to 10 points for interpolation), except when this leads to computing errors. We have used the same contour intervals (productivity levels) for each of the land maps so that they are directly comparable, both visibly and mathematically. Only the Ocean Productivity Map has different increments, as a result of the much lower productivity levels found in the oceans. For both land and ocean maps, the intervals are those of Lieth's original hand-drawn map (Lieth, 1964). Finally, although the maps are usually produced with legends (15° × 15° tic marks, the equator and tropics, etc.), special maps without legends have been produced for the MAPCOUNT evaluations so that no information would be obliterated by overprinting and thus lost to the evaluation.

The MAPCOUNT Program for the Quantification of Mapped Areas

The MAPCOUNT program, in combination with a user-supplied deformation routine to describe the map projection, is designed to assess quantitatively prestored, SYMAP-produced computer maps. It is designed for use with maps of straight, horizontal parallels and no more than 250 printed rows. It can be used on maps without straight, horizontal parallels only if no projection-related options are requested, that is, for a straight count of the symbols. If areas or area-based evaluations are desired, the requirement of a projection with straight, horizontal parallels cannot be circumvented.

In order to evaluate any map, MAPCOUNT must perform the following operations.

1. Position the map at its top border.
2. Determine for each map row the number of overprintable output records which constitute the row.
3. Count the number of occurrences per row of the requested symbols.
4. Pass these output records and sums individually to inner loops of the program for the processing required by the particular run.
5. Recognize and process the end of the map and subsequent legends.
6. Pass full control to sections of the program that compute the final results.

SYMAP maps are printed in vertical strips one page in width. If a map extends horizontally over more than one printer page, the intermediate SYMAP map borders are recognized and the line index is reset to line 1 so that sums are taken correctly over the entire horizontal extent of the map. Moreover, for area-dependent evaluations, MAPCOUNT must transform the sums into area values using the supplied scale and projection information, add the numbers of occurrences of symbols and the areas over the whole map, and print the resulting sums and areas for each symbol and for the entire map, optionally for each row of the map as well ("areas" option). If a thematic evaluation is requested, such as for productivity in our case, the evaluation is made for each thematic level based on the total area of the zone, and these values are added to give a thematic total for the whole map in the units requested ("evaluations" option). If desired, the counting and evaluation can be restricted to a spherically quadrilateral subsection of the map formed by user-specified parallels and meridians, such as the tropical zone delimited in MAPCOUNT notation by latitude 22.5° to −22.5° and longitude −180.0° to 180.0° ("sections" option). This restricted evaluation also may be repeated with changing geographic limits as often as desired, e.g., in order to evaluate a map in latitudinal belts. Furthermore the evaluation may be restricted to the interior of an arbitrarily shaped but totally convex region specified by the user in either SYMAP or geographic coordinates ("outline" option). The continent of Australia, for example, could be evaluated by specifying an outline that circumscribes Australia but that includes no other land. If a totally convex region, e.g. Europe or America, cannot be given, the evaluation may be performed piecewise. As an option, the map itself can be printed on a separate output file or storage device. As many maps or passes of the same map as desired can be processed in a single run, simply by stacking instruction packages, as the program terminates only when the input file becomes empty. Up to 16 symbols can be counted at one time. The use of MAPCOUNT generally requires the provision of a deformation routine that describes the map projection. The deformation routine must express the Earth's latitudinal, longitudinal, and areal deformation caused by the map projection, and it must be supplied to MAPCOUNT in the form of a subroutine DEFORM of standard format.

Evaluation of the Accuracy of Lieth's Land and Sea Outlines

We are now in a position to begin evaluating the Lieth maps. In order to interpret correctly the results of the MAPCOUNT evaluations, however, we first must have some idea of the accuracy of the areas rendered by the land and sea outlines used by Lieth in his maps. These outlines comprise the so-called UNC Biosphere model. The existence of the MAPCOUNT program (with the outline option) permits us to evaluate these areas for the first time, as given in Table 13–1. In order to evaluate the land outline, the land was first divided into 13 units, roughly corresponding to the continents and largest islands or island groups. The units are mutually exclusive but completely cover the Earth's land area. The "true area" values, therefore, add up to the Earth's total land area. The "computed area" values are those computed by MAPCOUNT for the land outline and also include the computed areas of the inland lakes appearing in the model in the cases of North America, Eurasia, and Africa. The computed areas reflect, of course, only those land areas included in the SYMAP model.

The SYMAP scheme for the representation of areas groups the land masses into three distinct categories, as shown in Table 13–1.

Table 13–1   Comparison of true areas of the world's land-masses and the effective areas rendered by the SYMAP–MAPCOUNT evaluation procedure[a]

Region	1 True area (km²)	2 Computed area (km²)	3 Discrep- ancy (km²)	4 Lati- tude	5 Error (%)
Iceland	103,000	141,318	38,318	64.5	37.2
W. Indies and					
Bahamas	235,858	265,794	29,936	20.0	12.7
Philippines	300,000	197,887	−102,113	11.0	−34.0
E. Indies +					
New Guinea	2,538,145	1,901,525	−636,620	−4.0	−25.1
Oceania excl. N.Z.	171,367	57,850	−113,517	—	−66.2
Greenland	2,175,600	1,891,850	−283,150	73.0	−13.0
N. American					
mainland	21,984,872	21,145,071	−839,801	45.0	−4.8
Eurasia	52,337,000	49,205,600	−3,131,400	45.0	−6.8
Africa +					
Madagascar	30,264,000	29,707,376	−556,624	7.5	−1.8
South America	17,793,000	17,599,508	−193,492	−15.0	−1.1
Australia	7,686,849	7,613,387	−73,462	−25.0	−1.0
New Zealand	268,676	256,650	−12,026	−42.0	−4.5
Antarctica	13,209,000	10,627,160	−2,581,840	−79.0	−16.5
Total land:	149,067,367	140,610,976	−8,455,791		

[a] True area" refers to the true land area of the Earth's surface according to figures given by the *Rand McNally International Atlas* (1969). The areas computed by MAPCOUNT for the SYMAP land outline are juxtaposed in column 2. They include computed areas for inland lakes appearing in the model in the cases of North America, Eurasia, and Africa. The "discrepancy" is obtained by subtracting computed area from the true area and changing the sign, i.e., a negative discrepancy indicates that the area has been underestimated by the SYMAP model. Latitude is the mean latitude of the land mass, with northern latitudes given as positive values. Error % (column 5) is obtained by dividing the discrepancy by the true area, so that underestimated area is represented by a negative error. The regions are defined as follows:

North America: All of North America except Greenland, the Bahamas, and the West Indies

Eurasia: All of the Eurasian land mass plus the British Isles, Faeroes, Spitzbergen, all Mediterranean islands, the Soviet Arctic Islands, the Japanese Archipelago, all Chinese offshore islands, Ceylon, and the South Asian Islands in the Indian Ocean.

Africa: All of the African land mass plus Madagascar, the Mascarene, Comores, and other islands in the Indian Ocean, and the Madeira, Canary, Azores, and Cape Verde Archipelagos in the Atlantic Ocean.

East Indies: All the islands of the Republic of Indonesia, including the entirety of the Islands of Timor, Borneo, and New Guinea

Antarctica: All of the Anarctic land mass plus the South Sandwich, South Georgia, South Shetland, South Orkney, and Falkland Islands and the French (Kerguelen) and British Antarctic and South Indian Ocean islands

Oceania: All remaining Pacific Islands (except New Zealand and its dependencies), primarily Polynesia (including the Hawaiian Islands), Melanesia (including all of the Bismarck Archipelago), and Micronesia

1. Islands and island groups the areas of which are overestimated in the model because of their relatively small sizes relative to the areas of single print positions (Iceland, West Indies, Bahamas)
2. Groups of islands, the areas of which are underestimated as a result of the omission of some constituent islands (Philippines, East Indies, Oceania)
3. The larger land masses, which show a truer picture of the pattern of area representation by the model (the continents, plus Greenland and New Zealand)

In designing the land units to be used, a special attempt was made to separate islands and island groups that could be badly over- or underestimated from their adjacent continents and thereby to evaluate the effect of poor representation of smaller land masses separately. Therefore, the West Indies and Bahamas are evaluated separately from North America, and so on. The total true area of all the islands in the first two categories (3,348,370 km²), however, is only 2.25% of the world's true land area. This entire area is given a net 25% underestimation (2,564,374 km²) by the SYMAP–MAPCOUNT evaluation procedure, and this underestimation (784,000 km²) is only 0.52% of the earth's total land area. The total effect of land omitted from the model, which we might estimate at about 1,000,000 km² (based on the 852,250 km² of the second category of land units), is larger but still < 1%. The over-all effect of the islands in the first two categories thus plays a relatively minor role. Looking at the larger land masses, however, one sees immediately that all areas are underestimated by the combined SYMAP–MAPCOUNT mapping procedure, with the magnitude of the area underestimation increasing unmistakably toward the poles. There seem to be five possible sources of error.

1. Overall scale error within the MAPCOUNT program
2. Error in the specification of projection characteristics
3. Approximation of continuous surfaces by discrete print positions
4. Error in Lieth's outlines, including discrepancy between the true mathematical specifications of the Robinson projection and the wall map from which Lieth's outlines and distances were taken
5. Approximation by MAPCOUNT of convex curves (all meridians) by straight secant lines at 5° intervals.

There is surely a certain amount of error from each of these sources, but it is difficult to say which are most significant. Probably more important is the fact that the Robinson projection exaggerates each of these five types of errors toward the poles, making all but the second item potentially critical in polar regions. The relatively greater distortion in the comparatively unproductive polar regions, however, is an advantage for our purpose, the evaluation of plant production. Even with an overall area underestimation of nearly 6% for the land surfaces, the corresponding underestimation of production does not exceed ~ 4%. Considering the problems involved and the expenditure of time and effort required to produce accurate SYMAP outlines on a world scale, this degree of error, which is probably even less in both the SYMAP outlines and in the

MAPCOUNT procedure individually, is acceptable. Therefore, the final production results, which appear in Table 13–9, are the raw figures from the MAPCOUNT evaluations, with no adjustment for the error in area, although with notation of the areas on which the production estimates are based. A short analysis of the distribution of the area underestimation over the various productivity zones is provided in the Conclusions section of this chapter.

The Primary Production of the Earth

As a development from the first world productivity map (Lieth, 1965), the total world primary production was estimated by Lieth (1971, 1973) at the Miami symposium to about 100.2×10^9 t/year of dry matter on land and about 55.0×10^9 t/year in the oceans.[2] These values were based on area estimates for each of 20 different vegetation types, obtained from the planimetry of a number of world vegetation maps prepared by Lieth's students. The areas were then multiplied by characteristic productivity values for the respective vegetation types using the estimates of Whittaker and Likens (Whittaker and Woodwell, 1969) and Lieth's own experience. Lieth's 1971 total for world primary production (155.2×10^9 t/year) and the somewhat higher estimate given in Chapters 10 and 15 (116.8×10^9 t/year continental and 55×10^9 t/year marine, for a world total of 171×10^9 t/year) may be compared with the computerized evaluations that follow.

The Innsbruck Productivity Map

In 1971–1972, the productivity map of 1965 was simulated by Lieth on the computer as two separate statistical maps, the Innsbruck Productivity Map (Lieth, 1972) for the land portions and the Ocean Productivity Map (Lieth and Box, 1972). The former is based on 930 datum points spaced as evenly as possible over the land masses. The values of these datum points were not productivity values but rather the integers 0 through 6 to represent the seven productivity levels that Lieth had used in his 1964 map. Because this was Lieth's first computer map, and because the productivity figures are so approximate, no further effort was made to break down the values within productivity levels. The results of the MAPCOUNT evaluation by productivity level of the Innsbruck Productivity Map are shown in Table 13–2. The total land production, as represented by the Innsbruck Productivity Map, is evaluated as 104.9×10^9 dry matter/year, a figure that is nearly 5% above Lieth's 1971 estimate, despite the 6% underestimation of the Earth's land area. Dividing this final production figure by the Earth's total (modeled) land area, we also can get an average productivity figure for the land: 747.7t/km² or g/m²/year. (This figure is provided also by MAPCOUNT.) This corresponds roughly to the productivity level of a very good grassland or of a woodland or poorer temperate forest, and the figure is also somewhat higher (by ~ 8.5%) than Lieth's 1971 estimate based on the same data. The significance of worldwide averages is limited by the fact that local

---

[2] Metric ton (10^6g) is abbreviated as t.

Table 13–2   Production evaluation of the Innsbruck Productivity Map[a]

Level	Productivity range (g/m²/year)	Symbol (over- (print)	Number of Occur- rences	Area estimates (10⁶ km²)	Average prod. (g/m²/ year)	Produc- tion (10⁹ t/year)
1	0–100	•	4,715	15.16	10.0	0.152
2	100–250	'	6,087	32.34	175.0	5.659
3	250–500	''	5,665	30.42	375.0	11.406
4	500–1,000	X	3,032	18.62	750.0	13.964
5	1,000–1,500	(O–)	2,896	19.64	1,250.0	24.554
6	1,500–2,000	(OHX)	2,035	14.69	1,750.0	25.711
7	2,000–3,000	(OXAV)	1,264	9.36	2,500.0	23.405
Total:			25,694	140.23		104.851

[a] All values represent amounts per year. Productivity values are in grams dry matter per square meter per year, or metric tons per square kilometer per year. The average productivity figures (column [6]) are the midpoints of the productivity intervals on the map, with the exception of that for level one, which was reduced to reflect the large sizes of the totally unproductive ice caps. The final production figure (column [7]) is obtained by multiplying the area estimate (column [5]) by the average productivity value (column [6]). The symbols actually appearing on the Innsbruck map are given in column [3], along with their counted totals in column [4].

productivity values (for areas not covered permanently by ice) range over about two orders of magnitude, from ~ 30 to ~ 3000 t/km²/year. If we keep this range in mind as a rough constant, however, the worldwide average is perhaps a useful figure as well.

A second version of the Innsbruck Productivity Map was produced by Lieth for use as the land portion of his combined land–sea Seattle Productivity Map. In this second version, only six levels appear, the last two having been combined into a single interval from 1500 to 3000 t/km²/year. The areas of the other levels are also slightly different, as a result of slightly differing data. Using an average figure of 2000 t/km² annually for this combined level 6 (based on the relative areas of levels 6 and 7 in Table 13–3), the values obtained from the MAPCOUNT evaluation of this second Innsbruck Map were only 92.6 × 10⁹ t/year for the whole earth and only 660.3 g/m²/year as a worldwide average.

The Miami Model

Additional estimates of terrestrial production are available from the predictive climate-based models of potential productivity, the Miami Model based on temperature and precipitation (Lieth, 1971, 1973), and the Montreal Model based on actual evapotranspiration (Lieth and Box, 1972). The Miami Model is based on data for mean annual temperature and average annual precipitation for the world's land areas, as read from the *Climate-Diagram World Atlas* (Walter and Lieth, 1960–1967). A single data set of 1001 data points was used, each point having both a temperature and a precipitation value associated with it. Separate production estimates based on temperature and precipitation thus

Table 13–3   Production evaluation of the Miami Model
Productivity Map[a]

Level	Produc-tivity range (g/m²/year)	Symbol (over-print)	Number of occur-rences	Area estimate (10⁶ km²)	Average prod. (g/m²/year)	Produc-tion (10⁹ t/year)
1	0–100	•	4,531	15.39	34.9	0.536
2	100–250	ı	2,958	14.15	170.0	2.406
3	250–500	ı ı	4,505	23.49	380.7	8.943
4	500–1,000	X	6,122	34.87	727.2	25.357
5	1,000–1,500	(O–)	3,126	20.43	1,222.2	24.969
6	1,500–2,000	(OHX)	2,884	20.36	1,743.2	35.489
7	2,000–3,000	(OXAV)	1,569	11.58	2,313.4	26.778
Total:			25,695	140.22		124.478

[a] All values represent amounts per year. Productivity values are in grams dry matter per square meter per year, or metric tons per square kilometer per year. The average productivity figures (column [6]) are the arithmetic averages of the values of all the datum points in the respective intervals. The final production figure (column [7]) is obtained by multiplying the area estimate (column [5]) by the average productivity value column [6]. The symbols actually appearing on the Miami map are given in column [3], along with their counted totals in column [4].

are directly comparable. Two productivty estimates were made, both as a function of temperature alone and as a function of precipitation alone, using least-squares formulas developed from a set of about 50 reliable productivity measurements from five continents, as described in Chapter 12. The lower of the two predicted productivity values was chosen as the productivity value for the Miami Model, reflecting Liebig's law of the minimum. The Miami Model map was then produced as a SYMAP statistical map of these 1001 predicted productivity values. The data points were so chosen that there are always at least three datum points per $15° \times 15°$ quadrat on the earth's surface, wherever three climate values were available in the atlas. The productivity intervals used in the Miami map are the same as those of the Innsbruck map. The results of the MAPCOUNT evaluation of the Miami Model map are shown in Table 13–3 above. The total production of the land, as represented by the Miami Model, is evaluated by MAPCOUNT to be 124.5×10^9 t dry matter/year, with a worldwide average of 887.7 g/m²/year. These figures are ~ 19% higher than those of the Innsbruck map.

The C. W. Thornthwaite Memorial (Montreal) Model

The Thornthwaite Memorial (or Montreal 2) Model (Lieth and Box, 1972) is based on a statistical map simulation of a single parameter, actual evapotranspiration, which, in a sense, combines the most important effects of temperature and precipitation, not as separate quantities, but rather as they simultaneously affect the plant. The data, which are much more difficult to obtain

Table 13–4 Production evaluation of the Thornthwaite Memorial
Model Productivity Map[a]

Level	Productivity range (g/m²/year)	Symbol (over-print)	Number of occur-rences	Area estimate (10⁶ km²)	Average Prod. (g/m²/year)	Production (10⁹ t/year)
1	0–100	•	5,728	19.68	5.7	0.112
2	100–250	ı	1,775	9.79	195.7	1.916
3	250–500	ıı	4,930	26.51	410.1	10.875
4	500–1,000	X	5,711	32.34	736.7	23.827
5	1,000–1,500	(O–)	3,225	20.72	1,205.4	24.973
6	1,500–2,000	(OHX)	3,488	24.95	1,741.9	43.707
7	2,000–3,000	(OXAV)	844	6.27	2,120.0	13.297
Total:			25,701	140.26		118.706

[a] All values represent amounts per year. Productivity values are in grams dry matter per square meter per year, or metric tons per square kilometer per year. The average productivity figures (column [6]) are the arithmetic means of the values of all the datum points in the respective intervals. The final production figure (column [7]) is obtained by multiplying the area estimate (column [5]) by the average productivity value (column [6]). The symbols actually appearing on the Thornthwaite Memorial Map are given in column [3], along with their counted totals in column [4].

and necessarily somewhat less accurate, are taken from a world map by Geiger (Geiger, 1965). The base map of evapotranspiration was produced from a data set of 562 datum points by the author. An initial data set of about 360 datum points spaced in a grid fashion over the land areas was supplemented by an additional 200 datum points where they appeared to be most needed to improve the contour fit to the Geiger map, generally in areas of high topographic complexity. The Thornthwaite Memorial Model was derived, as was the Miami Model, by a least-squares fit between reliable productivity measurements and corresponding climatic values, in this case, evapotranspiration values read also from the Geiger map. The productivity values used were the same approximately 50 that were used for the Miami Model. The productivity intervals on the Thornthwaite Memorial Map are again the same as those of the previous maps. The results of the evaluation of the Thornthwaite Memorial Map by MAPCOUNT are shown in Table 13–4. The total production of the land, as represented by the Thornthwaite Memorial Model, is evaluated by MAPCOUNT to be 118.7×10^9 t/year, with a worldwide average of 846.3 g/m²/year. The figures are ~ 13% higher than those of the Innsbruck map but ~ 5% lower than those of the Miami Model.

The Ocean Productivity Map

The Ocean Productivity Map is based on 1621 datum points spaced in a ~ 5° × 5° grid over the oceans, as read from Lieth's 1965 map by two students, E. Hsiao and P. van Wyck. Unlike the statistical simulation of the land

Table 13–5   Production evaluation of the Ocean Productivity Map[a]

Level	Productivity range (g/m²/year)	Symbol (overprint)	Number of Occurrences	Area estimate (10⁶ km²)	Average prod. (g/m²/year)	World ocean production (10⁹ t/year)
1	0–50	•	21,184	130.48	26.2	3.419
2	50–100	'	15,464	96.38	92.9	8.951
3	100–200	:	11,318	63.66	183.7	11.693
4	200–400	O	6,657	34.77	309.4	10.757
5	> 400	(O+)	1,268	6.51	508.6	3.310
Total			55,891	331.80		38.130
Adjusted for actual ocean area (Table 13–7)				361.0		43.763

[a] All values represent amounts per year. Productivity values are in grams dry matter per square meter per year, or metric tons per square kilometer per year. The average productivity figures (column [6]) are the arithmetic means of the values of all the datum points in the respective intervals. The final production figure (column [7]) is obtained by multiplying the area estimate (column [5]) by the average productivity value (column [6]). The symbols actually appearing on the Ocean Productivity Map are given in column [3], along with their counted totals in column [4].

portion of the 1964 map (in the Innsbruck map), the data for the Ocean Productivity Map are estimated productivity values, as opposed to indices of productivity level. Many points were given the values 50, 100, 200, or 400 g/m²/year, that is, exactly on boundaries between levels; the result was that the contours do not fit those of the 1964 map as well as possible. The Western Hemisphere is especially undervalued and, according to the 1964 map, should have the same highly productive algae belt at 60°S, which is shown in the Eastern Hemisphere. This inclusion would probably raise the total production estimate to a figure nearer Lieth's prediction of 55.0 × 10⁹ t/year. The results of the MAPCOUNT evaluation of the Ocean Productivity Map are shown in Table 13–5. The total production of the ocean is evaluated to be 38.1 × 10⁹ dry matter/year, with a worldwide average of ∼ 114.9 g/m²/year. These levels represent only ∼ 70% of Lieth's 1971 estimates, but we notice immediately that the area of the oceans has been underestimated not by ∼ 5.7%, as was the land, but by ∼ 8.1%, based on the figure 361 × 10⁹ km² for the total ocean area of the Earth. The worldwide average for the oceans is perhaps somewhat more significant than the terrestrial averages because the range of local productivity in the oceans (excluding the arctic ice cap) is only about one order of magnitude instead of two. This oceanic average approximates the productivity level of a continental shelf area without upwelling or reefs, or of the most productive parts of the open ocean with favorable currents.

Geographic Distribution of the Error

Although the SYMAP outlines and the MAPCOUNT evaluation procedure together underestimate the land area by ~ 6%, the consequent production underestimation, as mentioned previously, should be lower because much of the area
underestimation occurs in the highly unproductive polar regions. It is worthwhile to examine the distribution of the area underestimation and its effect on
the production values in greater detail, partly to convince ourselves that the
production error is, in fact, no greater than the error in the area, but even more
importantly to examine the nature of the error. Most of the error can probably
be removed from future models by modifying the land and ocean outlines and
by adjusting the scale in the MAPCOUNT evaluation. The Innsbruck and Ocean
maps are examined in detail. In order to examine the terrestrial case, we
grouped the 13 land units of Table 13–1 into five groups as follows, based on
similarity of productivity level and pattern.

Polar areas:	Greenland, Antarctica
Boreal areas:	North America, Eurasia, Iceland
Wet tropical areas:	Philippines, Oceania, West Indies and Bahamas, East Indies
Large diverse areas:	Africa, South America, Australia
Temperate insular area:	New Zealand

The area-underestimation in each group is then determined from the "discrepancy" values in Table 13–1, and average productivity figures are assigned
to each group, based on the values in Table 10–1, which reflect those on which
the Innsbruck map is based. The large northern continents are especially important. Neither North America nor Eurasia is dominated by boreal forests, but
because of the shapes and locations of these two continents and because of the
fact that the area underestimation increases toward the poles, most of the
underestimated regions of these continents are dominated by either boreal
forests or by biomes of a similar productivity level (e.g., grasslands). The distribution of area underestimation and its effect on production estimates of the
Innsbruck Productivity Map are summarized in Table 13–6. The results show
that the production underestimation of about 4×10^9 t/year is indeed only
~ 3.8% of the raw figure of 104.85×10^9 t, because about one-third of the
underestimation area is permanently covered by ice. The Miami and Montreal
maps can be examined similarly by using average productivity values based on
these models, but the overall patterns are roughly the same, and the production
underestimations are also ~ 4%.

Whereas the underestimation regions on land were fairly well dominated by
permanent ice or boreal forests, the situation in the oceans is somewhat more
complicated. The area underestimation increases toward the poles with most of
the ocean area comparatively uniform at a low productivity level. The general
pattern is complicated, however, by the following two factors.

1. The presence of bands of comparatively high productivity near the polar
   regions (45°–60°), especially in the Southern Hemisphere

2. The poor fit of the ocean outline to the land outline, resulting in the coastal gaps roughly one column wide on the western and southern sides of most land masses, strips that contain regions of high productivity as a result of the continental shelf and coastal upwelling zones

Because the projectional distortion is the same over both land and ocean, we assume that ~ 4.7% of the 8.1% underestimation is caused by distortion, and the remaining 3.4% by the omitted areas, primarily the coastal gaps. For the underestimation caused by distortion, we simply take the average productivity figure for the entire ocean area as being the best available estimate: 115 g/m^2/year. For the coastal gaps, we take a weighted average (based on total areas) of the average productivity figures (Chapters 8 and 15) for reefs and estuaries, for continental shelf, and for upwelling zones. This figure must be reduced, however, because these productive coastal regions have not been ignored completely by the poor coastal fit of the outlines. To some extent, the coastal productive areas have merely been displaced, and the unproductive open ocean has had its area underestimated near these coastal areas as well. A reasonable overall figure seems to be ~ 300 g/m^2/year. The resulting distribution of oceanic production underestimation is shown in Table 13–7.

As seen in the Table 13–7, the pattern of error in the oceans is somewhat different from that on land. The consequent additional production (5.645 × 10^9 t/year), rather than being less than the 8.1% area underestimation, amounts to

Table 13–6   Approximate distribution of terrestrial area underestimation and its effect on the evaluation of the Innsbruck Productivity Map[a]

Underestimated productivity level (biome types)	Area under-estimation (10^3 km^2)	Percent of under-estimation	Percent of total land	Average produc-tivity (g/m^2/year)	Additional produc-tion (10^9 t/year)
Polar areas	2,865	33.9	1.92	5	0.014
Boreal areas	3,933	46.5	2.64	500	2.282
Wet tropical areas	822	9.7	0.55	1,800	1.479
Large diverse areas	824	9.7	0.55	650	0.537
Temperate areas	12	0.1	0.008	800	0.010
Totals:	8,456	100.0	5.67		4.006

[a] The five classes of land units, grouped according to similar productivity level, are listed in column [1], with SYMAP–MAPCOUNT area underestimation in thousands of square kilometers in column [2]. The fraction of the total area underestimation which each of these areas represents is given as a percentage of the total underestimation in column [3], whereas column [4] shows the percentage of the Earth's total land area represented by each underestimation figure. The average productivity figures (column [5]) are rough averages for the Innsbruck Productivity Map based on the values in Table 10–1 and the known vegetation cover. The additional production figures (column [6]) are obtained by multiplying the area underestimation (column [2]) by the average productivity (column [5]).

Part 4: Utilizing the Knowledge of Primary Productivity

Table 13–7   Rough distribution of oceanic area underestimation and its effect on the evaluation of the Ocean Productivity Map[a]

Source of under-estimation of area	Area under-estimation (10^6 km²)	Percent of under-estimation	Percent of total ocean	Average produc-tivity (g/m²/ year)	Additional produc-tion (10^9 t/year)
Distortion	16.97	58	4.7	115	1.952
Gaps in outline	12.27	42	3.4	300	3.681
Totals:	29.24		8.1		5.633

[a] The two sources of area underestimation are listed in column [1] with the rough underestimation percentages (based on total oceanic area of 361 × 10^6 km², see discussion preceding Table 13–6) owing to each source in column [4]. The amount of underestimated area due to each source (column [2]) is calculated by multiplying the percentage in column [4] by the total oceanic area, and the percentages of the underestimation due to each source are given in column [3]. The average productivity figures (column [5]) are rough averages based on the values in Lieth and Box (1972), and the known geographic distribution of marine vegetation formations. Additional production (column [6]) is obtained by multiplying the area underestimation (column [2]) by the average productivity (column [5]).

~ 14.8% of the raw production figure of 38.1 × 10^9 t/year. These additional 5.6 × 10^9 t, plus the addition of the productive band in the South Pacific, would bring the total estimate significantly closer to Lieth's 1971 estimate of 55.0 × 10^9 t/year.

Observations and Conclusions

The MAPCOUNT evaluation of the Lieth productivity maps provides us at once with a large number of new values for various aspects of the Earth's production. The most important, of course, are the overall land and sea production estimates for the whole Earth, which are listed again in Table 13–8 together with the other production estimates contained in this book. From these overall values and the information on geographic distribution of productivity contained in Tables 13–2 through 13–5, we make the following major observations:

1. The overall values and the productivity distribution patterns for the three land maps generally agree with each other and with production estimates made by other scientists as summarized in Lieth (1973), Whittaker and Likens (1973a, b), and Chapter 2.
2. The value 104.85 × 10^9 t/year for the Innsbruck map agrees acceptably with Lieth's 1971, 1972 figure (100.2 × 10^9 t), even when one considers that the area underestimation would raise the Innsbruck figure to nearly 109 × 10^9 t. The value of 43.8 × 10^9 t/year for the oceans also agrees acceptably with the figures presented in Chapter 8.
3. The Miami and Thornthwaite Memorial climate-based models show higher overall production levels than does the Innsbruck map of actual

production, and the amounts of the difference (19% and 13%, respectively) are in surprising agreement with the revised world estimate in Chapter 15, and Lieth (1975). Furthermore, the fact that the Miami figure is slightly higher than the Thornthwaite Memorial figure, whether artifact or not, is also expected and reasonable, as the Miami Model does not include the effect of water lost to production due to runoff.

4. The most striking difference in production distribution between any of the

Table 13–8   Comparison of the various production estimates
for the earth[a]

1 Area (10^6 km²)	2 NPP est. (10^9 t/year)	3 Attempt	4 Source
Land			
140	96 (38.4×10^9 t C)	Planimetering Lieth's 1964 productivity map and checking it against annual, global CO_2 fluctuation	Junge and Czeplak (1968)
149	109.0	Sum of estimates by means for major vegetation types	Whittaker and Likens (1969) in Whittaker and Woodwell (1971)
149	100.2	Sum of estimates by means for major vegetation types	Lieth (1971, 1973)
149	116.8	Sum of estimates by means for major vegetation types	Chapter 15
140.2	104.9	Evaluation of Innsbruck productivity map (Lieth, 1972)	See Table 13–2 (Box)
140.2	124.5	Evaluation of Miami Model (Lieth, 1972)	See Table 13–3 (Box)
140.3	118.7	Evaluation of Montreal Model (Lieth and Box, 1973)	See Table 13–4 (Box)
149	121.7	Sum of estimates by means for major vegetation types	Lieth (1975)
Ocean			
332	46–51 (23×10^9 t C)	Sum of estimates by means for major zones	Chapter 8
361	55.0	Evaluation of major zones	Chapter 15
361	43.8	Evaluation of Oceans Productivity Map (Lieth and Box, 1972)	See Table 13–5 and 13–7 (Box)
361	55.0	Sum of estimates of major zones	Lieth (1975)

[a] The figure for land or ocean area on which each production estimate is based is given in column [1], along with the production estimate (net primary production) in column [2]. A short description of each attempt is given in column [3], and the author and location of his results are shown in column [4].

sets of results is the shift of the most common productivity level from level 4 (temperate forests and richer grasslands) in the climate-based models to level 2 (woodlands, poor grasslands, and semideserts) in the Innsbruck map of actual productivity.

Due to the nature of the data and the modeling and evaluation procedure, we cannot consider these results to be final. It is hoped, however, that the results underscore the usefulness of the MAPCOUNT procedure.

The evaluation of the Earth's production certainly does not end with the computation of figures for the entire Earth alone. The sections and outline options of MAPCOUNT also allow us to evaluate such important quantities as

1. Total and average production values for latitudinal belts around the Earth
2. The percentages of the earth's total production produced by the tropics, and by the northern and southern temperate zones, frigid zones, by the hemispheres, etc.
3. Production figures for individual land masses of the Earth, such as for the individual continents and larger islands
4. Production values for individual countries

These evaluations require some additional refinements to the MAPCOUNT program, as well as the provision of new outlines, and are therefore reserved for a future paper.

Acknowledgments

This research was supported by and performed at the Institut für physikalische Chemie, Kernforschungsanlage Jülich GmbH, Jülich, Federal Republic of Germany. Computer time was also received from the Computation Center of The University of North Carolina.

References

Box, E. 1975. MAPCOUNT und andere Programme zur Herstellung, Weiterverarbeitung, und quantitativen Auswertung statistischer, vom Computer gedruckter Landkarten. Special Publ. of the Kernforschungsanlage Jülich GmbH, (*In press.*)

Geiger, R. 1965. The atmosphere of the earth, Map No. WA6. *Annual Effective Evapotranspiration.* Darmstadt: Justus Perthes.

Harvard Graduate School of Design. SYMAP Program and SYMAP Reference Manual for Synagraphic Computer Mapping. Cambridge, Massachusetts: Harvard Univ.

Junge, C. E., and Czeplak, G. 1968. Some aspects of seasonal variation of carbon dioxide and ozone. *Tellus* 20:422–433.

Lieth, H. 1964. Versuch einer kartographischen Darstellung der Produktivität der Pflanzendecke auf der Erde. In *Geographisches Taschenbuch* 1964/65, pp. 72–80. Wiesbaden: Steiner.

——. 1971. The net primary productivity of the earth with special emphasis on the land areas. In *Perspectives on Primary Productivity of the Earth*, R. Whittaker, ed. (*Symp. AIBS 2nd Natl. Congr., Miami, Florida, October 1971.*)

283

13. Quantitative Evaluation of Global Primary Productivity Models

——. 1972. Über die Primärproduktion der Pflanzendecke der Erde. (*Symp. Deut. Bot. Gesell., Innsbruck, Austria, Sept. 1971.*) *Z. Angew. Bot.* 46:1–37.

——. 1973. Primary production: Terrestrial ecosystems. *Human Ecol.* 1:303–332.

——. 1975. The primary productivity in ecosystems. Comparative analysis of global patterns. In *Unifying Concepts in Ecology*, W. H. Van Dobben and R. H. Lowe-McConnel, eds. The Hague: Dr. W. Junk. (*In press.*)

——, and E. Box. 1972. Evapotranspiration and primary productivity; C. W. Thornthwaite Memorial Model. In *Papers on Selected Topics in Climatology*, J. R. Mather, ed., Thornthwaite Memorial Vol. 2 (*Publications in Climatology* Vol. 25(2):37–46). Elmer, New Jersey: C. W. Thornthwaite Associates.

Rand-McNally. 1969. *Rand-McNally International Atlas.* Chicago: Illinois: Rand-McNally.

——. *Standard outline map of the World*, Robinson Propection. 1″ = 500 miles equatorial scale. Chicago, Ill.: Rand-McNally and Co.

Reader, R., and UNC Applied Programming Group. 1972. SYMAP (Version 5.16A) *Instruction Manual*, 65 pp. Document No. LSR-139-0. Research Triangle Park, North Carolina: Triangle Universities Computation Center.

Robinson, A. H. 1974. A new map projection: Its development and characteristics. *Intl. Yearbook of Cartography*, 14:145–155.

Walter, H., and H. Lieth. 1960–1967. *Klimadiagramm-Weltatlas.* Jena: VEB Fischer.

Whittaker, R. H., and G. E. Likens. 1973a. Primary production: The biosphere and man. *Human Ecol.* 1:357–369.

——, and G. E. Likens. 1973b. Carbon in the biota. In *Carbon and the Biosphere*, G. M. Woodwell and E. V. Pecan, eds. *Brookhaven Symp. Biol.* 24:281–302. Springfield, Virginia: Natn. Tech. Inform. Serv. (CONF-720510).

——, and G. M. Woodwell. 1969. Measurement of net primary production of forests. (*IBP-UNESCO Conf. Brussels, 1969.*) Brookhaven, New York: Brookhaven National Laboratory Document No. 14056.

——, and G. M. Woodwell. 1971. Measurement of net primary production of forests. (French summ.) In *Productivity of Forest Ecosystems: Proc. Brussels Symp. 1969*, P. Duvigneaud, ed. *Ecology and Conservation*, Vol. 4:159–175. Paris: UNESCO.

14

Some Prospects beyond Production Measurement

Helmut Lieth

We have been asked what insight into an ecosystem can be gained if one knows its dry matter production. In some discussions of primary productivity I have heard such figures likened to the relatively insignificant house numbers along streets. The comparison is inept in at least one fundamental respect: the primary productivity of an ecosystem is its most essential resource base and the working income on which the development of community structure and function depends. Deeper insight may be gained from the study of productivity when that productivity is viewed as an approach to analysis of the structure and composition of communities. With the help of a few examples, this chapter investigates the expanded perspective one may develop by partitioning productivity into four aspects: stratal productivity, productivity of individual species, production and diversity, and the chemical composition of productivity.

Some of these have been considered in seminar discussions held with my students at the University of North Carolina; credit is given to the cooperating students in the individual sections.

Stratal Productivity

Two relationships described in Lieth (1953) and considered further in chapter 12 should be noted in connection with one another: the diagram relating formation types to climate (Fig. 12–1 this volume), and the curves relating productivity to temperature and precipitation (Fig. 12–3 and 12–4). There is

KEYWORDS: Primary productivity; species diversity; stratal productivity; chemical differences; ecosystems; trophic levels.

Primary Productivity of the Biosphere, edited by Helmut Lieth and Robert H. Whittaker.
© 1975 by Springer-Verlag New York Inc.

no doubt that productivity is a major, underlying determinant of vegetation structure. As one proceeds along either the moisture or the temperature gradient, and as productivity decreases, there is a general tendency for the community supported by that productivity to become simpler and lower in structure. Therefore, productivities control, in part, the ranges of climates occupied by different formation types shown in Figure 12–1, and the related diagrams of Holdridge (1967) and Whittaker (1970a, Fig. 3–8). However, community structure cannot be predicted from productivity or, except within broad limits, from mean annual temperature and precipitation. For certain combinations of temperature and precipitation it may be predicted within reason that the climax vegetation of an area will be forest, desert, or tundra. For climates that are intermediate to these, however, other environmental factors—notably seasonal distribution of rainfall, fire frequency, and soil characteristics—may determine whether the vegetation supported is woodland, shrubland, or grassland. Some of these relationships are also relatively predictable; for example, a given combination of precipitation and temperature will support sclerophyll shrubland in a maritime, dry-summer climate, but grassland in a climate with summer rain in the plains of a continental interior.

Despite the relative flexibility of vegetational expression of climate, it may be desirable to look further into the structural design of communities, as regards the stratal distribution of primary productivity. Certain stratal relationships are evident. As productivities decrease along either the temperature or the moisture gradient, forest canopies open and undergrowth productivities increase, until the community is dominated by shrubs or herbs. Broadly speaking, there is a complementary relationship between the productivity of the tree stratum and that of the undergrowth. In some areas as tree coverage and productivity decrease shrubs become dominant, and as shrub coverage and productivity decrease toward still drier environments, grasses become dominant. Thus there is a progressive movement of dominance downward to the lower strata. These trends correspond to the effects of stresses, such as irradiation, upon communities (Woodwell, 1967). The effects of these stresses also may cause decreases in productivity. In other areas, forests open to woodlands and these give way to grasslands, until, in still drier climates, the grasslands are replaced by desert scrub. In some cases the shrub and herb strata bear a complementary relationship to one another, one increasing as the other decreases; in other cases these strata decrease together as the tree stratum decreases.

Figure 14–1 illustrates two cases in which productivities of strata can be followed along the moisture gradient. In the Great Smoky Mountains of Tennessee (Whittaker, 1966), as the tree-stratum productivity decreases toward drier environments, the shrub stratum increases to share dominance with it. The response of the herb stratum is bimodal, with one peak in response to the most favorable moisture conditions at the mesic extreme of the gradient, and a second peak at the xeric extreme where the limited productivity and coverage of the tree and shrub strata combine to give the herb stratum relatively full exposure to light. In the Santa Catalina Mountains, Arizona, undergrowth productivity increases as tree productivity decreases (Whittaker and Niering 1975). The shrub and herb strata at first increase in parallel, but toward more xeric environ-

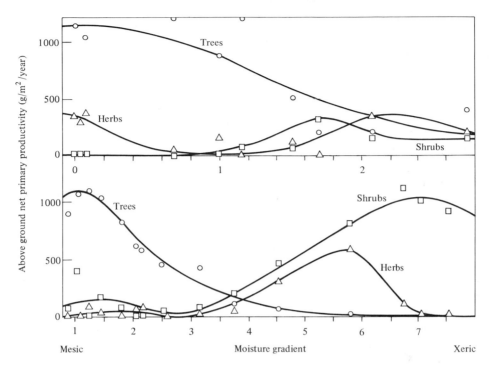

FIGURE 14–1. Trends in stratal productivity along moisture gradients.
○, Tree stratum; □, shrub stratum; △, herb stratum; all values dry
matter aboveground. Weighted-average indices of relative position
of samples along moisture gradient are indicated on horizontal axes.
(*Above*) Great Smoky Mountains, Tennessee, Cove forest (mesic)
through oak forests to pine heaths (xeric). (Herb stratum values
multiplied by 10 to show these on same scale.) Data of Whittaker
(1966). (*Below*) Santa Catalina Mountains, Arizona. High-elevation
fir forest (mesic) through pine forest, woodlands, and desert
grassland to Sonoran semidesert (xeric). (Both shrub and herb
strata × 10; the arborescent shrubs or small trees of the semideserts
have been treated as shrubs.) Data of Whittaker and Niering (1975).

ments herb productivity decreases while shrub productivity increases. (It decreases again in still more arid climates.) Both the herb and the shrub strata
show evidence of a minor, secondary peak of productivity in the most mesic
forests.

These examples show that no simply consistent and predictable patterns
should be expected. Stratal relationships differ in different kinds of climates, and
they may well differ in vegetations with different evolutionary histories—in
Australia, say, compared with the Northern Hemisphere. Nonetheless, it would
be interesting to try first to determine the boundaries that productivity sets on
different kinds of community structure, second to characterize major types of
communities in terms of the patterns or spectra of productivity in different strata
and growth-forms, and third to look for such regularities as may be found in
the relationships of stratal productivities to kinds of environment.

Species Productivity

The productivity in any ecosystem is shared among species. If one analyzes the relative production profile of species in natural communities, one observes that two conditions are almost nonexistent: (1) that all species in a community produce equal amounts and (2) that a single species carries out the primary productivity for a whole community. (There are communities, such as salt marshes, in which a single vascular plant species is present in samples from limited areas, but even in these communities other photosynthetic species are present.)

In an attempt to compare human sociologic systems with natural ecosystems on the productive level, Petrall (1972) analyzed many phytosociologic tables with regard to the relative importances of species in a given type of community. He evaluated phytosociologic tables that expressed the importance values of species as the percentages of cover. These numbers were used to construct summation curves as shown in Figures 14–2 to 4; the relative importances of species are accumulated from the least to the most important species. In such a diagram the equal importance case would yield a 45° linear curve between the 0,0 and 100,100% coordinates. The case of a monoculture would result in a straight line parallel to the ordinate, with no abscissa. The more realistic example for such a case is an agricultural field in which massive weeding results in a dominant stand of the desired crop species and a variable number of weak species with small productivities. Such a community would yield a sharply concave curve, with one arm of the curve (for the weak species) almost parallel to the abscissa, and the other arm of the curve (for the dominant) steeply rising to the 100,100% corner.

The following three figures show some results of this graphic analysis of species importance. Figure 14–2 shows three examples from different forest ecosystems. The curves are hyperbolic with almost equal shanks and the inflection point of the hyperbolas moving along the −45° line that divides the x,y field in half.

Figure 14–3 shows grassland and tundra communities that we consider under natural environmental stress either for water or temperature. This stress appears to move the curves toward the ideal, more equitable (egalitarian) situation. The inflection points tend to drop below the −45° line. We have included in this figure as curve 5 the data set presented in Table 14–2, columns 3 and 4. These data represent biomass values rather than phytosociologic estimates.

Figure 14–4 shows communities under severe management stress. The managed grassland communities fall within the same pattern described for Figure 14–3. Extreme management stress that seriously afflicts the number of persisting species results in unbalanced composition and pushes the inflection point for the curves to the extreme right (curve 1).

The significance of this investigation lies in its possible importance for the understanding of ecosystems. We assume that the curves in Figures 14–2 and 14–3 demonstrate the situation in stable communities. To the decreased external stress the communities respond with higher equitability—increased similarity

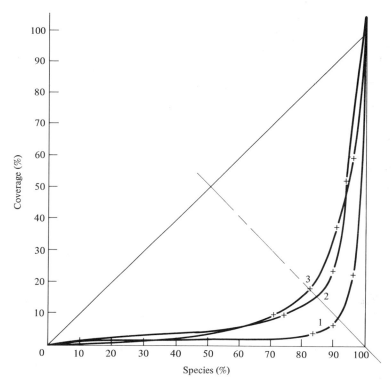

FIGURE 14–2. Plant communities without noticeable stress. Curve 1, pine forest in pioneer stage (Duke Forest, North Carolina); curve 2, old pine forest with climax species invaded (Duke Forest, North Carolina); curve 3, tropical rain forest. Graphs constructed by Petrall (1972) using data from ARPA (1968) and Whigham (1971).

of the importance values of adjacent species in the importance-value sequence. Heavy, unnatural management stress results not only in a sharp reduction of species, but also in the tendency of one or a very few species to be clearly separated from the subordinate species as in Figure 14–4, curves 1 and 2.

Different kinds of importance values—productivity, biomass, coverage, and species leaf-area index—can be used for this treatrent; of these productivity would appear to give the most fundamental expression of community function. The curves shown in Figures 14–2 to 14–4 are transformations of the dominance–diversity curves that have been studied in other forms by Williams (1964), MacArthur (1960), Whittaker (1965, 1970a, 1972), and others. The interpretations that can be drawn from these curves are limited, for a curve of a given form can be generated by more than one hypothesis on underlying community processes (Cohen, 1968; Whittaker, 1969). Nevertheless, when these curves differ markedly in form, differences in underlying community characteristics are expressed by these forms.

This consideration may be of real practical importance. Our present agricul-

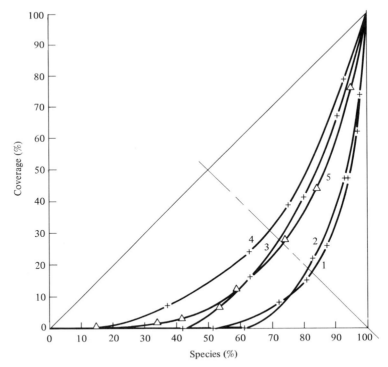

FIGURE 14–3. Plant communities under natural stress. Curve 1, Tropical grassland (Ivory Coast); curve 2, alpine tundra (Alps); curve 3, dry grassland (Morocco); curve 4, Arctic tundra (Baffin Island) [graphs constructed by Petrall (1972) using data from Lemée (1952), Dansereau (1954), Reisigl and Pitschmann (1958–1959), and Adjanohoun (1962)]; curve 5, grassland biome, U.S.I.B.P. aboveground biomass data.

tural and forestry systems apply an enormous force to maintain the situation shown by the curves in Figure 14–3. It might be worthwhile to consider future agricultural systems that follow the pattern of Figures 14–1 and 14–2, which might save much cost and labor and provide a better protection for the landscape than is afforded under present management schemes.

Primary Production and Species Diversity

One of the most important generalizations of ecology is the increase in species diversity from the poles (and high elevations) toward the lowland tropics. Net primary productivity (NPP) increases in the same direction, as shown in Chapters 10 and 11. It is natural to suppose that NPP and species diversity should be correlated both because of this observed relationship and because the NPP provides the working income for the species of the community. Presumably, the larger the NPP, the larger the number of species that NPP, subdivided among

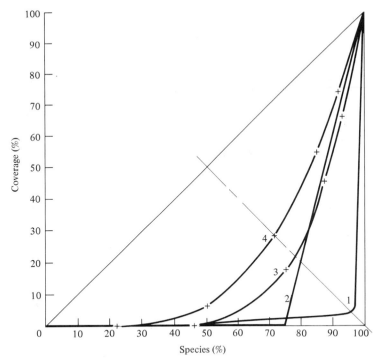

FIGURE 14–4. Plant communities under unnatural stress. (Examples taken from Lieth, 1953). Curve 1, Grain crop with weeds; curve 2, heavy trampled pasture; curve 3, wet lowland trampled path; curve 4, dry upland trampled path.

those species, can support. The dependence of diversity on productivity has been proposed as part of Connell and Orias' (1964) hypothesis on the regulation of species diversity and has been suggested also by MacArthur (1969); Whittaker (1965, 1969), in contrast, has denied the occurrence of a general relationship between diversity and productivity.

A serious obstacle exists for the accurate quantification and comparison of species diversity on land because no basic reference area or sample by which communities can be compared has been found. The thousand-count procedure of aquatic ecosystems (Sanders 1968) does not work on land because of the enormous size differences among higher plant species that range over three to four orders of magnitude.

In a class project Steve Leonard[1] attempted an evaluation of numerous florae of various sizes in one area to determine whether the concept of the minimum area commonly used in phytosociology would also be applicable in this context. Table 14–1 and Figure 14–5 illustrate his results for the state of North Carolina. Figure 14–5 shows that the number of species within a given biome begins to

[1] S. Leonard. Preliminary preparation toward a computer model of species diversity. Geobotany class project, 1972–1973. Chapel Hill, North Carolina: Univ. of North Carolina.

Table 14–1   Number of species in regional florae of different sizes within North Carolina

Region and county	Area (km²)	Number of species
Bullhead Mt., Allegheny	1.3	334
Island Creek, Jones	4	614
Deep River, Chatham and Lee	6	710
Umstead State Park, Wake	15	734
Bluff Mt., Ashe	16	680
Morrow Mt. State Park, Stanley	17	532
Uwharrie National Forest (part Montgomery)	44	694
Deep River, Randolph and Moore Counties; Neuse River, Wake, Durham, Granville	129	985
Sandhills	251	675
New Hanover	583	885
Jones	1213	987
Rowan	1365	787
Robeson	2461	810

level off at a certain area. The species number at that size area may be described as the species-saturation level. The area at which this level is reached is different but probably typical for each biome. The species-saturation level could be estimated with our library resources for North Carolina, New York State, northern Canada, and arctic Siberia with sufficient accuracy. The results are compiled in Figure 14–6, which also includes less accurate information for other areas. In Figure 14–6 the values for species-saturation levels are plotted against the NPP levels, which can easily be obtained for the individual regions from the tables and maps printed elsewhere in this volume, for example, Table 10–1, Fig. 10–1.

With the limited information at our disposal it appears that some correlation

FIGURE 14–5. Species/area curve to determine species-saturation level. Data from North Carolina. Abscissa; area size of flora in km²; ordinate; number of species/flora.

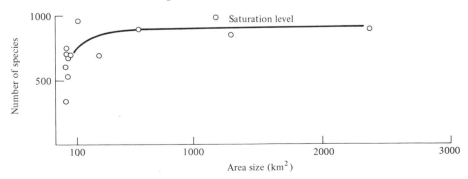

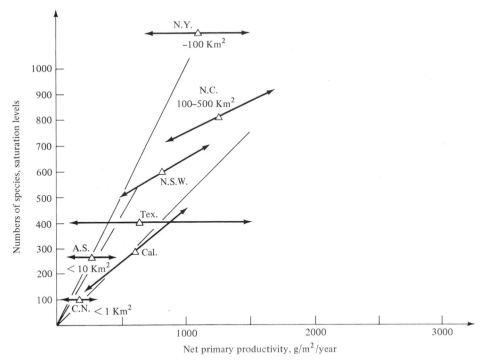

FIGURE 14–6. Species-saturation level versus net primary productivity. Areas shown: N.Y. = New York; N.C. = North Carolina; Tex. = Texas; Cal. = California; A.S. = Arctic Siberia; C.N. = Northern Canada; N.S.W. = New South Wales. Graph constructed from data by Blake and Atwood (1961), Williams (1943, 1964), Good (1964), Frodin (1964), and Van Balgooy (1969).

between dry-matter production of an area and its species-saturation level does exist. With the present accuracy, only widely diverse regions—such as different biomes—appear to be effectively distinguishable. Our preliminary investigations suggest, however, that it may be possible to use this relationship to describe the "maximum species load" of major community types. Figure 14–6 suggests that it takes 1-1.5 g/m^2/year regional NPP level for each species supported in its area. Such a relationship may not appear when diversity and productivity are compared for individual, local communities. Such relationship of species number (S) to area (A) may be reasonably fitted by $S = a + d \log A$, or $S = CA^z$ (MacArthur and Wilson, 1967). Major types of communities—biomes or formations—may differ and may be characterizable by the coefficients (d) relating S and A.

One other relationship that may add to the complexity should be suggested. It is possible to define for a given type of community not only a saturation area, but a threshold area, a mean area per individual for all species, $A_o = A/N = \exp[(1 - a)/d]$, where N is the number of individuals of all species in a sample of size A (Whittaker, 1972). The smaller the threshold area, the larger the number of individuals for a given area A, and consequently the larger the number of species (S) in that area (for a given value of d). Threshold area appears to be a variable affecting species richness in samples that may be at least partly independent of d, which is a rate of increase in species number with increase in area. We may return to Whittaker's (1969, 1972) observation that species diversity is higher on the average in woodlands with well-developed undergrowth and in grasslands than in closed forests. In these woodlands and grasslands mean plant size and individual productivity are lower than in forests, and threshold area may be small compared with both forests and deserts. May not the observed diversities of land plant communities then be interpretable as resultants of (1) the productivity correlation of Figure 14–6 as a broadest relationship, modified by (2) mean size, productivity, or area per individual, permitting higher sample species diversities in many nonforest communities, plus the further modifying effects, for particular communities, of (3) dominance and allelopathy (Muller, 1966; Whittaker, 1970b) and (4) successional status (Loucks, 1970; Auclair and Goff, 1971).

Chemical Differences within and among Ecosystems

The previous discussion centered around the classical species-diversity concept. Diversity of numbers of primary producer species is no doubt reflected in diversity of consumer species. Most consumer species have preferences for certain species of the primary producers, or certain tissues of these species. This consumer specificity with regard to the food (energy) source is well recognized in animal ecology and physiology, but it is not yet sufficiently incorporated into ecosystems models.

Food, once ingested, is acted on by digestive enzyme systems. The enzyme systems of individual species are sufficiently different that consumer species can only digest specific chemical compounds or can only tolerate or inactivate certain of the secondary substances in plant tissues. For those species, chemical energy has to be available in specific forms, which suggests that we look at the ecosystem not only for species diversity, but also from the viewpoint of chemical diversity or chemical differences in the different trophic levels. Many of the facts described in this section are well recognized in human and animal food technology. The knowledge gained from these fields may well be employed in the future for ecosystems research. We would like to illustrate how one may assess chemical composition of ecosystems and compare different ecosystems, as well as show that such comparisons may be significant. For the present we must limit our discrimination of chemicals to generalized categories as they are presently used in food technology.

The procedure to convert detailed dry-matter productivity data into chemical composition is demonstrated in Table 14–2. We have taken the information given by Risser (1971), which states the percentage composition in each species of the total productivity at the end of the vegetation period for some U.S.I.B.P. grassland sites. Almost all the tribes or families listed in Table 14–2 are cov-

Table 14–2   Chemical composition of aboveground Grassland biomass; classes of chemicals distinguished in nutritional studies, entered separately for individual species[a,b]

1	2	3	4	5	6	7	8	9	10	11	12	13	14
		Peak live biomass composition		Ash		Crude fiber		Ether extract		N-free extract		Protein	
Family or tribe	Species	%	g/m²	%	g/m²	%	g/m²	%	g/m²	%	g/m²	%	g/m²
Andropogoneae	Bluestem	24.2	384.3	6.6	25.4	34.3	131.8	2.3	8.8	50.2	192.9	6.6	25.4
Chlorideae	Grama	15.9	253.7	9.1	23.1	32.0	81.2	1.6	4.0	50.3	127.6	6.9	17.5
Compositae	Aster sp.	15.7	250.1	7.6	19.0	21.1	53.5	4.1	10.4	58.8	149.2	8.4	21.3
Festuceae	Fescue	9.5	151.7	7.5	11.4	31.9	48.4	2.9	4.4	48.1	72.9	9.6	14.6
Leguminosae	Lespedeza	6.7	107.2	7.4	7.2	44.9	48.1	2.1	2.2	32.8	35.2	12.8	13.7
Cactaceae	Opuntia	5.4	85.8	16.4	4.6	13.7	11.7	1.8	1.5	63.9	54.8	4.2	3.6
Stipeae	Needlegrass	5.4	85.4	9.9	8.4	36.7	31.3	1.9	1.6	43.2	36.9	8.3	7.1
Sporoboleae	Prairie dropseed	4.2	67.5	10.8	7.3	31.8	21.5	1.8	1.2	43.4	29.3	12.2	8.2
Triticeae	Wheat	3.9	62.4	6.5	4.0	27.9	17.4	3.8	2.4	49.8	31.1	12.0	7.5
Liliaceae	Yucca	2.3	36.9	7.1	0.5	42.7	15.7	2.0	0.7	40.5	14.9	7.7	2.8
Chenopodiaceae	Goosefoot	1.4	22.7	6.4	1.5	19.3	8.4	7.2	2.6	52.0	19.2	15.1	5.6
Paniceae	Panicum	1.4	21.4	13.1	2.8	38.3	8.2	1.3	0.3	38.8	8.3	2.5	1.8
Boraginaceae	Stickseed	0.8	12.5	11.6	1.4	10.6	1.3	3.3	0.4	51.7	6.5	22.8	2.8
Cyperaceae	Sedge	0.6	9.5	7.7	0.7	30.2	2.8	3.2	0.3	49.3	4.7	9.6	0.9
Caryophyllaceae	Silene	0.6	9.1	16.7	1.5	19.2	1.7	—	—	—	—	10.6	0.9
Danthonieae	Danthonia	0.5	7.6	4.6	0.3	30.2	2.3	3.6	0.3	52.1	3.9	9.5	0.7
Malvaceae	Mallow	0.3	5.6	16.8	0.9	21.2	1.2	2.7	0.2	36.9	2.1	22.4	1.2
Rosaceae	Avens	0.3	5.3	3.9	0.2	21.2	1.1	17.9	0.9	36.9	1.9	17.4	0.9
Aveneae	Oats	0.3	4.8	7.2	0.3	34.6	1.6	2.3	0.1	46.7	2.2	9.2	0.4
Euphorbiaceae	Euphorbia	0.3	4.3	11.0	0.4	13.0	0.5	4.1	0.2	57.5	2.5	14.4	0.6
Onagraceae	Evening primrose	0.3	4.0	7.9	0.3	20.0	0.8	3.0	0.1	45.7	2.2	14.4	0.6
Total or means:		100.0	1591.8	7.6	121.2	30.8	490.5	2.7	42.6	50.2	798.3	8.7	138.1

[a] Sources: Risser (1971); Atlas of Nutritional Data on United States and Canadian Feeds (1971). Table compiled by R. J. Reader.
[b] Columns 1–14: Family or tribes; [2] Representative plant listed as present in IBP/Grassland comprehensive sites as represented by Risser, in French (1971); [3] % composition represented by family or tribe; [4] Aboveground peak live biomass for all members in family or tribe as reported by Risser (1971); [5, 7, 9, 11, 13] = % constituents reported in Atlas of Nutritional Data on United States and Canadian Feeds (1971) for above ground portions of plants in column 2. [6, 8, 12, 13] = % × column 4.

ered by routine chemical analyses published in the *Atlas of Nutritional Data on United States and Canadian Feeds* (1971). Analyses of the particular materials from a given ecosystem would be preferable, but in the absence of such data, Table 14–2 is a feasible basis for a preliminary treatment.

The biomass distribution among the species is in accordance with the curves of Figure 14–3 (see curves 5 and 3). Seventy-five percent of the total biomass is concentrated in 30% of the species. The dominant species have, in general, a high level of crude fiber (structural carbohydrates) and a lower than average percentage of protein. The tail-end species in Table 14–2 have the reversed condition (high protein levels and low crude fiber values). It is no surprise that the dominant species, with a high productivity from operation in full sunlight in the canopy, invest much of their productive profit in structural cellulose and hemicellulose and the lignins by which this structure is defended against decomposition, whereas the subordinate species economize on structural tissues and concentrate on the essential photosynthetic and root tissues.

Looking at the chemical composition of the primary producers from the viewpoint of a consumer, we see that a species able to exploit soluble carbohydrates (N-free extract) usually has only 40–60% of the biomass for its use, whereas a species capable of using protein alone rarely finds, in this particular grassland, a food species containing 20% protein. With consumers demanding lipids, the situation would be even more critical, because the percentage for ether extract lies usually under 5% of the total biomass. Some relationships of interest are suggested: (1) The proportion of structural tissue in the community should correlate with the biomass accumulation ratio (BAR) (Whittaker, 1966; see also Chapter 15, this volume) and with the proportions of plant tissues harvested by animals (Whittaker and Likens 1973; see also Chapter 15, this volume). In temperate-zone forests with high structural content and BAR animals probably harvest less than 10% of NPP in most cases; in grasslands with lower structural content and BAR, animals may harvest 10–30% of NPP, aboveground at least; in phytoplankton communities with no structural content except cell walls and BAR's much below 1.0, animals may harvest 20–60% of NPP. (2) The significance in communities of organisms, notably fungi, that are able to digest cellulose, hemicellulose, and lignin should be directly related to BAR and inversely related to animal harvest percentage. (3) The productivities of individual heterotroph species should depend on the fraction of autotrophic productivity in a community they can use, times the fraction of the chemicals in that productivity they can digest and assimilate, times the fractions of their carrying capacities to which population controls limit them. (4) The diversity of the heterotroph species should depend on the chemical diversity of the autotrophs. The community has a primary chemical nucleus of structural and protoplasmic types of compounds common to all autotroph species, and a range of secondary compounds distinctive to individual autotroph species. These secondary compounds are believed to be primarily defensive against heterotrophs (Fraenkel, 1959; Whittaker and Feeny, 1971). Since a given heterotroph species can tolerate only a limited range of these in its food, the total community range of these should be expressed in heterotroph species diversity.

Further questions invite research: What fractions of NPP are invested in allelochemic substances in different communities? How do these fractions relate to environment, community structure, and perhaps to herbivore pressures in different communities? Do long-lived species and tissues need heavier expenditure of NPP in secondary substances (including the lignins of wood and the tannins and resins of evergreen leaves) than do short-lived species and tissues? Are there significant differences in kinds of allelochemics prevalent in different communities? Muller (1970) suggests that volatile allelochemics, particularly terpenoids, are more prevalent in some arid or semiarid climates, water-soluble allelochemicals, particularly phenolic compounds, in more humid climates. In several ecosystems like grasslands, tropical forests, and diatom communities a defense against herbivores is provided by an inorganic substance, silica. The energetic expense to a grassland community of concentrating silica particles into plant tissues, with the effect of reducing intensity of grazing, is quite unknown.

Data are not available to answer these and related questions. However, our search was for a very preliminary and relatively crude compilation of chemical characteristics of communities, using data extracted from the literature during a seminar at the University of North Carolina.[2] In order to gain some insight into the possible range of differences, we chose as biome types a freshwater lake, temperate grassland, summergreen deciduous forest, and coniferous forest. For each type we also tried to collate information for the three main trophic levels: primary producers, consumers, and reducers. We were forced to combine data from different sources and to calculate chemical compositions from similar but not strictly comparable species categories. Table 14–3 must be regarded as a demonstration rather than as a conclusive presentation. Nevertheless, the numbers in it are so contrasting as to deserve attention and comment.

The percentage figures for the different nutritional categories are interesting for their demonstration of contrast in allocation of photosynthates to different chemical categories in the different communities. From water to herbaceous plants (grassland type as representative) to forest, the amount of ash decreases about one-half for each step. In contrast, the allocation of structural carbohydrates about doubles in each step. Nonstructural carbohydrates are similar in water and herbaceous terrestrial ecosystems but are reduced by half in a forest. The protein level is cut in about half for each step. The lipid components are lower in the water than in grassland and deciduous forest, and are in marked contrast between deciduous and coniferous forest. The low lipid content shown in this table seems to contradict the high energy values listed for plankton ecosystems in Table 3–1. Only high ether-extract contents can bring the combustion values of plant material close to 5 kcal/g; further investigations are needed. Of interest is the high ether-extract content of coniferous wood that markedly increases the combustion value of the coniferous forest. We discussed in Chap-

---

[2] The information was compiled from the sources listed in Table 14–3 by R. J. Reader, P. Carlson, W. Martin, R. Kneib, G. P. Doyle, and B. Katz (students at the University of North Carolina, Chapel Hill). In computer searches for more references, we were assisted by P. Pineo (Duke University), N. Ferguson, and J. Gillett (both of Oak Ridge National Laboratory).

Table 14–3 Chemical differences in % of dry weight of selected ecosystem types[a-c]

1	2	3	4 Structural carbohydrates (crude fiber)			5 Nonstructural carbohydrates	6 Protein	7 Ether extract or equiv. (fat)	8 Total (%)	9 Sources[d]
Ecosystem	Species and comments	Ash	Cl	Hcl	Lgn[e]					
A. Primary Producers										
Coniferous forest	Biomass, mature stand	0.3	43.5	14.5	30	1.1	1.3	7.7	98.4	1,2,3,4
	Productivity	4.2	44.1	8.7	18	15.5	4.0	5.8	100.3	1,2,3,4
Summergreen deciduous forest	Standing biomass	0.3	46.6	24	20	0.8	2.5	1.8	96	1,2,3,4
	Productivity	4.2	37	14.4	12	22.5	6.4	2.8	99.3	1,2,3,4
Temperate grassland	Aboveground peak Biomass	7.6		30.8		50.2	8.7	2.7	100	Table 14–1 4,5
Freshwater lake	Plankton	14		18		50	17	1.5	100	6
	Macrophytes	—		14–20		43–60	8–19	1–2.5	100	7
	Benthic algae	—		9–17		36–44	5–18	0.7–2	—	7
	Typha	—		30–39		38–48	7–12	1.5–3.5	—	8
	Scirpus	osc. 6.5		33		53	7	0.5	100	9

[a] This table was compiled as a seminar project at the Univ. of North Carolina, Chapel Hill. Participants in this project were: H. Lieth, J. R. Reader, W. Martin, P. Carlson, G. Doyle, and R. Kneib.

[b] Synthesis table taken from H. Lieth (1973), Chemical differences of contrasting ecosystems and their trophic levels. An exploration of a new viewpoint in systems ecology. *US IBP EDF Biome Memo Rep.* 73–6.

[c] *Parts A–C:* Numbers are given in percentage averages as extracted from data in the literature.

[d] 1, Wenzel (1970); 2, Kollmann (1968); 3, Hillis (1962); 4, *Atlas of Nutritional Data* (1971); 5, Risser, in French (1971); 6, Birge and Juday (1922); Pourriot and Lebourgne (1970); 7, Welch (1953); 8, Hegnauer (1963); 9, Seidel (1955); 10, Stiven (1961); 11, Bump (1947); 12, Davis (1970); 13, Jordon (1912); 14, Kleiber (1961); 15, Heen and Kreutzer (1962); 16, Boyd (1970), Borgstrom (1962); 17, French et al. (1957); 18, Cochrane (1958).

[e] Cl, cellulose; Hcl, hemicellulose; Lgn, lignin.

1	2	3	4 (Chitin)	5	6	7	8	9
B. Consumers								
Temperate forest	Isopodae	6	9	19	63	3	—	10
	Homopterae	3	10	10	73	4	—	10
	Coleopterae	3	4	18	71	5	—	10
	Dipterae	6	6	24	61	3	—	10
	Formicidae	2.6	6	24	67	3	—	10
	Arachnida	2.4	3	17	74	3	—	10
	Camponotus herculeanus	28.5		} 36	30	5.5	100	11
	Turkey (edible parts)	2		?	63	35	100	12
	Ox (total)	17		?	56.4	25.6	—	13
Temperate grassland	Insect	3	6	24	67	3	103	10 Average
	Chicken (edible parts)	8.6	?	?	49	38	95.6	4
	Cattle	11		?	41	47	99	14
	Sheep	8–10.5	?		33–47	42–57	—	13,14
Lake	Zooplankton	15	9	16	50	10	100	6
	Carp ⎱ (edible	8.4	?	?	89.5	3.1	100	15
	Salmon ⎰ parts)	3.4	?	?	71.4	25.2	100	12
	Fish (total)	4–16.3	?	19	50–64	1.2–13	—	4,12,15,16
C. Decomposers								
Summergreen deciduous forest	*Lumbricus*	15–23		16–18	53–64	5–6	100	17
	Eisenia rosea	16		19	61	4		17
	Fungi mycelia:							
	Gliocladium	?	8.4	34.4	21	22.4	86.2	18
	Rhizopus	5.5	2	2	39	9.7		18
	Penicillium	?	5.5–10.6	45–60	12–16	1.4–34.5		18
	Phymatotrichum	6.2	?	32.4	24	4.5		18

ter 10 the possibility that this might have been a reason angiosperm species have eradicated the gymnosperms from most parts of the world where they once prevailed. It has been suggested also that high lipoid and resin content and high combustibility may be a desirable fire adaptation for some coniferous and eucalyptus species (Mutch, 1970).

The consumer part of Table 14–3 shows little consistency in assimilate allocation in any specific chemical category for the three ecosystem types. The differences occur between taxonomic groups rather than between ecosystems. The vertebrates apparently have no structural carbohydrates, whereas the insects contain 5–10% (chitin); ash and protein levels are similar; the ether-extract levels (lipids), however, are up to 10 times higher in vertebrates than in invertebrates. Future comparative studies should reveal more details of ecologic significance.

The comparison between the primary producers and the consumers shows that, for all ecosystems, the consumers have about 10 times the protein concentrations, half or less the structural carbohydrate concentrations, similar to half the nonstructural carbohydrate concentrations, and similar or higher concentrations of ash. The latter is true also for vertebrates, because they maintain their body structure primarily with inorganic reinforcements of protein structures. Table 14–3 shows 4–16% ash for fish, 2–9% for birds, and 8–17% for mammals. The respective values for primary producers are $\sim$ 8% for grassland, 14% for plankton, and 4% for forests, which contrasts against the 30–75% structural carbohydrates for plants in the terrestrial ecosystem. Inorganic components as structural building material appear to be less costly in energy than structural carbohydrates, but the balance between these shifts toward the carbohydrates, proceeding from aquatic through grassland community to forest, and from consumers to producers.

The decomposer portion of Table 14–3 is even less documented in the literature than is the consumer part. We have therefore restricted our comparison to one ecosystem, the deciduous forest, Table 14–4. The values in this Table show trends in a few categories in contrast to the primary producer and consumer

Table 14–4   Comparison of the chemical composition of trophic levels within a forest ecosystem[a]

Trophic level	Ash	Structural carbo-hydrates	Non-structural carbo-hydrates	Protein	Ether extract
Primary producers	4	75	15–22	4–6	3–6
Consumers	2–28	3–10	10–24	30–75	3–5 (I)
					25–35 (V)
Decomposers	5–23	2–11	2–30–60	50–65 (I)	4–6 (I)
				10–40 (F)	2–23 (F)

[a] I, invertebrates; V, vertebrate; F, fungi. Averages extracted from Table 14–3.

groups of the same ecosystem. Animals e.g., (*Lumbricus* and *Eisenia*), that we have grouped with the decomposers, because they are part of detritus chains, have of course a chemical composition similar to the consumers. The fungal reducers differ; their protein levels are cut in half compared with those of the animals, but are higher than those of the producers. Concentrations of the nonstructural carbohydrates seem to be dispersed in the fungi, but are quite high in some of these.

Conclusion

This chapter is intended to offer projections rather than conclusions in relationship to production research. The limitations both of our data and the inferences drawn from them should be evident. But the measurement of productivity is not in itself a sufficient goal; this measurement can be the means to a broad spectrum of characterizations and interpretations of communities and their relationships to environment. A measurement of primary productivity should be compared not with a street address but with map coordinates. The measurement first locates a given community in relationship to other communities in its most important characteristic, second, it alerts the investigator to the further detail that is recorded on the map but not initially observed, and which awaits our research and interpretation.

Acknowledgments

Part of the work reported in this chapter was funded by the National Science Foundation under Interagency Agreement AG-199, 40-193-69 with the Atomic Energy Commission, ORNL, Eastern Deciduous Forest Biome Project, US-IBP. The part of the manuscript dealing with species diversity was much improved by R. H. Whittaker.

References

Adjanohoun, E. J. 1962. Étude phytosociologique des savanes de Basse Côte d'Ivoire (savanes lagunaires), Table II. *Vegetatio* 11:1–38.

Advanced Research Projects Agency (ARPA). U.S. Department of Defense, Cooperative Research Programme No. 27, Tropical Environmental Data (TREND), Table 7, p. 42. Bangkok: ASRCT.

Atlas of Nutritional Data on United States and Canadian Feeds. 1971. Washington, D.C.: National Academy of Science.

Auclair, A. N., and F. G. Goff. 1971. Diversity relations of upland forests in the western Great Lakes area. *Amer. Nat.* 105:459–528.

Birge, E. A., and C. Juday. 1922. *The Inland Lakes of Wisconsin. The Plankton: I. Its Quantity and Chemical Composition. Wisc. Geol. Nat. Hist. Serv. Bull.* 64 Sci. Ser. 13:1–222.

Blake, S. F., and A. C. Atwood. 1967. *Geographical Guide to the Floras of the World,* Part I. New York: Hafner.

————, and A. C. Atwood. 1961. *Geographical Guide to the Floras of the World*, Part II. (USDA Misc. Publ. No. 797.) Washington, D.C.: U.S. Government Printing Office.

Borgstrom, Georg (ed.). 1962. *Fish as Food*, Vol. II: *Nutrition, Sanitation and Utilization*. New York: Academic Press.

Boyd, C. E. 1970. Nutrient content of offal and small fish from some freshwater fish cultures. *Trans. Amer. Fish. Soc.* 99:809–811.

Bump, G. 1947. *The Ruffed Grouse: Life History, Propagation, Management*. New York State Conservation Dept.

Cochrane, V. W. 1958. *Physiology of Fungi*, 524 pp. New York: Wiley.

Cohen, J. E. 1968. Alternate derivations of a species-abundance relation. *Amer. Nat.* 102:165–172.

Connell, J. H., and E. Orias. 1964. The ecological regulation of species diversity. *Amer. Nat.* 98:399–414.

Dansereau, P. 1954. Studies on central Baffin vegetation, L. Bray Island, Table II. *Vegetatio* 5–6:329–339.

Davis, A. 1970. *Let's Eat Right to Keep Fit*. New York: New America Library.

Fraenkel, G. S. 1959. The raison d'être of secondary plant substances. *Science* 129: 1466–1470.

French, C. E., S. A. Liscinsky, and D. R. Miller. 1957. Nutrient composition of earthworms. *J. Wildlife Mgmt.* 21:348.

Frodin, D. G. 1964. *Guide to the Standard Floras of the World*. 59 p. mimeo. Knoxville, Tennesee: Univ. of Tennessee.

Good, R. 1964. *Geographical Guide to the Floras of the World*. London: Longmans.

Heen, E., and R. Kreuzer. 1962. Fish in nutrition. *Fishing News*.

Hegnauer, R. 1963. *Chemotaxonomie der Pflanzen*, Vol. 1. Basel: Birkhäuser.

Hillis, W. E. 1962. *Wood Extractives*, pp. 332–333. New York: Academic Press.

Holdridge, L. R. 1967. *Life Zone Ecology*, rev. ed. San Jose, California: Tropical Science Center.

Jordan, W. H. 1912. *The Feeding of Animals*, 9th ed. New York: Macmillan.

Kleiber, M. 1961. *The Fire of Life. An Introduction to Animal Energetics*, 454 pp. New York: Wiley.

Kollmann, W. A. 1968. *Principles of Wood Science and Technology*, p. 56. New York: Springer-Verlag.

Lemée, G. 1952. Contribution à la connaissance phytosociologique des confins Saharo-Marocains, Table III. *Vegetatio* 4:137–154.

Lieth, H. 1953. *Untersuchungen über die Bodenstruktur und andere vom Tritt abhängende Faktoren in den Rasengessellschaften des Rheinisch-Bergischen Kreises*. Ph.D. dissertation, Table III. Cologne: Univ. of Cologne.

Loucks, O. L. 1970. Evolution of diversity, efficiency, and community stability. *Amer. Zool.* 10:17–25.

MacArthur, R. H. 1960. On the relative abundance of species. *Amer. Nat.* 94:25–36.

————. 1969. Patterns of communities in the tropics. *Biol. J. Linn. Soc. London* 1:19–30.

————, and E. O. Wilson. 1967. *The Theory of Island Biogeography*, 203 pp. Princeton, New Jersey: Princeton Univ.

McNaughton, S. J. 1966. Ecotypic function in the *Typha* community. *Ecol. Monogr.* 36:297–325.

Muller, C. H. 1966. The role of chemical inhibition (allelopathy) in vegetational composition. *Bull. Torr. Bot. Club* 93:332–351.

———. 1970. The role of allelopathy in the evolution of vegetation. In *Biochemical Coevolution*, K. L. Chambers, ed. *Corvallis: Oregon State Univ. Biology Colloq.* 29:13–31.

Mutch, R. W. 1970. Wildland fires and ecosystems—a hypothesis. *Ecology* 5:1046–1051.

Petrall, P. 1972. A comparative analysis of income (productivity) distribution in plant communities and human societies. In *Papers on Productivity and Succession in Ecosystems*, H. Lieth, ed., pp. 19–51. EDF Biome Memo Report 72-10.

Pourriot, R., and L. Leborgne. 1970. Teneus en proteines, lipides, et glucides de zooplanctons d'eau douce. *Ann. Hydrobiol.* 1(2):171–178.

Reisigl, H., and H. Pitschmann. 1958–1959. Obere Grenzen von Flora und Vegetation in der Nivalstufe der Zentralen Ötzaler Alpen (Tirol), Table 14. *Vegetatio* 8:93–129.

Risser, P. G. 1971. Composition of the primary producers on the Pawnee site. In *The Grassland Ecosystem. A Synthesis Volume*, N. French, ed. Fort Collins, Colorado: Range Science Dept., Colorado State Univ.

Sanders, H. L. 1968. Marine benthic diversity: A comparative study. *Amer. Nat.* 102:243–282.

Seidel, K. 1955. Die Flechtbinse. In *Die Binnengewässer*, A. Thienemann, ed., Vol. 21, 216 pp. Stuttgart: Schweizerbart.

Stiven, A. E. 1961. Food energy available for and required by the blue grouse chick. *Ecology* 42:547–553.

Van Balgooy, M. M. J. 1969. A study on the diversity of island floras. *Blumea 17* 1:139–178.

Welch, P. S. 1953. *Limnology*, pp. 273–276; 301–304. New York: McGraw-Hill Book Co.

Wenzel, H. F. J. 1970. *The Chemical Technology of Wood*, pp. 97–100. New York: Academic Press.

Whigham, D. F. 1971. An ecological study of *Uvularia perfoliata* in the Blackwood Division of Duke Forest, North Carolina. Ph.D. dissertation. Chapel Hill, North Carolina: Univ. of North Carolina.

Whittaker, R. H. 1965. Dominance and diversity in land plant communities. *Science* 147:250–260.

———. 1966. Forest dimensions and production in the Great Smoky Mountains. *Ecology* 47:103–121.

———. 1969. Evolution of diversity in plant communities. *Brookhaven Symp. Biol.* 22:178–196.

———. 1970a. *Communities and Ecosystems*, 162 pp. New York: Macmillan.

———. 1970b. Biochemical ecology of higher plants. In *Chemical Ecology*, E. Sondheimer and J. B. Simeone, eds., pp. 43–70. New York: Academic Press.

———. 1972. Evolution and measurement of species diversity. *Taxon* 21:213–251.

———, and P. P. Feeny. 1971. Allelochemics: Chemical interactions between species. *Science* 171:757–770.

————, and G. E. Likens. 1973. Carbon in the biota. In *Carbon and the Biosphere*, G. M. Woodwell and E. V. Pecan, eds. *Brookhaven Symp. Biol.* 24:281–302.

————, and W. A. Niering 1975. Vegetation of the Santa Catalina Mountains, Arizona 5. Biomass, production, and diversity along the elevation gradient. *Ecology* (in press).

Williams, C. B. 1943. Area and number of species. *Nature (London)* 152:264–267.

————. 1964. *Patterns in the Balance of Nature*, 324 pp. London: Academic Press.

Woodwell, G. M. 1967. Radiation and the patterns of nature. *Science* 156:461–470.

15 The Biosphere and Man

Robert H. Whittaker and Gene E. Likens

Preceding chapters in this volume have dealt with the history of productivity study, methods of measurement, patterns of productivity in different kinds of communities, and some applications in research. Two topics remain: the characterization of the biosphere as a whole in terms of productivity and related properties, and consideration of man's relationship to the biosphere. The first topic is the focus of the book as a whole, and it is summarized here as well as in Chapters 10 and 13. The second topic is inescapably problematic; we can offer only a viewpoint on it.

Much of the concern of ecologists is with the diversity of the biosphere—the differences in structure and function of communities in adaptation to different environments. Considering production alone, the biosphere is a film of quite variable density from deserts and tropic seas to coral reefs and forests. It is best, in describing this complex and variable mantle, to distinguish at least its major ecosystem types or biome types. Some characteristics of interest are summarized for these ecosystem types in Table 15–1.

Biosphere Characteristics

Productivity

Columns 4 and 5 in Table 15–1 give estimates of net primary production for major kinds of ecosystems and the world. For land ecosystems we have largely followed Lieth's Table 10–1. Some changes from this and our 1969 estimates

KEYWORDS: Biomass; biosphere characteristics; chlorophyll; global carrying capacity; human populations; leaf surface areas; man's harvest; primary production; stability.

Primary Productivity of the Biosphere, edited by Helmut Lieth and Robert H. Whittaker.

Table 15–1 Net primary production and related characteristics of the biosphere

1	2	3	4	5	6	7	8	9	10	11	12
		Net primary production (dry matter)			Biomass (dry matter)			Chlorophyll		Leaf-surface area	
Ecosystem type	Area (10^6 km²)	Normal range (g/m²/year)	Mean (g/m²/year)	Total (10^9 t/year)	Normal range (kg/m²)	Mean (kg/m²)	Total (10^9 t)	Mean (g/m²)	Total (10^6 t)	Mean (m²/m²)	Total (10^6 km²)
Tropical rain forest	17.0	1,000–3,500	2,200	37.4	6–80	45	765	3.0	51.0	8	136
Tropical seasonal forest	7.5	1,000–2,500	1,600	12.0	6–60	35	260	2.5	18.8	5	38
Temperate forest:											
evergreen	5.0	600–2,500	1,300	6.5	6–200	35	175	3.5	17.5	12	60
deciduous	7.0	600–2,500	1,200	8.4	6–60	30	210	2.0	14.0	5	35
Boreal forest	12.0	400–2,000	800	9.6	6–40	20	240	3.0	36.0	12	144
Woodland and shrubland	8.5	250–1,200	700	6.0	2–20	6	50	1.6	13.6	4	34
Savanna	15.0	200–2,000	900	13.5	0.2–15	4	60	1.5	22.5	4	60
Temperate grassland	9.0	200–1,500	600	5.4	0.2–5	1.6	14	1.3	11.7	3.6	32
Tundra and alpine	8.0	10–400	140	1.1	0.1–3	0.6	5	0.5	4.0	2	16
Desert and semidesert scrub	18.0	10–250	90	1.6	0.1–4	0.7	13	0.5	9.0	1	18
Extreme desert— rock, sand, ice	24.0	0–10	3	0.07	0–0.2	0.02	0.5	0.02	0.5	0.05	1.2
Cultivated land	14.0	100–4,000	650	9.1	0.4–12	1	14	1.5	21.0	4	56
Swamp and marsh	2.0	800–6,000	3,000	6.0	3–50	15	30	3.0	6.0	7	14
Lake and stream	2.0	100–1,500	400	0.8	0–0.1	0.02	0.05	0.2	0.5	—	—
Total continental:	149		782	117.5		12.2	1837	1.5	226	4.3	644
Open ocean	332.0	2–400	125	41.5	0–0.005	0.003	1.0	0.03	10.0		
Upwelling zones	0.4	400–1,000	500	0.2	0.005–0.1	0.02	0.008	0.3	0.1		
Continental shelf	26.6	200–600	360	9.6	0.001–0.04	0.001	0.27	0.2	5.3		
Algal beds and reefs	0.6	500–4,000	2,500	1.6	0.04–4	2	1.2	2.0	1.2		
Estuaries (excluding marsh)	1.4	200–4,000	1,500	2.1	0.01–4	1	1.4	1.0	1.4		
Total marine:	361	—	155	55.0	—	0.01	3.9	0.05	18.0		
Full total:	510	—	336	172.5	—	3.6	1841	0.48	243		

(Whittaker, 1970; Whittaker and Woodwell, 1971) incorporate Murphy's means for tropical types (Chapter 11) and other data in this volume. The marine estimates are based on close agreement of Lieth's values and ours (Whittaker, 1970; Whittaker and Likens, 1973a,b), and the Koblentz-Mishke *et al.* (1970) summary discussed in Chapter 8. We have not for the present modified these estimates upward as suggested by Bunt, except that benthic as well as plankton estimates are included. The estuarine value is from Woodwell *et al.* (1973) and the freshwater value from Likens (Chapter 9).

From the combination of these values, as given in column 5, we estimate world net primary production as about 170×10^9 t/year. In this total, production on land predominates to a degree that was only recently recognized. The mean productivity of the oceans is about one-fifth that of the land; total production on land is consequently somewhat more than twice that of the oceans on somewhat less than half as large an area. A major source of this contrast is the difference in nutrient function of plankton and land communities. The nutrients available to a plankton community are cycled rather rapidly among its short-lived organisms. However, in stratified waters at a distance from continents the sinking of organisms and particles carries nutrients downward out of the lighted zone, thereby impoverishing the plankton. In contrast, the characteristics of land ecosystems, which have evolved in relation to stable land surfaces, tend to hold a larger capital of nutrients in plant tissues and soil at the lighted surface of the earth where they may support primary productivity.

Confidence limits cannot be set as yet for the estimates of world production and other biosphere characteristics. The means given in Table 15–1 are in some cases averages of published values; but in many cases they have been subjectively chosen as reasonable, intermediate values for a range indicated by a few field measurements. In general, more useful data are available for productivity than for the other characteristics. A number of recent American estimates of global net primary production (Whittaker, 1970; Olson, 1970; SCEP, 1970; Golley, 1972; Whittaker and Likens, 1973a; Ryther, 1969; see also Chapters 10 and 13), and the Russian marine estimates of Koblentz-Mishke *et al.* (1970) have converged toward the ranges of $90–120 \times 10^9$ t/year for the land, and $50–60 \times 10^9$ t/year for the sea. The convergence of these estimates suggests that they may not need major revision; unless human interference changes the NPP drastically (Lieth, 1975).

The convergence does not extend, however, to the recent Russian estimate (Bazilevich *et al.*, 1971; Rodin *et al.*, 1975) of 172×10^9 t/year for total continental productivity. The Russian estimate appears more detailed than ours, for production values are given for more than 100 land ecosystem types distinguished by climatic belt, physiognomy, and soil. It is not clear which of these many values are means, which are individual measurements, and which are estimates. Some of the values seem clearly too high to be taken as means. Rodin *et al.* give 100 g/m²/year for polar desert and 150 g/m²/year for mountainous polar deserts. A community with productivity over 100 g/m²/year is semidesert rather than desert, and such values can scarcely be means including the true arctic and alpine deserts. For subboreal sand deserts they give 500

g/m^2/year, a productivity sufficient for a fairly good grassland; for steppified desert on sierozem they give 1000 g/m^2/year, a value typical not of a desert but of a productive grassland. For subtropical and tropical bogs they give extraordinary values, 13,000 and 15,000 g/m^2/year, and for floodplains 4000–9000 g/m^2/year—values that, if justified, would appear to be extremes and not means. The largest single contribution to their total is for tropical humid ever-green forest on red-yellow ferallitic soils, 3000 g/m^2/year. Although some tropical secondary forests (and grasslands that have replaced them) are this highly productive, we think this value excessive as a mean for the climax forests to which the estimate of Bazilevich *et al.* applies (see also Chapter 11, this volume). In other values, too, the estimates by Bazilevich *et al.* seem to us biased on the high side.

Many of the early estimates of world production summarized by Lieth in Chapter 2 and by Whittaker and Likens (1973) were low for lack of informa-tion. More recently with reasonable although incomplete data available, esti-mates have diverged partly because of the ways in which values are chosen and means are derived. Given a range of values, there is often a tendency to prefer the higher of these as being, perhaps, more representative and more complete. However, given measurements that may be trusted, it is not the high values that should be observed but the ranges of values. Temperate grasslands, for example, have productivities mostly ranging (with marshes excluded) from 200–300 to 1200–1500 g/m^2/year. The high values of 1200–1500 g/m^2/year are as far from being typical of grasslands on a continental scale as are the low values. Furthermore, because the areas of dry grasslands in the interiors of continents are greater than the areas of more humid grasslands or prairies in near-forest climates, a realistic mean should be weighted on the low side—hence, 500 or 600 g/m^2/year, rather than a middle value of 800 or 900 g/m^2/year. With all respect to the contribution of the Russian students of productivity, we think that many of their individual values are too high, and that their over-all estimate does not reflect a careful consideration of ranges of values and most realistic means. We also think the realism of our estimates is supported by Lieth's use of alternative approaches (Chapters 12 and 13)—estimates based on ecosystem types, on correlations with precipitation and temperature, and on correlation with evapotranspiration. These estimates, which are independent of one another as means of averaging and summarizing productivity (although they may use many of the same measurements), converge with one another and the other estimates by Western authors.

Biomass

Land and sea communities are even more in contrast in their biomass values (Table 15–1, columns 7 and 8). In land communities that are not desert, biomass ranges mostly between 1 and 60 kg/m^2; plant biomass in aquatic com-munities that are not dominated by attached plants ranges mostly downward from 0.1 kg/m^2. The biomass of an ordinary forest is of the order of ten thousand or more times an average plant biomass for the open ocean. If the attached aquatic plants were set aside, the biomass on land would be more than one thousand times that of the seas. Stability of surface is again critical. Given stable surfaces, land plants have so evolved that long-lived plants are dominant. These plants use the biomass that is the accumulated profit of net productivity for their extensive root and aboveground structures. These structures are in turn part of the basis of high productivity through their support of photosynthetic surfaces and contribution to the pattern of nutrient use and retention. The nutrient function of the plankton is, in a sense (that we would not want taken literally), "primitive," based on rapid overturn of limited resources with little capital accumulation; and that of terrestrial communities is "advanced," with extensive accumulation of capital and long-term tangible assets.

These and other relationships form the basis for the correlation of production and biomass mean values shown in Table 15–1. The correlation, although significant, is loose (see Chapter 10 and Figure 4–3 in this volume) and is much affected by age of the dominant plants of communities. The relationship is expressed as the biomass accumulation ratio (biomass present/net annual primary production); and these ratios are typically from 2 to 10 in desert, 1.5 to 3 in perennial grasslands, 2–12 in shrublands, 10–30 in woodlands and young forests, and 20–50 in mature forests (Whittaker, 1966, 1970). For aquatic communities the biomass accumulation ratios are fractions; and the contrast of terrestrial and marine communities is expressed in their mean ratios: 15.7 and 0.07 (Table 15–1). Some other characteristics of terrestrial communities are related to their accumulation of biomass—the longer time scales through which their dominant populations may be stable, the variety of structure or physiognomy by which major types of communities may be recognized as in Table 15–1, and the diversity of intracommunity functional positions, or niches, offered microorganisms and small herbivorous animals, particularly insects, as a basis for the great species diversity of these groups (Whittaker and Woodwell, 1972).

The Russian biomass estimates of Rodin *et al.* (1975) give a considerably higher total for the land surface: 2.4×10^{12} t. The Russian estimate differs from our own in that it is intended to represent potential world biomass, or that of the climax vegetation before the effect of man; ours is an estimate for a world occupied by man (in 1950, before the recent accelerating destruction of natural vegetation). Even as a potential world biomass, the Russian estimate may be rather high. The largest contributions to it are from tropical forests, including humid evergreen forest on red–yellow ferallitic soils (650 t/ha), and humid tropical mountain forest on red–yellow ferallitic soils (700 t/ha). No doubt there are tropical forests of such magnitudes, but these values seem to be closer to maxima than to means. A preliminary estimate of world animal biomass (Whittaker and Likens, 1973a) gave roughly 1.0×10^9 t each for the continents and for the seas, excluding man ($\sim 0.05 \times 10^9$ t in 1970) and his livestock (0.26×10^9 t).

Leaf area, chlorophyll, gross production, and efficiency

Columns 9–12 in Table 15–1 give estimated means and totals for two indices of the photosynthetic equipment of communities—chlorophyll content and, for land communities, leaf-surface areas (see also Lieth's Table 10–2). Conservative values, including only leaf-blade chlorophyll and surface for land plants, have been preferred in both cases. As indicated by Lieth, chlorophyll is rather evenly distributed among the more productive land communities, but very unevenly distributed between terrestrial and aquatic communities. For a wide range of land communities (excluding the least and most productive), chlorophyll contents of 1–4 g/m^2 support net productivities of 200–2000 g/m^2/year. The mean productive efficiency of chlorophyll on land, expressed as net annual productivity in grams dry matter per gram of chlorophyll, is 518; expressed as energy it is 2200 kcal/g. Efficiencies are higher for forests (mostly 300–700 g/g, with lower values in temperate evergreen forests) than in deserts, tundra, and dry grasslands (100–300 g/g). For a wide range of phytoplankton communities chlorophyll contents of 0.002 to 0.1 g/m^2 support net productivities of 50–1000 g/m^2/year; for marine phytoplankton Table 15–1 suggests that mean net productive efficiencies are 3300 g/g and 16,300 kcal/g. The corresponding values for the biosphere are 705 g/g and 3100 kcal/g.

Most land communities of reasonably favorable environments intercept sunlight with 3–8 m^2 of leaf-surface area displayed above 1 m^2 of ground surface; higher values occur in some communities. The mean efficiencies for dry matter productivity and energy capture per unit leaf-surface area on land are 180 g/m^2/year and 7600 $kcal/m^2$/year. For land communities of favorable environments these efficiencies are generally 150 to 300 g/m^2/year, with lower values in evergreen communities; for many communities of arid and cold environments they are 50–150 g/m^2/year.

Lieth (1973, cf. Chapter 10) has given the efficiency of net primary production in relationship to total annual solar radiation for the earth as 0.13%,[1] with values of 0.07% for the ocean and 0.3% for the land. Lieth's 0.13% (for total radiation) is convergent with the 0.25% (for energy in the visible spectrum only) calculated earlier (Whittaker, 1970) for efficiency of net primary production. Information on respiration in plant communities is limited (Chapter 4). Plant respiration in forests appears to be 50–75% of gross primary productivity, with the higher values in tropical forests; respiration is probably 25–50% of gross primary productivity in most other land and water communities. For the biosphere as a whole the values above and below 50% tend to compensate for one another, so that world gross primary production is about twice world net primary production (Golley, 1972). Using the data of Table 15–1, we have recalculated efficiency of world net primary production in relation to sunlight energy in the visible spectrum as 0.27%. Using estimates of plant respiration that seem reasonable for different communities (based on Figure 4–6 for land communities, and from 30% respiration in the open seas to 50% in algal beds), we obtain mean ratios of gross to net primary production of 1.5 for the seas, and 2.7 for the continents. The efficiency of world gross primary productivity in relation to sunlight energy in the visible spectrum is then 0.6%.

---

[1] 0.16% in a revised assessment (Lieth 1975), with 67% of the energy bound on land and 33% in the ocean.

Man's Harvest

The energetic magnitude of world primary production, estimated as 6.9×10^{17} kcal/year by Lieth (1973, Chapter 10), much exceeds that of any of the works of man. Man's total use of fossil fuels and other industrial energy in 1970 was 4.7×10^{16} kcal/year (Cook, 1971), hence $\sim 7\%$ of net production and 3.5% of the gross primary production that supports the world's life. The recent doubling time for world consumption of industrial energy has been approximately 10 years, a rate of increase that is taxing presently available energy resources (Hubbert, 1969, 1971; Cook, 1972) and that suggests a formidable and accelerating rate of release of heat and materials from industry into environment.

Man's harvest of food is also small compared with biosphere production. The 14×10^6 km² of arable land produced in 1950 $\sim 8.5\%$ of land surface net production (Table 15–1); and $\sim 9\%$ of the total production of agricultural plants on land was available to man as harvested food. Production and the fraction harvested in 1970 were higher, probably 11×10^9 t/year and 12% to give a yield of 1200×10^6 t/year of cereal grains and 570×10^6 t/year of other food crops in fresh weights (FAO, 1971a), approximately 1000 and 220×10^6 t dry weights. A larger fraction of the land surface, about 30×10^6 km², is used as pasture and range land. World harvest of food from animals is important for its protein content but small in quantity compared to that from plants (Kovda, 1971); it includes, in millions of tons fresh weights (and approximate dry weights in parentheses), 80 (20) of meat, 20 (4.7) of eggs, and 400 (48) of milk (FAO, 1971a). World harvest of aquatic organisms for food was 69 (17) $\times 10^6$ t in 1970 (FAO, 1971b), with about 88% of this from the oceans (FAO, 1970). The marine yield to man of about 15×10^6 t/year dry matter is only 0.027% of total marine net primary production, but represents a much larger fraction of that production concentrated through animal food chains. Man's total food harvest of about 1.22×10^9 t/year of plant and 80×10^6 t/year of animal dry matter (of which some of the latter has been fed on the former) is $\sim 0.72\%$ of the energy of net primary production of the world. A very approximate estimate of the harvest of plant productivity by herbivorous animals other than man and his livestock is 7.2×10^9 t/year on land, 20.2×10^9 t in the seas, for 0.82 and 3.0×10^9 t/year secondary productivity by these animals (Whittaker and Likens, 1973).

It is easy to think that man's harvest should be much more. Although the amount can be increased, there are reasons it will not increase as much as simple arithmetic might suggest. These reasons include, first, the limitations on what portion of biosphere production can be used for human food, within the limits of acceptable diet and economics. Most food for man must include tissues high in food value—animal flesh and milk, and seeds and other plant food-storage organs. Lacking the means for conversion into suitable food by either animals or by industrial energy, the greater abundance of wood, grass tissues, and phytoplankton cells of the biosphere cannot be used effectively. The principal increase in food yield must be through conventional agriculture (FAO, 1970).

There are limits on the amount of land suitable for agriculture; although these limits are elastic, the rapid increase in cost of irrigation and other conversion to agriculture of less suitable land limits the expansion of farmlands (FAO, 1970; Ehrlich and Holdren, 1969; SCEP, 1970). Advances in agricultural technology, notably the new cereal crops of the "green revolution," have recently made possible increases in yield of 2–3%, and locally more, per year, mainly by greater yields from lands already in use for agriculture. It may be possible to increase world food production for some years; however, the present (1975) food supply and projections are not encouraging. We offer no prediction as to how long the increase can keep pace with the growth of population or when it will cease. The increase should be expected to end, however, not only because of the limits of agricultural land but for a second group of reasons involving the effects of man on the biosphere.

Man's Effects

Some of these effects include the consequences of overharvest. Some of the world's fishery resources have been depleted by excessive catch (Holt, 1969; TIE, 1972). Given fish populations that are limited, ever-increasing demands on these for food, and competing and largely unregulated national fisheries, excessive harvest is inescapable. The overharvest is, in fact, a paradigm of man's relationship to the biosphere, as it is an example of the principle stated by Hardin (1968) as the "tragedy of the commons." Agricultural production is less easily depleted, for removal of nutrients from the soil can be more than compensated for by fertilization. However, while population grows and the harvest to feed it increases in one area, the potential agricultural land available for the future is decreased in another area by erosion, and in a third area by urban growth. Available land has not in the past exerted a close limit on world food production. However, the convergence of accelerating demands upon the limited area suitable for agriculture, with reduction in that area by conversion of land to other uses and increasing costs of energy and other resources for intensive agriculture, may imply an unexpectedly early limit on the expansion of farm production (Meadows *et al.*, 1972)—even assuming that production is not affected by pollution, or climatic change resulting from man's effects on the atmosphere.

Current harvest of wood is 2.2×10^9 m³/year, of which 0.93×10^9 m³ is used for fuel (FAO, 1971c). The 2.2×10^9 m³ converts to about 2×10^9 t/year of aboveground dry matter in the trees cut down. This yield, which is not a large fraction of world forest production (Table 15–1), does not include forest clearing from which the wood is not used. The biomass estimates in Table 15–1 are for the year 1950, before recent acceleration of clearing in Amazonia and elsewhere in the tropics, and increased cutting for pulp and structural timber in the temperate zone. The biomass estimates can be modified for conditions before the effects of civilized man by assigning the agricultural land to other ecosystems and replacing the mean forest biomass values with others for wider occurrence of undisturbed forests. We estimate the resulting world biomass at 2.1–2.2 × 10¹² t, whereas Bazilevich *et al.* (1971) estimate 2.4 × 10¹² t. It seems clear also that our world biomass estimate of 1.83 × 10¹² t for 1950 is outdated, for the 90% of that biomass that was in forests (56% in tropical forests) has been affected by accelerating cutting. As old-growth forests are replaced by young stands subject to reharvest before they have reached mature ages, and as forests are converted to other land uses, forest biomass and therefore world biomass decrease. We are unable to estimate the rate of this decrease.

The biosphere is affected also by widespread pollution. In the past, industrial pollutants caused local damage to ecosystems and man's environment, but they were lost in the vast processes of dispersal, conversion, and deposition of the ecosphere. In recent decades, however, the release of pollutants has begun subtly to affect the chemistry of environment in areas remote from cities and including, in fact, the whole of the earth. Figure 15–1 attempts to characterize the pattern of pollutant increase in environment. Some actual pollution data and a larger number of hypothetical curves have been superimposed to illustrate the dual exponential increase in amounts of individual pollutants and in numbers of pollutant substances. Interactions between some of the increasing number of substances may be biologically significant. As some individual pollutants (e.g., DDT in the United States) slow or reverse their increase, others continue

FIGURE 15–1. Pattern of pollutant levels in environment in a wealthy society with expanding industry; some curves represent actual data, others are hypothetical. Broken line *P* represents level at which given pollutant is recognized as increasing in the environment (after a lag of some years). Although some pollutants are controlled or reversed, most increase exponentially; and numbers of new potential pollutants being introduced and of existing pollutants being recognized in environment also increase exponentially.

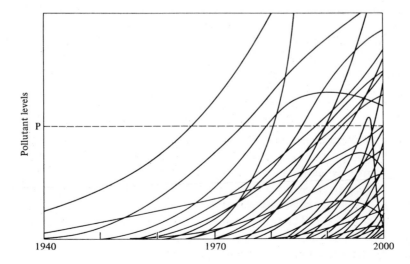

314

Part 4: Utilizing the Knowledge of Primary Productivity

to increase and are added to the ecosphere as new pollutants. The process is impelled by industrial growth and technological enterprise on a worldwide scale, and as a whole is not much altered by reductions or replacements of particular pollutants in individual countries.

Effects on the productivity of the biosphere are to be expected in due course. Persistent pesticides and heavy metals from technological agriculture and industry drain into and accumulate in fresh and coastal waters. These toxic materials may be expected to contribute to the decline of some food fish populations, and to make some of the remaining fish unsuitable for human food. On land, air pollution has reduced the growth of or killed plants of forests, deserts, and agricultural crops in a fairly wide area of southern California; and the geographic extent of the effects is increasing (Miller, 1969; Heggestad and Darley, 1969). Differences in plant growth in ambient (and polluted) versus charcoal-filtered air at Beltsville, Maryland, have been observed under experimental conditions; the differences cannot be applied directly to field conditions (Howell and Kremer, 1970) but are suggestive of effects to be expected on some more sensitive native species. Because of increasing pollutant sulfur dioxide and nitrogen oxides in the atmosphere, precipitation has become increasingly acid over wide areas of the eastern United States and western Europe (Likens *et al.*, 1972). Decreases in tree growth and forest production attributed to the effects of rainfall acidity have been observed in Sweden (Bolin, 1971) and in our study area at Hubbard Brook, New Hampshire (Whittaker *et al.*, 1974). Although the effect is not yet adequately understood, it may now be geographically widespread. These observations suggest that the modest increases in productivity of land vegetation that could, in principle, result from increased nutrient input (carbon, nitrogen, and sulfur oxides) from the atmosphere are more than canceled by the adverse effects of other pollutants.

We do not argue that increasing pollution threatens a short-term crisis for biosphere production, but emphasize the acceleration and difficulty of control inherent in pollution processes. Reduced productivities of some ecosystems are early signs of what must be expected to occur at an increasing rate, from the logic of Figure 15–1, if world-wide industrial growth continues. To longer-term projections of world food harvest should be added another consideration: the possibility that increasing pollution of the ecosphere will reduce biosphere production, and with this man's harvest from agriculture and the seas.

World carrying capacity for man

Both the production by the biosphere and man's effects on it bear on what ought to be a guiding question for man's policy in occupying and using the world. This question is the world's carrying capacity for man: the size of the human population that can be supported on a long-term, steady-state basis by the world's resources without detriment to the biosphere or exhaustion of non-renewable resources that are reasonably available (cf. Ehrlich and Ehrlich, 1970; Cloud, 1973). The question has not only been neglected in policy; it has proved, because of the effects of technology, to be almost unanswerable. Industrialization has permitted the population of Europe to grow much beyond the level at which Malthus would have expected disasters we would now term Malthusian. American farmlands now produce more food per acre, relying on heavy use of technology and expenditure of fossil-fuel energy, than was produced a century ago when the farm population was larger (Steinhart and Steinhart, 1974). Some would judge the United States is already overpopulated for its present standard of living, as indicated by the rate of exhaustion of resources, destruction of farmland and natural areas, urban and pollution problems, and heavy reliance on petroleum. Yet it seems that the North American continent itself is so provident that more than a billion Americans could be supported at a standard of living at least comparable to that of India or China. Neither for North America nor for the world can a carrying capacity be defined unless a standard of living and a role of technology are first specified. If the role of technology is large, a time scale for exhaustion of resources may need to be part of the definition; but this time scale itself is (given the uncertainties of substitution and feasible use of low-grade resources) almost indeterminable.

During favorable periods of history, certain happy notions of historic processes become popular. One of these is belief in the essential permanence of the favorable circumstances of a given period. For example, during the industrial growth of the West it was argued until recently that man, with his technological ingenuity, could continue to find resources to support population and industrial growth indefinitely. Such growth might lead, perhaps, to the creation of a single great world-city (Doxiadis, 1970). A further comforting belief was that the decline in birth rate experienced by societies that became industrialized and prosperous ("demographic transition") would be sufficient to relieve them of population problems. More recently, in response to the disturbing projections of the world model of Forrester (1970) and Meadows *et al.* (1972), it has been argued that the overshoot of which the model warns will not occur because increasing costs and shortages of materials will become self-limiting mechanisms for industrial growth and will produce a stabilization, rather than an overshoot. There has never been a world so industrialized before, but historic experiences of societies may offer comment on ideas of indefinite growth and automatic stabilization.

Figure 15–2 shows historic population levels in three civilized societies for which reasonably effective, long-term records and estimates are available. Egypt has long been supported by a great river that annually flooded and enriched the soils of its valley. The response to this relatively stable carrying capacity has been one of conspicuous population fluctuation, with periods of growth ending in drastic reductions by plague and conquest. Chinese civilization has experienced, through the period shown and history before it, a sawtooth alternation of population growth to overpopulation for the resources then available, with periods of famine and population decline, war and disorder, and dynastic change. Following each population descent, expansion of agricultural production with new crops or new land made possible a renewal of population growth to the next period of overpopulation at a higher level. The shorter Irish record shows population growth supported by grazing and limited use of the "Irish" potato to one period of population saturation, 1720–1780, that was punctuated by

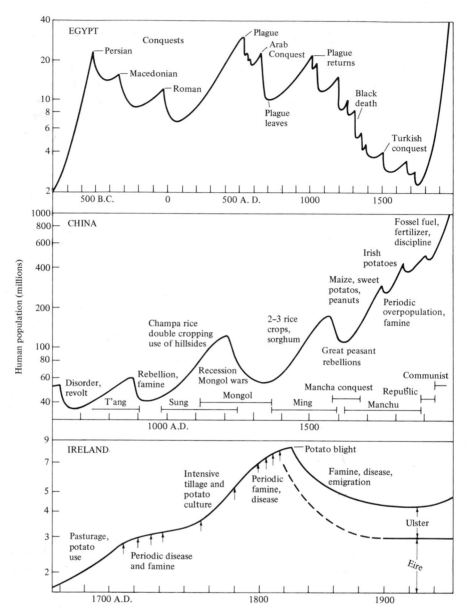

FIGURE 15–2. Historic populations of civilized societies. Estimates of population numbers, together with some events affecting them, are plotted for three nations; the population numbers are on logarithmic scales (*Top*) Egypt, 800 B.C. to the present, as interpreted from various historic records and estimates (Hollingsworth, 1969). (*Middle*) China, 600 A.D. to the present, based on censuses in some periods and estimates from other historic records (Cook, 1972; see also Durand, 1960; Ho, 1959; Clark, 1967). (*Bottom*) Ireland, 1650 to present, based on censuses from 1821 to present and interpretation of less reliable household counts before 1800 (Connell, 1941; Reinhard *et al.*, 1968).

famine and smallpox epidemics. After 1780 more intensive farming and heavy reliance on the highly productive potato permitted population growth to a new saturation about 1820. This population suffered periodic famine and disease until, in 1845, it was struck into disaster by the potato blight. Following the resulting population descent, Irish population became relatively stable in the period 1900–1970, apparently by means of a national consensus that population must be limited by delayed marriage, nonmarriage, small families, and emigration to that which the land could support with mixed agriculture.

We shall not try to calculate the incalculable world carrying capacity for man, but we offer these suggestions.

1. Chinese history reinforces the point that no carrying capacity can be defined independent of technology. Furthermore, at least two concepts should be distinguished: (a) a favorable or optimum population for an industrialized world with a standard of living corresponding, say, to that of Europe 20 years ago, before the recent industrial overgrowth of the West, and (b) a maximum population for a world predominantly nonindustrial.

2. A reasonable estimate for the first of these might be a world population of 1 billion at an American standard of living (Hulett, 1970) or 2 to 3 billion at a more frugal European standard. Such a population could be supported, if steady-state systems of resource use and cycling were established, for an extended period without either resource exhaustion or biosphere detriment. The suggested population may seem low, but realism must take into account the contradictory effects of technology. Industrial development has made possible major increases in population, but in the longer term the exponential growth of industry leads to exhaustion of resources and biosphere degradation. A very large, industrialized and wealthy world population, permitting itself exponential growth, would be unstable.

3. There appears to be little prospect that the world population as a whole will be raised to a Western European or American standard of living. The opportunity to create a world based on a favorable or optimum population may have passed irretrievably in the burst of growth of population and exploitative industry following World War II. The difficulty of bringing about population stabilization, as the necessary beginning of a rational use of the world's resources by man, suggests that creation of such a world will be very difficult if not impossible. Major international strains may result from the combination of aroused hopes for development to wealth in the poor countries, with the disappointment of these hopes (Heilbroner, 1974).

4. An agricultural world, in which most human beings are peasants, should be able to support several billion human beings, perhaps 5–7 billion, probably more if the large agricultural population were supported by industry promoting agricultural productivity. In theory a quite large human population could be supported for a fairly extended period if world industry were focused on relatively slow use of nonrenewable resources for

agriculture, with only limited use for luxuries and generally low standards of living. Considering, however, the extent of hunger, the apparent near-saturation of world food resources, and the increasing costs of technological agriculture at the time of this book in 1975, an estimate should not be too generous in assuming carrying capacity for additional billions. Some expansion of world population beyond its present 4 billion could be supported if more of the plant crops that are converted into animals for food were instead consumed directly. On the other hand, probably one fourth of the present world population is fed by the increase in farm crop productivity made possible by technological agriculture and fossil fuel use. The uncertain future role of technology and cost of energy to agriculture makes a much larger estimate of world carrying capacity seem questionable. Further uncertainty results from the likelihood that in a densely occupied world, war and disease would periodically reduce national, if not world, populations below the potential maximum in a pattern resembling that of Egypt in Figure 15–2.

5. It appears that fluctuation is normal to the long-term population behavior of civilized societies. No natural tendency toward stabilization without hardship is evident.

6. The Irish experience is a special case from which some encouragement may be derived. It should also be noted, however, that the Irish experience of population stabilization was painful—both in the tragedy that taught its necessity and in the restrictions on human life that became its means. Means of stabilization less restrictive of individual life should be feasible, but it need not be supposed that stabilization will occur without effective national decision, enforcement by consensus and moral force or by law, and restriction of individual rights to reproduce.

Unstable Systems

We therefore judge that the relationship of man, and especially of industrial society, to the biosphere is unstable (cf. Ehrlich and Holdren, 1969; Istock, 1969; Crowe, 1969; Forrester, 1970; Meadows *et al.*, 1972; Goldsmith *et al.*, 1972). We pass now to another mode of consideration, from direct comment on data, limited to some extent by the incompleteness of those data, to interpretation of the instability of man's societies, in which we are limited also by the complexity of the relationships.

Among living systems several states with respect to growth may be distinguished.

1. *Steady states:* Input and output of the system are in balance, and approximate constancy of the system is maintained, superimposed on the flow of matter and energy through it.

2. *Regulated growth:* Complex growth processes are controlled in relationship to one another. Regulated growth may be either determinate, subject to some limitation at a steady-state or mature condition, or indeterminate

without such limitation. Negative feedbacks acting on individual processes of growth control these processes, maintain the balances among them, and define the limits of growth.

3. *Unregulated growth:* The system expands, usually in an exponential manner, with the growth of its parts (if it is a complex system) accelerated by positive feedbacks and not or weakly controlled in relation to one another.

Biologic cases of these on the three levels of organisms, populations, and communities include: (1) the steady states of mature organisms with determinate growth, stable populations with equal birth and death rates, and the climax of natural communities, (2) the regulated growth of individual organisms, either determinate as in most higher animals or indeterminate as in most higher plants, the sigmoid stabilization of a population at its limit (Fig. 15–3a), and community succession to the climax, and (3) eruption with potential collapse of populations that are not stabilized by negative feedback as the carrying capacity is approached (Fig. 15–3c). Human societies do not fit closely into these categories, but: (1) A few primitive societies may have maintained population steady states by devices that brought births and deaths into balance. (2) The earlier and slower growth of many other societies, as complex systems expanding without reaching the limits of their resources, may be regarded as primarily regulated, but indeterminate growth. (3) The accelerating recent expansion of both developing and industrial societies, especially since World War II, has some of the characteristics of eruptive or explosive growth.

For human societies in a world of competing states, both growth and non-growth may be dangerous. The danger of not growing while a potential enemy does is evident to national leaders; but the dangers of growth are more complex, indirect, delayed, and pervasive. As we have observed, the optimum population for a society is very difficult to define and unlikely to be recognized until it has been exceeded. As suggested by Malthus (1798; Fig. 2d; cf. Whittaker, 1970; Meadows *et al.*, 1972) and by Fig. 15–2, human societies have an inherent tendency to overshoot the limits that should be set by their resources; the tendency appears in pollution processes, as well as population growth. The consequences of pollution, because they are mostly cumulative and indirect, are either not foreseeable or are discounted because they are uncertain, until the time comes when the effects are unmistakable. These effects also are delayed in expression; by the time a problem is recognized it is already the result of some years' development, and reduction of its industrial sources may then be difficult or time-consuming. As the causes are complex, industrial and public interests different, and the restraints that may permit solution unwelcome, agreements and decisions on sustained pollution control are difficult, particularly so if competing industries (or states) are responsible. Because solutions are difficult, false hopes of easy solution are often accepted as prospects for the future, while real efforts at solution fail. The negative feedbacks that would reduce pollution may be long delayed, and if they produce reversals of some pollutant increases, these reversals may be counteracted by other consequences of continued indus-

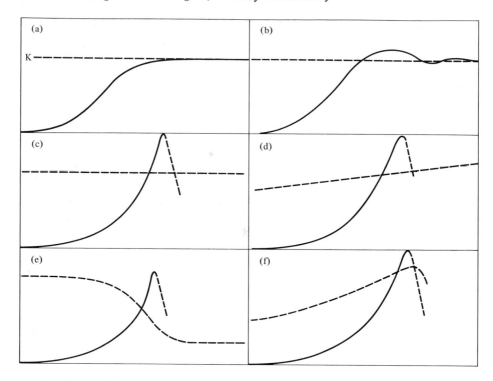

FIGURE 15–3. Patterns of stable and unstable growth (—) in relation to resource limit or carrying capacity *K* (- - -). (a) Regulated growth to steady-state limit at carrying capacity; (b) growth with limited overshoot, followed by stabilization; (c) eruptive growth beyond constant carrying capacity, and collapse; (d) exponential growth overtaking a linear increase in carrying capacity, Malthus' (1798) interpretation of population increase beyond food supply; (e)"convergent" pattern of exponential growth with reduction of resource by that growth; (f) modified Malthusian pattern of exponential growth overtaking slower exponential increase in resource.

trial growth. There is therefore a general tendency for the problems produced by population growth and industrial growth to reach an advanced stage before their seriousness is recognized and to continue to intensify after their seriousness is recognized.

These tendencies apply, although in different ways, to rich nations and poor. For the poor or "developing" countries growth leads ultimately toward over-population and a national life on the edge of hunger. No agricultural landscape is so provident that population growth, at current doubling times of 20–35 years, cannot exceed its production. Most poor countries have so far increased food production at paces that prevented massive famine; for many the consequence of population growth has been not sudden and massive, but chronic and marginal famine. It is a mistake, however, to identify the predicament of poor countries as a problem only of nutrition. Overpopulation works against the improvement of human conditions in a poor society in a variety of ways—by the unfavorable balance between population and industrial, as well as agricultural, resources; by the necessity of centering national effort on food; by the sheer numbers that make provision of reasonable education, employment, and public services difficult; by the difficulty of assembling the resources, capital, educated and dedicated personnel, and organized interrelationships among industries that would permit development commensurate with population size; and by the discouragement of leadership and deflection of effort toward problems other than population and its consequences. Overpopulation is thus self-intensifying. Not only is population growth exponential; the resulting numbers make more difficult the organization and education that might control those numbers. Hence, the implication of population growth for poor countries can be progressive entrapment in relative poverty—relative in the sense that per capita income may increase for a time, but at a low rate at which the gap between rich nations and poor widens and for many of the latter there is no real hope that they may join the former.

It is in this perspective that the effects of the new agricultural technology of the green revolution should be viewed. The success of this technology in increasing food production has been accepted as if it were the solution to the problems of the poor countries. The food is a great short-term benefit to the peoples of those countries, but the revolution may bring long-term intensification of their problems. Among environmental effects (Brown, 1970), the fertilizers and pesticides that increase food production on land are likely to decrease the smaller, but sometimes critical food production in coastal and inland waters. The food produced on land supports continued population growth, while application of capital and technology in the countryside disrupts traditional village life and displaces rural population to already overcrowded cities. Both farm and city populations become increasingly dependent on technology and energy expenditure for intensive agriculture and the transport of food, and therefore increasingly vulnerable to effects of increased energy costs or of social disorder. Even in the near future (to 1985) many of the poor countries face problems of unemployment and urban growth far more intractable than those of food supply (FAO, 1970). If not linked with population control, the green revolution does not solve the underlying problems, but implies a future encounter with acute or chronic famine by a larger, more heavily urbanized and more vulnerable population stripped of the sustaining psychological context of village life and traditional culture. History may cite no more painful combination of humanitarian achievement and potential for increased human suffering. It does not belie the achievement to observe that, if the green revolution should continue to be a substitute for, rather than an accompaniment to, real population control, then its effects will lead to magnification of tragedy for some of the poor countries.

Consequences work in different ways for rich, or overdeveloped countries. ("Overdeveloped" we apply to countries in which the combination of population and industrial wealth is producing accelerating environmental and psychological detriment, with the United States 1960–1975 as the type specimen. In the matter of wealth, as of many other human problems, there are detriments to the right and left and only an unstable balance between. It is a further irony that over-development can bring the problems of wealth without solving, for many of a country's people, those of poverty.) With the increase of pollution, industrial exploitation of environment, and urban spread, the rich country observes that the quality of its environment is declining. The awareness of environmental decline and increasing problems, congestion, visual and aural ugliness, sub-ordination to the power of technology, and erosion by rapid change of social values and sense of self, come to play more strongly upon the consciousness of many of its citizens than the dubious reassurances of consumer wealth. As the national wealth loses its luster, population growth continues and is sensed to be involving the society in further difficulties. The population of the United States in particular still grows, though more slowly than it did. If the reproductive rate, now at the replacement level, should remain at that balance for half a century or more, the United States population could eventually stabilize at a level around 270 million. Much of the lag in stabilization even with low birth rates is a consequence of the post-World War II baby boom, a national indulgence that casts a long shadow into the American future. Although the United States has now, in the view of some observers, more people and greater urban problems than it knows how to handle, it must live with the continued increase in both.

Growing population and wealth may therefore produce environmental and social problems that mount on all sides; not only do some problems intensify at exponential rates, but the rate of recognition of new problems increases (see Fig. 15–1). The result is that the society is overloaded with problems, and its means of attention, selection, decision, and effort at solution are overtaxed. Such a condition of overloading is likely to result in confused responses to symptoms and short-term pressures to the neglect of long-term policy, and devotion to gestures and statements that have only the appearance of solutions. It may also result in efforts to solve with money problems that are not thus soluble, until even a rich society may discover one of the patterns of overgrowth (Fig. 15–3f)—an expansion of committed governmental expenditures beyond the economic resources to support them.

We suggest (without wishing to overload our discussion with problems) further aspects of instability in an overdeveloped society. The most evident of these is the historic uncertainty of continued exponential increase in wealth and the possibility of depression. It would appear that the longer-term vulnerability of the society to economic disruption may be increased by the process of over-development. The post-World War II development of the United States has converted the economic basis of the society from one of moderate use of non-renewable resources to provide most citizens with a reasonable standard of living, to one of dependence on rapidly increasing use of resources for the creation of "wealth." Instability may result also from the commitment to reliance on technology for solutions of problems. Because of the past rewards of technology, it may seem that technology is the means of solving most problems, and that the net effects of further technological change must continue to be beneficial in the future. Technology too is subject to overshoot: an historic growth from a moderate use at lower population levels when its net effects seem conspicuously beneficial, to heavy reliance on increasingly powerful technology the effects of which (compounded by high and increasing population) produce increasing disadvantages. It is true both that our society is dependent on technology, and that for some years now the further expansion of technology has been producing environmental and social problems more rapidly than actual solutions for those problems. Furthermore, in a densely occupied world of interconnected problems, many short-sighted efforts at technological remedy are not solutions but are sources of new problems. There may be two implications. First, the problems of overdeveloped societies cannot now be solved by technology alone; they can be solved only by a coherent strategy (cf. Platt, 1969; Goldsmith *et al.*, 1972; Meadows *et al.*, 1972) including population stabilization, industrial restraints, advanced technology of pollution control and materials cycling—and, we suggest, some consideration of psychological problems. Second, the prospects of the future should not be discussed in terms of technological possibilities alone. It is the decision for such a strategy, and the psychological factors bearing on the decision, that are now crucial.

It is not within our province to analyze the psychological effects of over-development here. Increasing crime rates and drug use in the United States need only be mentioned as comments on these effects. Some current trends in psychological characteristics also suggest that the effects of heavy commercialization, and of wealth carried beyond sufficiency to luxury and the emphasis of comfort and passive pleasure, are detrimental. If the United States is something like a "mass aristocracy," the experience of past aristocracies may warn of weakness as the direction of growth in too easy and indulgent an environment. Wealth and passive entertainment can insulate human beings from the problems and challenges of life that, as they are faced, contribute to growth toward adult strength. Wealth may indirectly reduce the involvement of children with parents and other adults and thereby gravely weaken family structure and its role in the true education from childhood into effective and responsible adulthood (Bronfenbrenner, 1970). In a wealthy society of assured livelihood such qualities as trained competence, self-discipline, realism, devotion to work, and respect for institutions that serve others no longer seem so important to many individuals. These effects of wealth, and others that interact with them, may lead to progressive lowering of average standards in individual behavior, acceptance of work and quality of performance, means of recognizing and advancing the more competent, and willingness to accept sacrifice for the sake of long-term realism. The reduced acceptance of realism might apply not only to individual lives, but also to social policy in relation to economic limitations, and foreign policy in relation to a dangerous world. Yet the technological society is a complex, intricately functioning, sensitively balanced system that may depend on these qualities in individuals to make its successful function possible. A final convergence (Fig. 15–3e) affecting the overdeveloped society may deserve the attention of its leaders. On the one hand, expansion of the society's complexity and problems implies increasing need for effective leadership and competent and devoted service, whereas on the other hand indirect, psychological effects of wealth are contracting the availability these same qualities of leadership and service among the society's citizens.

It may thus be true that for both rich nations and poor, crucial effects of overgrowth appear in societal morale and quality of leadership. These effects not only result (in partial, complex, indirect ways) from overgrowth of population and industrial wealth, but imply that the longer the overgrowth is allowed to occur, the more difficult it becomes to set limits on that growth. It may no longer be true that time is on the side of solutions because the growth of problems eventually will compel societies to realism regarding the problems. After some point time may well be on the side of failure, in the sense that intensification of problems reduces the social and political feasibility of solutions. These factors of societal competence and leadership should be among the reasons for concern about the time available for solution of world problems (cf. Platt, 1969; Meadows *et al.*, 1972). In our view, they are reasons why a technological society is intrinsically unstable and cannot expect self-stabilization to occur in the pattern of Fig. 15–3a, but must seek the leadership and self-discipline to stabilize itself.

Conclusion

A degree of instability is inherent to human societies both in their relationship to environment and in their internal development. Population overgrowth is one source of both instabilities, and failure to control population is a major cause of historic tragedy. The population problem should not be regarded as one of food alone, critical as food may be for some countries; population increase beyond some ill-defined level produces a complex of interconnected and intensifying economic, environmental, and social detriments. Our theme may be simply stated. The primary production of the biosphere is immense, and could provide a long-term abundance of food for (without serious damage from pollution by) a human population stabilized at a favorable level. From man's failure to stabilize population may result increasing entrapments: of poor countries in poverty, of wealthy countries in environmental degradation, and of the world that includes both in trouble.

We wish to detach ourselves from one epithet often used against those concerned with population problems—"alarmist." The time for alarm is already past; alarm might have been appropriate at the end of World War II, when many of the implications of population growth could be discerned. We now live among the consequences, and it is not alarm but realism that is in order. The opportunity for the more favorable balance of population with environment that might then have been reached has passed, and a maximum effort of leadership that accepts present prospects and seeks a durable human future is needed. Some directions for that effort have been formulated by British scientists as a "blueprint for the future" (Goldsmith *et al.*, 1972), but the blueprint is for the present visionary: Solutions can be envisioned, but not the political means of achieving them. For the United States a new course should be sought toward a future that may be less wealthy but that will not, at least, be self-defeating.

It will be difficult either for individual nations to solve these problems for themselves while others seek maximal growth, or for a world of nations to agree on common policies of self-limitation. That the prospects are unpromising should not imply a counsel of defeat. The world is not so unitary as the Meadows *et al.* (1972) model; no result so singular as either continued world enrichment or simultaneous world collapse is to be expected. Not all nations can be spared the tragedies for which their population growths prepare them, but not all will suffer alike. Whatever can be salvaged for the improvement of man's future should be sought; encouragement may be taken from the movement of some nations, at least, toward awareness of the problems of exponential growth and toward policies for the future. Leadership is called for, in which major nations direct their efforts and those of others that can be influenced toward policies based on longer-term accommodation of population and industry to the limitations of the world. If such policies are not possible, if governments are still impelled to maximize their nations' wealth and power, the future may show little of what might have been hoped for, given human intelligence and technologic power, but will be more typical of history.

References

Bazilevich, N. I., L. Ye. Rodin, and N. N. Rozov. 1971. Geographical aspects of biological productivity. *Sov. Geogr. Rev. Transl.* 12:293–317. (Transl. from *Mat. V. Syezda Geogr. Obshch. USSR*, Leningrad, 1970.)

Bolin, B. (ed.). 1971. *Air Pollution Across National Boundaries: The Impact on the Environment of Sulfur in Air and Precipitation*, 96 pp. (Rep. of the Swedish Preparatory Committee for the U.N. Conference on Human Environment.) Stockholm: Norstedt.

Bronfenbrenner, U. 1970. *Two Worlds of Childhood, U.S. and U.S.S.R.*, 190 pp. New York: Sage Foundation.

Brown, L. R. 1970. Human food production as a process in the biosphere. *Scient. Amer.* 223(3):160–170.

Clark, Colin. 1967. *Population Growth and Land Use*, 406 pp. New York: Macmillan.

Cloud, P. 1973. Is there intelligent life on earth? In *Carbon and the Biosphere*, G. M. Woodwell and E. V. Pecan, eds. *Brookhaven Symp. Biol.* 24:265–280. Springfield, Virginia: Natl. Tech. Inform. Serv. (CONF–720510).

Connell, K. H. 1941. *The Population of Ireland 1750–1845*, 293 pp. Oxford: Clarendon.

Cook, E. 1971. The flow of energy in an industrial society. *Scient. Amer.* 224(3):134–144.

———. 1972. Energy for millenium three. *Technol. Rev.* 75(2):16–23.

Crowe, B. L. 1969. The tragedy of the commons revisited. *Science* 166:1103–1107.

Doxiadis, C. A. 1970. Ekistics, the science of human settlements. *Science* 170:393–404.

Durand, J. D. 1960. The population statistics of China, A.D. 2–1953. *Population Studies* 13:209–256.

Ehrlich, P. R., and A. H. Ehrlich. 1970. *Population, Resources, Environment: Issues in Human Ecology*, 383 pp. San Francisco, California: Freeman.

————, and J. P. Holdren. 1969. Populations and panaceas, a technological perspective. *BioScience* 19:1065–1071.

FAO. 1970. *Provisional Indicative World Plan for Agricultural Development*, 2 vols., 672 pp. Rome: Food and Agricultural Organization of the United Nations.

————. 1971a. *Production Yearbook, 1970*, Vol. 24, 822 pp. Rome: Food and Agricultural Organization of the United Nations.

————. 1971b. *Yearbook of Fishery Statistics, 1970*, Vol. 30: *Catches and Landings*, 469 pp. Rome: Food and Agricultural Organization of the United Nations.

————. 1971c. *Yearbook of Forest Products, 1969–70*, 230 pp. Rome: Food and Agricultural Organization of the United Nations.

Forrester, J. W. 1970. *World Dynamics*, 142 pp. Cambridge, Massachusetts: Wright Allen.

Goldsmith, E. R. D., R. Allen, M. Allaby, J. Davoll, and S. Lawrence. 1972. A blueprint for survival. *Ecologist* 2(1):1–43; also *Congressional Record* H209–232, Jan. 24, 1972.

Golley, F. B. 1972. Energy flux in ecosystems. In *Ecosystem Structure and Function*, J. A. Wiens, ed., Corvallis: *Oregon State Univ. Ann. Biol. Colloq.* 31:69–90.

Hardin, G. 1968. The tragedy of the commons. *Science* 162:1243–1248.

Heggestad, H. E., and E. F. Darley. 1969. Plants as indicators of the air pollutants ozone and PAN. (French and German summs.). In *Air Pollution: Proc. 1st European Congr. on Influence of Air Pollution on Plants and Animals, Wageningen 1968*, pp. 329–335. Wageningen: Centre for Agricultural Publishing and Documentation.

Heilbroner, R. L. 1974. *An Inquiry into the Human Prospect*, 150 pp. New York: Norton.

Hollingsworth, T. H. 1969. *Historical Demography*, 448 pp. Ithaca, New York: Cornell Univ.

Ho, Ping-Ti. 1959. *Studies on the Population of China, 1368–1953*, 341 pp. Cambridge, Massachusetts: Harvard Univ.

Holt, S. J. 1969. The food resources of the ocean. *Scient. Amer.* 221(3):178–194.

Howell, R. K., and D. F. Kremer. 1970. Alfalfa yields as influenced by air quality. *Phytopathology* 60:1297.

Hubbert, M. K. 1969. Energy resources. In *Resources and Man, a Study and Recommendations*, by the Committee on Resources and Man, P. Cloud, chairman, pp. 157–242. Washington: National Academy of Sciences; and San Francisco: Freeman.

————. 1971. The energy resources of the earth. *Scient. Amer.* 224(3):60–70.

Hulett, H. R. 1970. Optimum world population. *BioScience* 20:160–161.

Istock, C. 1969. A corollary to the dismal theorem. *BioScience* 19:1079–1081.

Koblentz-Mishke, O. J., V. V. Volkovinsky, and J. G. Kabanova. 1970. Plankton primary production of the world ocean. In *Scientific Exploration of the South Pacific*, W. S. Wooster, ed., pp. 183–193. Washington, D.C.: National Academy of Sciences.

Kovda, V. A. 1971. The problem of biological and economic productivity of the earth's land areas. *Sov. Geogr.: Rev. Transl.* 12:6–23.

Lieth, H. 1973. Primary production: terrestrial ecosystems. *Human Ecol.* 1:303–332.

———. 1975. The primary productivity in ecosystems. Comparative analysis of global patterns. In *Unifying Concepts in Ecology*, W. H. Van Dobben, and R. H. Lowe-McConnell, eds. The Hague: Dr. W. Junk (in press).

Likens, G. E., F. H. Bormann, and N. M. Johnson. 1972. Acid rain. *Environment* 14(2):33–40.

Malthus, T. R. 1798. *First Essay on Population*. Reprinted with notes by J. Bonar, 1926, 396 pp. London: Royal Economic Society and New York: Macmillan.

Meadows, D. H., D. L. Meadows, J. Randers, and W. W. Behrens III. 1972. *The Limits to Growth: A Report for the Club of Rome's Project on the Predicament of Mankind*, 205 pp. New York: Universe.

Miller, P. R. 1969. Air pollution and the forests of California. *Calif. Air Envir.* 1(4):1–3.

Olson, J. S. 1970. Carbon cycles and temperate woodlands. In *Analysis of Temperate Forest Ecosystems*, D. E. Reichle, ed., *Ecological Studies* 1:226–241. New York: Springer-Verlag.

Platt, J. 1969. What we must do. *Science* 166:1115–1121.

Reinhard, M., A. Armengaud, and J. Dupaquier. 1968. Histoire générale de la population mondiale, 708 pp. Paris: Montchrestien.

Rodin, L. E., N. I. Bazilevich, and N. N. Rozov. 1975. Productivity of the world's main ecosystems. In *Productivity of the World Ecosystems: Proc. Seattle Symp. 1974*, D. E. Reichle, ed. Washington, D.C.: National Academy of Sciences. (*In press*).

Ryther, J. H. 1969. Photosynthesis and fish production in the sea. *Science* 166:72–76.

SCEP. 1970. *Man's Impact on the Global Environment: Report of the Study of Critical Environmental Problems (SCEP)*, 319 pp. Cambridge and London: Massachusetts Institute of Technology.

Steinhart, J. S., and C. E. Steinhart. 1974. Energy use in the U.S. food system. *Science* 184:307–316.

TIE. 1972. *Man in the Living Environment: Report of the Workshop on Global Ecological Problems*, 267 pp. Madison, Wisconsin: The Institute of Ecology and Univ. Wisconsin.

Whittaker, R. H. 1966. Forest dimensions and production in the Great Smoky Mountains. *Ecology* 47:103–121.

———. 1970. *Communities and Ecosystems*, 162 pp. New York: Macmillan; see also 2nd ed., 1975, 385 pp.

———, and G. E. Likens. 1973a. Carbon in the biota. In *Carbon and the Biosphere*, G. M. Woodwell and E. V. Pecan, eds. *Brookhaven Symp. Biol.* 24:281–302. Springfield, Virginia: Natl. Tech. Inform. Serv. (CONF–720510).

———, and G. E. Likens. 1973b. Primary production: The biosphere and man. *Human Ecol.* 1:357–369.

———, and G. M. Woodwell. 1971. Measurement of net primary production of forests. (French summ.) In *Productivity of Forest Ecosystems: Proc. Brussels Symp. 1969*. P. Duvigneaud, ed., *Ecology and Conservation*, Vol. 4: 159–175. Paris: UNESCO.

———, and G. M. Woodwell. 1972. Evolution of natural communities. In *Ecosys-*

tem *Structure and Function*, J. A. Wiens, ed. Corvallis: *Oregon State Univ. Ann. Biol. Colloq.* 31:137–159.

———, F. H. Bormann, G. E. Likens, and T. G. Siccama. 1974. The Hubbard Brook ecosystem study: Forest biomass and production. *Ecol. Monogr.* 44:233–254.

Woodwell, G. M., P. H. Rich, and C. A. S. Hall. 1973. Carbon in estuaries. In *Carbon and the Biosphere*, G. M. Woodwell and E. V. Pecan, eds. *Brookhaven Symp. Biol.* 24:221–240. Springfield, Virginia: Natl. Tech. Inform. Serv. (CONF–720510).

Index

Abies fraseri (Fraser fir), 107
Abies balsamea (balsam fir), 60
Abies (Japanese fir), 72–75
Acer saccharum (sugar maple), 70, 75, 90, 93, 94
Acer spicatum (mountain maple), 70, 93
Adiabatic calorimeter, 123
Africa, 159, 195, 196, 218, 224, 238, 270, 278
Agarum (sea colander), 178
Ages of trees and shrubs, 62, 67, 70, 80
Agricultural land availability, 312, 313
Alaskan lakes, 194
Algae, 172, 178, 187–188, 193. *See also* Macrophyte
Algae, inorganic and organic needs, 173
Algal beds and reefs, 178, 306
Algal mass culture, 207
Alkalinity calculation, 45
Allochthonous carbon, 188–189
Allometric approach in production measurement, 66, 67–76
Allometric relations, 66, 76, 83, 149, 151, 152
Alpine lakes, 190, 192, 194
Amazon River, 191
Amazon rainforest, 265, 312
Ambrosia artemisifolia (ragweed), 91
Andropogon scoparius (bluestem grass, prairie), 91
Angiosperms, caloric values and evolution, 206, 300
Antarctic diatoms, measured growth of, 173

Antarctic lakes, 192, 194
Antarctica, 194, 271, 278
Appalachian Mountains, 154, 161
Aquatic communities, biomass, 306, 308
Aquatic communities, chlorophyll content of, 207–208, 306, 310
Aquatic primary production, 176–181, 195, 276–278, 306
Aquatic primary productivity, measurement of, 19–53, 189–191
Aquatic productivity, 169–202
Aquatic productivity, modeling, 43–44, 172, 178–180
Arctic lakes, 192, 194
Areas of land masses, 271
Aristotle, 8, 12
Arrhenius, 10
Ascophyllum nodosum (brown alga), 176
Asia Minor, 239
Australia, 194, 195, 218, 224, 271, 278, 287
Autochthonous production, 188–189
Autotroph species, 296

Bacteria (chemosynthetic), 188, 189
Baby boom, 322
Bahamas, 271, 272, 278
Bamboo Brake, 223, 226, 227
Basal area and increment, 67, 80, 84, 103–104, 152–153
Beckmann thermometer, 123
Beltsville, Maryland, 314
Benthic algae. *See* Macrophytes

Benthic biomass, method of estimation, 42
Benthic microflora, 178
Benthic production, 42–43, 176, 180, 181, 186–88
Betula alleghenienses, Betula lutea (yellow birch), 75, 94
Beta sp. (beets), 92
Biomass, 151–154, 206, 306, 308–310
Biomass accumulation ratios, 55, 62, 71, 80, 82, 99, 100, 149, 309
Biomass
  Definition, 4
  Distribution among species, 296
  Percents in tissues, 62, 70, 81
  Ranges, 207, 306
  Relation to elevation, 105–106
  Versus production, 82, 309
  World, 306, 313
Biome types, 204
Biomes, 148, 278
Biosphere, 3, 10, 305, 310, 313
  Characteristics, 305–307
  Man's relation to, 212, 312, 318
  Production, 306, 311, 314
  Production for human food, 311
Blanket bog productivity, 59
Bomb calorimeter, 119–125
Boreal areas, 278, 279
Boreal forests (taiga, spruce-fir), 80, 107, 205, 206, 207, 239, 278, 306
Bottle effects, 23
Bottle incubations, 36
Branch measurements, 60, 67–71, 73–82
Branch surfaces, 68, 80
Brockmann–Jerosch, 11
Brookhaven forest, 72–81, 90, 97, 100
Brookhaven inversion approach to measuring productivity, 97
Brookhaven National Laboratory, 66, 75, 76, 89, 96, 97
Burma, 226

^{14}C method, 33–39, 43, 171, 190, 191
C_4 plants, 20
Calabozo Plains, Venezuela, 226
Calcareous red algae, 178
California, 194, 314
Calluna vulgaris (heather), 93
Caloric content of chemical compounds, 126–127
Caloric contents in net primary productivity, 125, 204, 206
Caloric measurement, 119–129
  Calculations, 124–126

Milling samples, 121
Drying samples, 121
Preparation of tablets, 121, 122
Preparation of bomb and water bath, 122, 123
Caloric value
  Calculating from chemical analyses, 127
  Definition, 120
Calorie equivalents in plant materials, 124–127
Canada, 190, 292
Canary Islands, 178
Carbon, allochthonous, 188–189
Carbon content of organic matter, 4, 11, 22, 175, 181, 187
Carbon dioxide, 9, 10, 11, 30, 31, 32, 36, 37, 95–97, 148
Carbon dioxide, measurements, 21–24, 95–99
Carbon fixation, 189
Carbon productivity, 172–181, 186–198
Carbonic acid, 30
Carex lacustris (lake-margin sedge), 91
Carrying capacity, world, for man, 314–318
Cassia fasciculata (prairie senna), 91
Cimacian, 10
Circaea quadrisulcata (enchanter's night shade), 92
Chad, 225
Chaparral, 205, 207, 239
Chemical composition of heterotrophs, 299–301
Chemical composition of primary producers, 294–301
Chemical difference between ecosystems, 298–301
Cheyenne River, 194
China, 315, 316
Chlorophyll
  Content, 62–65, 80, 193, 195, 206–208, 301, 306, 309–310
  Determination, 39, 68
  Relation to productivity, 39–42, 62, 64–65, 80–81, 102, 208, 306, 310
  World amount, 306
Classification of communities, 204, 205, 306
Clear Lake, California, 194
Clethra acuminata (white alder), 63, 64, 90, 93
Climate
  Productivity, 105, 237–264
  Vegetation formations, 239

Climatic maps, 245, 247, 253, 255
Climax, 82, 100, 104
Clipping dry weight for production
    estimate, 61, 101
Codium fragile (sea staghorn), 178
Coefficients of variation, 83
Combustability of trees, 300
Community production, 57
Community properties, 206–209, 306–
    310
Computer analysis of maps, 265–283
Computer maps, 237–264, 266
Congo, 191, 225
Conic surface area, 67, 70, 80
Consumers, chemical composition of,
    299–300
Consumption
    Of food by man, 311–314
    Of industrial energy, 311
Coniferous forests, 80, 105, 107, 136,
    152–153, 297, 298
Conifers. *See Picea, Pinus, Tsuga*
Continental shelf, 207, 306
Continuous forest inventories, 150
Conversion factors, 22, 132–138, 155–
    156
Coral reef, tropical Pacific, 176
Corn (*Zea mays*), 88, 91, 98, 99, 125,
    162
Corrections for
    Daytime respiration, 29
    Diffusion, 26–29, 33
    Large trees, 85
    Logarithmic transformation, 83–84
    Loss of plant parts, 57, 60, 79
        Root loss, 89
        Scintillation counting, 38–39
        Temperature compensation, 25
Correlation models, 239
Corylus avellana (hazel), 93
Costa Rica, 224, 226
Creosote bush (*Larrea divaricata*), 58
Crop conversion factors, 136, 155–166
Crop yield statistics, 131–136, 149,
    155–156
Cultivated land, 205, 207, 306, 311–
    312
Cumberland Plateau, 154
Cultural eutrophication, 196–198
Cuvettes, 95, 96
Cytoseira (chain bladder), 178

DDT, 313
Deciduous forests, 80, 100, 105, 107, 205,
    207, 221, 239, 298, 306
Deciduous trees. *See Acer, Betula,*

Corylus, Fagus, Liriodendron,
    Populus, Prunus, Quercus
Decomposers, chemical composition of,
    299–301
Demographic transition, 315
Denmark, 194
Depth
    Mean, of ocean and lakes, 169, 186
    Productivity in relation to, 190–191
Desert, 79, 159, 205, 207, 219, 239, 286,
    306, 307, 309, 310
Desert shrubs, 207, 209, 306
Diffusion, 26–29, 33
Diffusion corrections, 26, 28, 33
Dimension analysis, 58–90
    Application to forests and woodlands,
        78, 80, 81
    Calculations, 67–69
    Field analysis of sample trees, 67, 68
    Summary tables, 62, 64, 70, 72, 80
Dissolved organic carbon, 186, 187
Dissolved organic loss from productivity,
    171
Diurnal curves, 24, 27
Diversity in relation to productivity,
    290–294
Dogwood (*Cornus florida*), 256, 258
Drude, 11
Dry matter
    Calorie values of, 119, 125, 127, 205
    Carbon content of, 4, 11, 22, 175, 181,
        187
    Chemical composition of, 294–301
    Net primary productivity, 4, 98, 204,
        205, 306
Dystrophic lakes, 193

East Asia, 239
East Indies, 271, 272, 278
Ebermayer, 10, 12
Ecosphere, 3, 148, 314
Ecosystems, 97, 148, 190, 307
    Chemical difference, 294–301
    Classifications, 4, 168, 176, 177, 192,
        193, 204, 205, 306
    Reduced productivities, 314
    Stress effects on, 288
Efficiency of productivity relative to
    Biomass. *See* Biomass accumulation
        ratio
    Chlorophyll, 42, 62, 64–65, 80–81,
        102, 208, 306, 310
    Leaf surface area, 62, 64–65, 71, 80–
        81, 103, 306, 310
    Light, 179, 206, 310

Egypt, 316
Eisenia rosea (earthworm), 299, 301
Elevation, relation to productivity and
    biomass, 105–107
Energy fixation, estimate of total, 204–
    206, 310
Energy from chemical analyses, calcula-
    tion of, 126–127
Energy industrial release of, 311
Energy of productivity, 3–4, 19–22, 98,
    119–129, 204–206
Eniwetok, 178
Environmental correlations of produc-
    tivity, 105–107, 158–161, 173–177,
    192–195, 224–227, 237–264
Environmental maps, 245, 247, 253,
    255
Error, sources of, 22–23, 83–86, 190
Estimate of relative error, 69, 75, 83
Estimated volume increment (EVI), 61–
    63, 67, 70, 74–76, 80, 83, 86, 103
Estuaries, 306, 307
Ethiopia, 195, 196
Eurasia, 159, 270, 271, 278,
Europe, 211, 238, 265, 270, 314
Eutrophic lakes, 192–195, 197
Eutrophication, cultural, 196–199
Evapotranspiration map, global, 252
Evapotranspiration models, 212, 254, 256
Evapotranspiration, relation to produc-
    tion, 105, 158–161, 250–254
Evergreen forest, 205, 207, 222, 226,
    239, 306, 308. *See also* Coniferous
    forests
Evergreen rain forest, primary produc-
    tivity, 205, 226, 306
Excavation of roots, 87–89
Experimental lakes area, 190

Fagus grandifolia (American beech), 75,
    94
Famine, 315, 317
Festuca ovina (sheep's fescue grass), 92
Florida, 161, 178
Flower and bud scale production, 79
Foliage biomass and production, 60, 62,
    64, 69, 70, 76–81, 85, 87, 92, 101,
    151
Forests. *See* Deciduous forests, Ever-
    green forests, Boreal forests, Rain
    forests
Forest biomass, 80–86, 107, 154, 207,
    306
Forest dimension analysis, 58–90
Forest inventory plot data, 149–152

Forest production, 205, 306
Forest productivity, 78–82, 98–107, 133–
    137, 139, 140, 150–158, 205, 306
Forest stand measurements, 66–69, 80, 81
Forest yields, 133–137, 155
Forms for production measurement, 68,
    134
Free-water methods, 21–24, 30
Freshwater ecosystems
    Production, 195, 306
    Productivity, 19–53, 185–202, 306
Fruit production, 59–62, 71, 79, 81
Fucus vesiculosus (rockweed), 171

Galium aparine (bedstraw), 91
Galvanic probe, 25
Gas exchange methods, 24–33, 56, 95–
    99, 170–172
Gaylussacia baccata (black huckleberry),
    62, 64, 65, 76, 89, 92
Geiger map, 161, 250, 252, 276
Geoecology, 267
Georgia, 151, 161, 171
Germany, 132
Giant cylinder, 97
Global net primary production estimates,
    11, 12, 204–206, 281, 305–308
Global productivity modeling, 237–283
Grasslands, 57, 82, 91–92, 150, 207, 209,
    239, 286, 288, 289, 294, 295, 297,
    298, 299, 306, 307, 309, 310
    Biomass, 207, 295, 306
    Productivity, 79, 105, 158, 205, 218–
    220, 224, 306
    Tropical, 218–220, 224
Great Barrier Reef, 171
Great Lakes, Laurentian, 186, 194, 197
Great Smoky Mountains, Tenn., 61–65,
    82, 83, 90, 103, 105, 107, 159, 162,
    286, 287
Great Valley, 154
Green Lake, 198
Greenland, 268, 271, 272, 278
Green revolution, 321
Grid system, 150
Gross photosynthesis, 24
Gross plant respiration, 21
Gross primary productivity
    Concept, 4, 20, 100, 147
    Measurement, 21, 98–101
    World, 310
Growing season, 138–139, 146, 254–259
Growth
    Allometric, 63, 66
    Patterns of, 318–320

Gymnosperms, caloric values and
evolution, 206, 300

Harvest of food, 311, 312, 314
Harvest of wood, 312
Harvest techniques, 57, 59, 172
Heath balds, 79
Heights of trees, 62–63, 66–67, 70, 76,
78, 80, 102–104
Helianthus annuus (sunflower), 91, 125
van Helmont, 8, 9, 12
Heteropogon contortus, 224
Heterotrophs, chemical composition of,
299–301
Hierarchy of regions, 148, 163
Highland Rim, 154
Himalayas, Nepalese, 106
Historical survey, 7–15
van't Hoff relation, 243
Hubbard Brook, New Hampshire, 66, 72–
75, 89, 90, 314
Hyperbolic equation, 85

Ibadadan, Nigeria, 225
Iceland, 271, 278
India, 217, 218, 224, 315
Industrial energy release, 311
Industrial growth, 315, 319, 320
Infrared gas analyzer, 148
Ingenhousz, 8, 9, 12
Inland water bodies, 185–202
Innsbruck productivity map, 11, 261,
261, 266, 273–280, 282
Inorganic carbon, determination of
total, 31
Insect consumption, 60, 79
In situ methods, 21–24, 172
Instability in society, 322
International biological program (IBP),
15, 147, 149, 162, 163, 203, 204, 211
International decade of ocean explora-
tion, 175
Interspecies regressions, 72–79, 89, 90,
151
Intertidal blue-greens, 178
Intertidal seaweeds, 178
Inversion measurement of respiration, 97
Ireland, 315, 316, 317

Japanese beech forest, 105
Japanese tree regressions, 72, 76
Jhansi, India, 224
Jodhpur, India, 218

Kalmia latifolia (mountain laurel), 64,
65, 92
Komor, 11

Laccadives, 178
Lake, 299
Lake and stream production, 195, 205,
207, 306
Lake and stream productivity, 185–202
Lake Aranquadi, 195, 196
Lake Ashtabula, North Dakota, 194
Lake Ashtabula Reservoir, North Dakota,
192
Lake Baikal, 194
Lake Constance, 191, 194
Lake Erken, 40
Lake Erie, 194, 197, 198
Lake Huron, 195, 197
Lake Huron's Saginaw Bay, 197
Lake Kilotes, 195
Lake Klopein, 191
Lake Mariut, Egypt, 192, 196
Lake Michigan, 195, 197
Lake Michigan's Green Bay, 197
Lake Millstatt, 191
Lake Nakuru, 195, 196
Lake Ontario, 195, 197
Lake Superior, 195, 197, 198
Lake Suwa, 192
Lake Tahoe, 194
Lake Vanda, 194
Lake Washington, 198
Lake Werowrap, 194
Lake Wörth, 191
Laminaria (brown alga, wrack), 178
Landscape productivity patterns, 131
Land use categories, 132–135, 137,
143, 154–158, 162
Large diverse areas, 278, 279
Large tree effects on estimates, 84–85
Larrea divaricata (creosote bush), 58, 93
Law of the minimum (Liebig), 9, 10,
250, 275
Law of yield (Mitscherlich), 10, 161,
262
Lawrence Lake, Michigan, 188, 189
Leaf
Production and biomass, 60, 62, 64,
69, 70, 76–81, 85, 87, 92, 101, 151
Area efficiency, 62, 64–65, 71, 80–81,
103, 306, 310
Area ratio and index, 62, 64, 80, 101,
103, 207, 209, 306, 310
Surface area, world, 306
Lespedeza cuneata (bush-clover), 92

Liebig, 9, 12, 15
Liebig's law of the minimum, 9, 10, 250, 275
Lieth–Box Model, 159, 160, 161
Lieth–Box production map, 162
Light–dark bottle method, 24, 36
Light effects on productivity, 173
Light penetration, 66, 80, 101, 171, 190–191, 193
Lilac (*Syringa*), 256
Liriodendron tulipifera (yellow poplar, tulip tree), 76, 79, 100
Litter fall, 66, 225
Litter fall as index for production estimate, 101
Logarithmic regressions, 63, 66–78, 83–86, 151
Long Island Sound, 178
Low energy beta particles, 38
Lumbricus (earthworm), 299, 301
Lyonia ligustrina (maleberry), 64, 65

Macrophytes, 42, 43, 176, 186, 187, 188, 192, 194, 198, 298
Madison, Wisc., 157
Maize (*Zea mays*), 88, 91, 98, 99, 125, 162
Malthus, 315, 319
Management stress, 288
Mangrove, 178, 223, 227
Map evaluation, 267
MAPCOUNT, 267–271, 273–276, 278, 280, 282
MAPMATH, 267
MAPMERGE, 267
Mapping of productivity, 13, 14, 140, 237–283
Maps
  Evapotranspiration, world, 253, 255
  Precipitation, world, 247
  Productivity, Mozambique, 210
  Productivity, North Carolina, 138, 142
  Productivity, oceans only, 174
  Productivity, U.S., 158, 160
  Productivity, world, 209–212, 249, 251, 257, 259, 261
  Stations for productivity measurements, 242
  Temperatures, world, 245
Mapzones, 267
Massachusetts, 154–157, 212
Marine
  Environment, 172, 173
  Production, 176–177, 180–181, 306
  Productivity, 169–184
  Productivity measurement, 170–181

Mean tree approaches, 58–60, 84
Measurement of caloric values, 119–129
Measurements, productivity, 17–166. *See also* Methods and Productivity measurement
  Benthos, 42–43
  Crops and fields, 57, 59, 132, 136, 155–157
  Forests, 58–60, 63–76, 82–87, 95–97, 101–103, 136–138, 150–154
  Grasslands, 57, 59, 94
  Macrophytes, 42–43, 172
  Periphyton, 42
  Plankton, 24, 27–42, 45, 170–172
  Plantations, 58, 60
  Shrublands, 57, 58, 61–65
  Soil respiration, 97–98
  Streams, 26–27
  Undergrowth, 61–63
Metabolism, 19–22, 24, 26, 28, 30, 32, 189
Methods, assessing regional production, 149–157
Methods, productivity, 17–166. *See also* Measurements and Productivity measurement
  Fresh water, 19–53, 189–191
  Marine, 19–53, 170–172
  Regional, 149–157
  Streams, 26–27
  Terrestrial, 55–118
Miami model, 210, 212, 237, 238, 244, 245, 246, 247, 248, 250, 252, 258, 261, 266, 274–276, 278, 280, 281
Micrometeorological approach to measuring productivity, 96, 97
Minimum area, concept of, 291
Minimum, Liebig's law of, 10, 250, 275
Mirror Lake, New Hampshire, 188, 189
Mitscherlich's yield law, 10, 161, 262
Mississippi River, 191
Mixed dry forest, 221
Models
  Of production process, 43–44, 101, 172, 178–180
  Of production response to climate, 105, 237–264
  Of regional and world production, 158–163, 178–180, 237–283
Montreal model, 237, 250, 266, 267, 274, 275, 278
Mozambique, 208, 210

Nashville Basin, 154
Net community growth, 56

Net ecosystem production, 56, 100, 226
Net photosynthesis, 24
Net primary productivity, definition, 3, 4, 21, 56, 147
Net primary productivity measurement. *See* Methods, Measurements, and Primary productivity measurement
New agricultural technology, effects of, 321
New York, 154, 155, 156, 157, 212
New York State, 292
New Zealand, 271, 272, 278
Nicolai de Cusa, 8, 12
Nigeria, 217
Nile, 191
Noddack, 11, 12
North America, 156, 186, 238, 270–272, 278
North Carolina, 28, 131, 138–144, 151, 154, 155, 156, 157, 211, 291, 292
North Carolina productivity profile, 134, 140–142
Nova Scotia, 176, 178
Nutrient effects on productivity, 170, 173, 175–177, 196–197
Nutrient pools, 206, 209
Nutrients, 307
Nyquist analysis, 241–242

Oak heath, 79
Oak–pine forest (Brookhaven), 72–81, 90, 97, 100
Oak Ridge National Laboratory, 96, 100
Oak Ridge, Tennessee, 72–76
Oaks. *See Quercus*
Ocean, 207, 306
Ocean environment, 169
Ocean primary production, 176–177, 180–181, 306
Ocean productivity, 170–181
Oceania, 271, 272, 278
Oceans productivity maps, 266, 269, 273, 276, 277, 278, 280
Oligotrophic lakes, 192–194
Overdeveloped societies, 321–323
Overgrowth, effects of, 324
Overharvest, 311, 312
Overpopulation, 315, 320, 321
Overshoot, 319, 323
Oxygen analysis, 25–26
Oxygen bomb, preparation of, 122, 123
Oxygen methods of production measurement, 24–30, 170

Pacific, western region, 217
Panicum aciculare (panic grass, old-field), 91
Parabolic volume (VP), 61, 63, 67, 70, 72–76, 80, 85, 86
Particulate organic carbon, 186, 187
Pederborgsø, 194
Periphyton, 42, 186, 188, 198
pH change, production measurement by, 31
Phenology, 149, 163
Philippines, 271, 272, 278
Phlox caespitosa (alpine phlox), 92
Photic zone, 179, 180
Photorespiration, 20, 21, 29, 30, 95
Photosynthesis, 7, 8, 20, 22
  Measurement of, 30–32, 95–96, 101, 171, 189–190
  Prediction from chlorophyll data, 41–42
  Relation to primary productivity, 7–9
Photosynthetic quotient, 21
Photosynthetic surfaces, 306, 309. *See also* Leaf area
Phragmites communis (reed grass, marsh grass), 188
Phyllostachys bambusoides (bamboo), 93
Phytoplankton, 170, 172, 175, 180, 186, 187, 188, 197, 198
  Production, 171–181, 276–278, 306
  Productivity, 169–202, 306
Phytorespiration, 21
Picea rubens (red spruce), 70, 72–75, 89, 93, 107
Pine forests, 107, 136–137, 152, 289
Pine heaths, 107
Pinus echinata (short-leaved pine), 136
Pinus elliotii (slash pine), 136
Pinus rigida (pitch pine), 70, 83, 89, 93
Pinus taeda (loblolly pine), 136, 137
Pinus virginia (scrub pine), 152
Plankton, 171, 173, 186, 209, 243, 298, 310
Plankton productivity, measurement, 24, 27–42, 45, 55, 170–172
Planktonic production, 176–181, 276–278, 306
Planktonic productivity, 169–202, 306
Plant production, historical, 9
Plant respiration, percentage of gross productivity, 98–100, 191–192, 310
Plant-water relationships, 8, 159–161, 237–254
Polar areas, 278, 279
Polar waters, 170

Poletimber, 150–155
Pollution, 181, 197, 198, 313, 314, 319
Population growth, 315, 317, 320, 321, 322, 324, 325
Population stabilization, 316–318
Populus tremuloides (aspen), 90, 94
Precipitation
    Average annual, 238, 239, 240, 247
    Relation to primary productivity, 105, 224, 243, 246, 248
Priestley, 8, 9, 12
Primary production
    Freshwater, 195, 306
    Marine, 176–181, 276–278, 306
    Regional, 131–166
    Terrestrial, 204–206, 306–308
    World, 204, 273–281, 306
Primary productivity
    As different from photosynthesis, 7–9
    Correlation to length of vegetation period, 138–139, 146, 254, 258
    Data, 64, 80–81, 176–178, 188, 196, 219–223, 240–241
    Effects of rainfall and temperature on, 105, 241–248, 285–287
    Equation and balance, 8, 15, 100
    Freshwater, 185–202
    Maps, 140, 144, 174, 210, 212, 237–283
    Marine, 169–184
    Modeling. *See* Models
    Predicted from evapotranspiration, 159–161, 250–257
    Regional, 131–166
    Terrestrial, 100–107, 203–231
Producer, 300
Production
    Balances, 100
    Below-ground, 90–95, 152
    Estimates. *See* Primary production
    Ratios, 61–62, 103, 132–138, 155–156
Productivity
    Assessments, 11
    Benthic, 42–43, 176, 178, 180
    Mapping, 134, 209–212. *See also* Maps
    Measurement, aquatic, 18–47. *See also* Measurements and Methods
        "Allen curve" method, 43
        Bottle methods of productivity measurement, 22, 24
        Carbon 14 method, 33–40, 170–171, 190
            Equipment used, 34
            Advantage of, 36, 171
            Drawbacks of, 39, 171

        Chlorophyll method, 39–42
        Diurnal curve, 24
        Diurnal pH method, 30
        Free water oxygen technique of measurement, 19, 21, 24, 30, 36
        Gas exchange methods, 24–33, 170–172
        In situ method, 22, 23
        Macrophyte methods, 42, 171–172
        Models, 43–44, 178–180
        Periphyton productivity methods, 42
        Single curve method, 27
        Two-station method, 27
        Winkler method, 170
        Winkler oxygen determinations, 21, 25
    Measurement, terrestrial, 55–107. *See also* Measurements and Methods
        Allometric approach, 66
        Caloric value measurement, 123–124
        Dimension analyses of forests, 61, 62, 63, 66–69, 76, 78, 80–87
        Gas exchange measurement, 55, 58, 95–107, 149
        Harvest techniques, 57–59
        Mean tree approach, 58, 60, 61
        Micrometeorological approach, 96, 97
        Models, 101, 105, 158–163, 237–264
        Plantation approach, 58, 60
        Root biomass and production, 87–95
        Undergrowth, 61–63
    Pattern, 15
    Percents in tissues, 62, 71, 79, 81
    Profiles, 154–157
    Regional, 131–166
    Relations to environment
        Elevation, 105, 107
        Evapotranspiration, 105, 158–161, 250–254
        Nutrients, 170–177, 196–197
        Precipitation, 105, 224, 243, 246, 248
        Temperature, 107, 173, 241–243
    Species, 288–294
    Stratal, 285–287
Profiles of productivity, 154–157
Prunus pensylvanica (pin cherry), 90, 93, 94
Psychological effects of overdevelopment, 323, 324
Puerto Rico, 226
Puerto Rican rain forest, 97

Pygmy conifer-oak scrub, 80
Pyrus melanocarpa (black chokeberry), 65

Quenching, 38, 39
Quercus alba (white oak), 70, 76, 83, 90, 93
Quercus coccinea (scarlet oak), 90, 93
Quercus ilicifolia (shrubby, bear oak), 89, 90, 92
Quercus prinus (chestnut oak), 64
Quercus robur (oak), 70, 93

Radial increments, 62, 67, 70, 80
Radioactive carbon, 33–39
Radioactivity, determination of, 37
Radiocarbon method, 33–39, 43, 171, 190, 191
Rain forest, 205, 206, 207, 306, 308, 309
Raingreen and tropical seasonal forest, 205, 207, 225, 239, 306
Rate-of-change curve, 27, 29, 30
Red Rock Tarn, 196
Redbud (*Cercis canadensis*), 256
Red maple (*Acer rubrum*), 258
Reefs and estuaries, 178, 207, 306
Regional productivity, 131–166
    History, 132
    Methods, 132–133, 147–166
    Models relating to climate, 158–163
    North Carolina, 131–146
    Sources of data, 133–136, 149–158
    United States, 147–163
Regulated growth, 318
Regression, 63, 76, 86
    Correction for logarithmic transformation, 83–84
    Deviations from, 83
    Equations, 63, 83, 258
    For roots, 73, 89, 90
Regressions, interspecies, 72–79, 89, 90, 151
Reliability of production estimates, 82–86
Respiration, 4, 20, 24, 29, 96, 97, 310
    Heterotroph, 100
    Light and dark, 29–33, 171
    Percents of gross productivity, 4, 98–100, 191–192, 310
    Photorespiration, 20, 21, 29, 30, 95
    Relation to pH, 30
    Soil, 97–98
Rhizophora mangle (mangrove), 223, 227
Rhododendron catawbiense (mountain rosebay), 62, 65

Rhododendron maximum (great rosebay), 62, 64, 90, 92
Rivers, productivity of, 191, 194
Root and shoot relations, 87–95
Root mass, 73, 88, 89
Root production, 79, 88, 90, 152,
Root regressions, 73, 89, 90
Root/shoot ratios, 89–96
Rosenzweig model, 105, 159, 160, 161
Russia, 238

Saline lakes, 186, 194, 196
Sample plots, 61, 66, 69, 76, 150–154
Sample trees, 67–69
Santa Catalina Mountains, Arizona, 286, 287
Sapling, 150, 151, 152
Sarawak, 226
Sargasso Sea, 171
Saturation curve, 160, 161
de Saussure, 8, 12
Savanna, 217, 220, 306
    Definition, 224
    Primary productivity, 205, 220–225, 306
Sawtimber, 150–155
Scales of production study, 148–149, 163
Scheele, 8, 12
Schroeder, 10, 11, 12, 15
Scintillation counting, 37–39
Sea grasses, 178
Seattle productivity map, 209, 211, 250, 261, 265, 266, 274
Seasonal forests, tropical, 205, 207, 225, 226, 239, 306
Secale cereale (rye), 91
Semidesert, 239, 306
Senegal, 225
Sequential harvesting, 42
Shrub dimensions, 61
Shrublands, 64, 79, 82, 205, 207, 239, 286, 306, 309
Shipboard productivity estimates, 23, 172
Siberia, 292
Sierra Nevada Mountains, 194
Silica, 297
Soil respiration, 97, 98
Solanum lycopersicum (tomato), 91
Solanum tuberosum (potato), 92
Solar radiation, relation to productivity, 179, 206, 310
Søllerod Sø, Denmark, 194
Sorghastrum nutans (Indian grass, old-field), 91
South America, 159, 238, 271, 278

Southern Appalachians, 63
South Pacific, 278
Southeast Asia, 239
Southwest Africa, 239, 246
Spartina (marsh-grass), 43
Species–area relations, 292–294
Species diversity, relation to primary
    productivity, 290–294
Species productivity, 288–290
Species-saturation level, 292
Spruce–fir forests. *See* Boreal forests
Statistical mapping, 267
Statistical regressions, 44
Statistics of production measurements,
    82–86
Steady states, 318
Stem surface, 70, 80. *See also* Conic
    surface
Stem volume, 72, 80. *See also* Parabolic
    volume
Steppe, arid, 92
Stored energy, calculation of, 125, 126
Stratal productivity, 285–287
Stratification, thermal, in water bodies,
    170, 175
Stream metabolism, 27
Stress effects, 288–290
Structural carbohydrate concentrations,
    300
Stocking, 152, 153, 154
Succession, productivity in, 79–80, 82,
    87, 104, 163
Summergreen (temperate deciduous)
    forest, 80, 100, 105, 107, 205, 207,
    221, 239, 298, 306
Surfaces
    Branch, 68, 80
    Conic, 67, 70, 80
    Leaf, 62, 64, 80, 101, 103, 209
    Mapping, 246
    Stem, 70, 80
Swamp and marsh, 195, 205, 207, 306
Sweden, 314
SYMAP, 248, 252, 254, 265–267, 270,
    271, 275, 278
SYMAP Earth models, 268, 269
SYMAP-MAPCOUNT evaluation
    procedure, 271, 272
Syringe–gas chromatograph technique, 31

Taiga. *See* Boreal forests
Technology, problems of, 323–324
Temperate forests. *See* Deciduous, Ever-
    green, and Boreal forests
Temperate lakes, 192

Temperate rivers, 192
Temperate-zone waters, 170, 278, 279
Temperature
    Compensation, correction of, 25
    Probes, 25
    Relation to productivity, 107, 173,
        241–243
    Values, average annual, 238–243
Tennessee, 140, 150, 151–159, 162,
    212
Tennessee Valley Authority (TVA), 150,
    151, 152, 154
Terrestrial communities
    Biomass ranges, 207, 306
    Chlorophyll, content of, 207, 306
    Leaf areas, 207, 306
Terrestrial primary production, 204–206,
    306–308
Terrestrial productivity, 100–107, 131–
    166, 203–231, 237–264, 306
    Methods, 55–118
Thematic evaluation of maps, 270
Thalassia testudinum (eelgrass), 172
Theoretic treatment of productivity,
    178–180
Thornthwaite memorial model, 237, 238,
    248, 254, 256, 257, 258, 261, 266,
    267, 275, 276, 278, 280
Thornthwaite method, 160, 161
Threshold area, 294
Time scales, 148–149, 163
Tree dimensions, relations between, 63,
    66, 70–71, 77–78
Trifolium parryi (alpine clover), 92
Trophic status of lakes, 192–193
Tropical aquatic productivity, 176–178,
    192, 195, 306
Tropical areas, wet, 278, 279
Tropical deciduous (raingreen) forest,
    93, 205, 207, 225–226, 239, 306
Tropical forests, 159, 160, 206, 239, 289,
    306, 310
    Biomass, 205, 306, 309
    Method of production measurement,
        87
    Production balance, 100
    Productivity, 85, 87, 205, 221–223,
        225–227, 306
Tropical grassland, 218–220, 224
Tropical rivers, 192
Tropical savanna, 205, 217, 220–225,
    306
Tropical terrestrial productivity, 205,
    217–231, 306
Tsuga canadensis (eastern hemlock), 107

Tundra, 159, 205, 207, 209, 239, 286, 288, 306, 310
Twig and leaf measurements, 59–61, 67–69, 75–81
Two-station analysis, 26
Typha (cattails), 188
Typha latifolia (broadleaved cattail), 92

Undergrowth production, 58, 61, 79, 81, 151, 152
Unit definitions, 4
United States, 147, 150, 159, 161, 197, 198, 212, 313–315, 322, 323, 325
U.S.-IBP eastern deciduous forest biome, 254
UNC biosphere model, 248, 252, 258
Unregulated growth, 318–320
Unstable systems, 318–325
Upwelling zones, 175, 207, 306

Vaccinium constablaei (highbush blueberry), 64, 65
Vaccinium vacillans (lowbush blueberry), 76, 89, 92
Varanasi, India, 218, 224
Vegetation formation types, 11, 204, 205, 239, 306
Vegetation period, 144, 254, 258, 261
Venezuela, 224

Viburnum alnifolium (hobble bush), 62, 90, 93
Virginia, 27
Volume, stem, 72, 76, 80. *See also* Parabolic volume

Walter's ratio, 246
Water value, calculation of, 124, 125
Wealth, effects of, 323
Weather records, 158, 240–241
West Indies, 271, 272, 278
Wisconsin, 154–157, 162, 211
Wood and bark measurements, 60–63, 67–82, 100
Woodlands, 179, 205, 207, 286, 306, 308
World animal biomass, 309
World food production, 311
World primary production estimates, 12, 206, 306, 310
World productivity maps, 11, 13, 15, 209–212, 249, 251, 257, 259, 261
World vegetation maps, 11, 273

Yellow poplar (*Liriodendron tulipifera*), 258
Yield, 9, 100, 155
Mitscherlich's law of, 10, 161, 262

Zea mays (maize), 88, 91, 125, 162